Veterinary Molecular Immunology

Shijun J. Zheng

Veterinary Molecular Immunology

Springer

中国农业出版社
CHINA AGRICULTURE PRESS

Shijun J. Zheng
College of Veterinary Medicine
China Agricultural University
Beijing, China

ISBN 978-981-99-8928-7 ISBN 978-981-99-8929-4 (eBook)
https://doi.org/10.1007/978-981-99-8929-4

Translation from the Chinese language edition: "Veterinary Molecular Immunology" by Shijun J. Zheng, Published by China Agriculture Press. All Rights Reserved.

Jointly published with China Agriculture Press
The print edition is not for sale in China (Mainland). Customers from China (Mainland) please order the print book from: China Agriculture Press.

This Springer imprint is published by the registered company Springer Nature Singapore Pte Ltd.
The registered company address is: 152 Beach Road, #21-01/04 Gateway East, Singapore 189721, Singapore

Preface to the Second Edition

Since its publication, *Veterinary Molecular Immunology* has been well received by readers because of its novel contents, easy-to-understand and up-to-date knowledge of immunology. The book is not only of high academic value but also of significant social value. The academic value is embodied by its profound understanding and exposition of new knowledge in the field of modern immunology; the social value is reflected by the fact that the book makes immunology, which contains profound truths and theories, easy to understand and interesting to read, especially with the author's own color illustrations, and makes the readers more interested in immunology and looking forward to its future prospects. The book also contains inspiring stories of immunologists, which have led many students to become interested in research and to take up immunology research as a lifelong pursuit. The fact that students have set their lifelong goals and taken up research because of reading the book fully illustrates the social value of this book! In addition, all of the hundreds of reader reviews after the book was sold online were five stars, which also reflects the overall quality and value of the book from another perspective. Modern science advances rapidly, and the contents of books should be timely updated and renewed as time goes on, providing readers with more accurate, advanced, easy-to-learn, and easy-to-understand knowledge in disciplines of sciences. The second edition of *Veterinary Molecular Immunology* (hereinafter referred to as "the second edition"), while retaining the illustrations of the first edition, is more advanced and complete, especially in the chapters on innate immunity and lymphocyte maturation and activation, covering the latest knowledge and research results in the field of immunology in recent years. The book is equipped with 146 illustrations to help readers learn and understand the contents. In addition, the appendices have been updated in real time, and new contents have been added, such as information on the homology of some avian and porcine cytokines with human and mouse (Appendix VIII–XI) and important information on the Nobel Prizes in immunological research (Appendix XII), for the convenience of readers. The second edition retains the audio (QR code) of the inspiring stories of the Nobel Laureates in Immunology from the first edition, which serve as a beacon of light for readers, especially young researchers. Innovation requires the courage to encounter difficulties and not to be afraid of failure, and to see the rainbow only through storms. All the achievements of Nobel Prize laureates resulted from their continuous efforts and persistence in their research. In the last half century, immunological research in animals has relatively fallen behind, and much knowledge is borrowed from human or mouse immunology, but the immune systems of different species differ, for example, the cytokine homology between avian and mammalian (see the appendix of this book) varies greatly, whereas the specific functions of some animal cytokines are still not clear; in addition, the proportion of CD4CD8DP T-lymphocytes in peripheral blood is high in chickens and pigs, accounting for about 30%, while in humans and mice they are less than 1%. Animal immunity differs significantly from that of humans and mice in some aspects and needs to be further studied. However, studies on Veterinary Immunology fall relatively behind, lacking talented people devoted to this research field, and facing with a lot of difficulties, and has become a weak spot in research on effective measures for prevention and control of infectious diseases in animals, limiting the effective prevention and control of infectious diseases in animals, allowing some disease prevention and control problems unsolved for a long time. For instance, porcine reproductive and respiratory syndrome (PRRS), also called blue ear disease, occurs here and there, and it is difficult to obtain an optimal control result using conventional vaccines. Thus, there is an urgent need to develop novel vaccines and control measures. In clinical practice, there is also an urgent need for an immunology book that integrates theories with practice. The second edition not only presents the latest theoretical knowledge and research achievements of immunology but also integrates theories with practice, explaining how to use the theoretical knowledge of immunology to serve prevention and control of infectious diseases, analyzing the key issues in clinical practice and providing guidance to clinical immunization in control of diseases, so that immunology is no longer a profound knowledge restricted in the ivory tower, but returns to its original position as a thorough, practical and applied discipline of science. For the reader, a good book is a mentor, and a good book requires a high sense of responsibility and dedication from the author. The second edition was written and edited by myself alone, including all the illustrations (146) in the book. I sincerely hope that readers will make critical comments and correct errors or mistakes in the book, and I also hope that colleagues and like-minded people will cooperate sincerely to keep the book updated and enriched, promoting the flourishing advance in life sciences.

Beijing, China
January 2021

Shijun J. Zheng

Preface to the First Edition

The rapid development of immunology, with fast knowledge update and renewal, of especially the rapid progresses in the field of innate immunity research in recent years, have led to a qualitative leap in the understanding of the physiological functions of the immune system. However, students who are new to immunology often feel overwhelmed by the profound knowledge contained in it, while students with some basic knowledge of immunology find it difficult to update their knowledge in a short period of time. Since I started to learn immunology in 1986, I have been fascinated by many scientific issues revealed in the course, and taught "Veterinary Immunology" to undergraduate students in Veterinary Medicine from 1990 to 1996. In 1996, I obtained Research Assistantship from University of Massachusetts (UMass), USA, to pursue a PhD with a thesis on the regulation of the reproductive cycle of horses by immunological techniques. During my PhD study, I learnt from immunologists Dr. Richard Goldsby (Editor-in-Chief of the *Kuby Immunology* textbook), Dr. Barbara Osborne (Editor-in-Chief of the *Kuby Immunology* textbook), Dr. Samuel Black, and Dr. Cynthia Baldwin. I was able to listen to the teachings of these masters and touched by their deep love and dedication to immunology. I still recall their humble words and amiable smiles, especially when Professor Goldsby gave lectures, his knowledge and elegance set a shining example for the students. During my research, I came to understand and appreciate that the pathogenesis of autoimmune diseases is one of the most elusive topics in immunology research. After completing my PhD study in UMass, I went to the University of Pennsylvania School of Medicine to work on the pathogenesis of autoimmune diseases and the mechanisms of infection and immunity under the supervision of Dr. Youhai Chen, a molecular immunologist, using mice as a model. During this time, I really experienced the wonders and joys of immunological experiments, and since then I have become a real practitioner of immunological research. In the summer of 2005, I returned to my alma mater, College of Veterinary Medicine, China Agricultural University, to teach and conduct research in immunology. In the same year, I contacted the publisher and planned to translate the textbook *Kuby Immunology* from English to Chinese. However, due to the copyright of the illustrations, no agreement could be reached, which was a great pity. The desire to write a book on an up-to-date Veterinary Molecular Immunology emerged therefrom. It was indeed a great test of will to sit down and write a book, and after many years of hard work, this book is finally completed! This book provides a focused description of the core elements of immunology. The core of immunology is T and B lymphocyte development, maturation and activation (Chaps. 4 and 5), which is the root and backbone of immunology. If this part of the text is fully and deeply understood, problems encountered will be readily solved, and the rest of the text is just a matter of minutiae. Each chapter is devoted to the latest research findings in immunology and the most popular areas of research, with a "reading tips" space at the end of the chapter where readers can summarize their reading. The book focuses on the core theories of immunology and disease prevention and control in a thematic manner and can be used by researchers in the field of biomedicine, especially veterinary medicine. The book was written soly by me, and all 125 color illustrations in the text were done by me alone with a mouse through Power Point, except for Figure 11-3, which is a citation from the literature. I sincerely hope that readers will make critics and corrections, and that like-minded people will cooperate to keep this book updated and enriched!

April 2015 Shijun J. Zheng

Contents

Chapter 1
Introduction

Overview:
The immune system is a multifunctional defense system that has evolved over long time in animals, and has the function of protecting animals against pathogenic infections, preventing tumor formation and maintaining homeostasis. Immunology is a biological discipline that studies the composition and function of the immune system in animals and reveals its mechanism of action. With the application of cell biology and molecular biology techniques to immunological research, the mechanism of action of the immune system has been explored and discovered in depth, thus ushering in the era of molecular immunology. The special techniques and methods in immunological research have also greatly promoted the advance in life sciences, molecular medicine and other related research fields, and molecular immunology has become an important core content in the field of biomedical research.

1.1 The Birth of Immunology and Its Development

The phenomenon of immunization has been known for more than 2000 years, dating back to a war in 430 BC (Peloponnesian War-430 BC), when the Athenian historian Thucydides described a plague among soldiers, and only those who recovered could care for the sick for the company without further illness. In the eleventh century, folk physicians in the Song Dynasty of China baked and ground the scabs of blisters of smallpox patients into powder and blew them into the nose of people through a bamboo pipe to prevent smallpox (Fig. 1.1), an immunization method that later spread to Turkey along the Silk Road. Mary Wortley Montagu, the wife of the British ambassador in Constantinople (nowadays called Istanbul), observed that this method was effective in preventing smallpox, and she inoculated her own children with the expected effect. Soon afterward, this method was introduced in Britain. As this method posed a high risk to the public due to the use of a virulent virus, it was

S. J. Zheng, *Veterinary Molecular Immunology*,
https://doi.org/10.1007/978-981-99-8929-4_1

Fig. 1.1 The folk physician in the Song dynasty of ancient China prevented against smallpox by blowing smallpox-scab powders into the nose with a bamboo pipe. In ancient China, during the Song Dynasty (960–1279 AD), folk doctors ground the scabs of smallpox patients into powder and inoculated against smallpox by blowing the nose through a bamboo pipe, which was the first record of inoculation against pathogenic infection, around 700 years earlier than that of Jenner's contribution to the control of smallpox with cowpox

later terminated. At the end of eighteen century, Edward Jenner, a British physician, observed that the milkmaids had smooth faces free of smallpox infection except for a small number of blisters on their hands. Recognizing that people infected with the cowpox virus might develop resistance to smallpox, he assumed that injection of people with cowpox pustules might provide protection against smallpox infection. In 1798, to test this hypothesis, he injected an 8-year-old boy with cowpox pustule and then infected the boy with the smallpox virus. As a result, the boy did not develop the disease. This was the first scientific evaluation of the protective efficacy of a vaccine in human history. Jenner's method of injecting people with cowpox pustules to prevent smallpox soon spread throughout Europe, making an outstanding contribution to the prevention and control of smallpox epidemics. However, for more than 100 years after the invention of the technique, it was not extended and used in the prevention and control of other infectious diseases.

Scientific discovery depends on keen observation and deep thought, and the development of immunology follows the same pattern. In the late nineteenth century, Louis Pasteur, a famous French scientist, succeeded in isolating the pathogenic bacterium while studying avian cholera and found that infection of chickens with the isolated bacterium could cause their death. Pasteur instructed his assistant, Charles Chamberland, to infect the chickens with a culture of the pathogenic bacteria and then left the laboratory for a summer vacation. The assistant watched Pasteur leave on vacation and secretly left the laboratory for a vacation of his own. One month later, Pasteur returned from vacation, and so did the assistant. They

Fig. 1.2 Inoculation of chickens with an out-of-date bacterial culture did not cause any clinical signs (schematic). Pasteur instructed his assistant Charles Chamberland to infect the chickens with a culture of pathogenic bacteria and left for a vacation. One month later when Pasteur returned from vacation, his assistant infected the chickens with an out-of-date culture of bacteria that had been in place for a month and caused no disease in inoculated chickens. Pasteur was surprised when the chickens remained unharmed after infection with a fresh-culture bacteria again, a phenomenon that attracted his attention (40 × 50 canvas, titled "Pasteur's Surprise")

inoculated the chickens with the out-of-date bacterial culture that had been sitting there for a month, and all the chickens inoculated with the old bacterial culture survived healthy (see Fig. 1.2). The assistant thought he had made a serious mistake and wanted to discard the rest of the old bacterial cultures, while Pasteur stopped the assistant and kept the cultures. Considering that the old bacterial cultures might fail, Pasteur prepared freshly cultured pathogenic bacteria and infected the chickens with them. Because of the limited number of chickens in the test, he had to use chickens left over from the previous test. To his surprise, those chickens that had previously been inoculated with the old bacterial culture remained unharmed, a phenomenon that caught his attention. Further replications of the test confirmed that inoculation of chickens with the old pathogenic culture indeed generated resistance against the pathogen. Pasteur proposed that the old bacterial culture had become so avirulent that inoculation of chickens caused only mild symptoms, followed by rapid recovery, and that the recovered chickens would resist against the challenge of virulent bacteria. He called this attenuated strain "vaccine", a word originating from the Latin vacca, meaning "cow", in honor of Jenner's outstanding contribution to the prevention of smallpox by inoculation with cowpox.

Pasteur extended these findings to research on the prevention and control of other epidemics. In 1881, in Pouilly-le-Fort, France, Pasteur conducted a classic vaccination and challenge experiment: 48 sheep, 2 horses, and 10 cattle were divided half into two groups. One group with half number of animals was immunized with attenuated Bacillus anthracis vaccine, followed by a booster 12 days later and a booster two weeks later, and the other were used as unimmunized controls. Two weeks after the last immunization, all animals were challenged with virulent Bacillus anthracis. All animals in the unimmunized control group died, while all animals in the immunized group survived in good health. This was the first time in human history that humans employed intelligence to attenuate the virulence of a pathogen and use the attenuated pathogen for the prevention and control of an infectious animal disease. The success of this experiment marked the birth of Immunology. In addition, Pasteur developed a rabies vaccine by serial passages of rabies virus in rabbits. In 1885, Pasteur had the audacity to immunize Joseph Meister, a 9-year-old boy who had been severely bitten by a rabid dog, against rabies. At that time, as Pasteur was a scientist rather than a licensed doctor, he risked being prosecuted for the treatment of Joseph with the vaccine. However, if Joseph was not immunized, he would surely die. After serious consideration, Pasteur bravely immunized Joseph with the rabies vaccine he developed. Fortunately, Joseph survived and the treatment was very successful. Pasteur was hailed as a "hero" for his success in saving Joseph's life, and won acclaim and generous donations from the community, which led to the birth of the Pasteur Institute. After Pasteur passed away (1895), his body was kept in a secret chamber in the basement of the Pasteur Institute, which Joseph always guarded. During the Second World War, when Hitler attacked France, Joseph was the only one who insisted on staying behind to guard the Pasteur Institute. When the Germans tried to enter the chamber, Joseph committed suicide at the entrance to the chamber to show his loyalty, and Joseph's courage and sense of duty were equally celebrated.

Although Pasteur achieved a great success in developing and using vaccines, the rationale was not understood at the time. In 1883, the Russian scientist Elie Metchnikoff observed that certain white blood cells (which he named phagocytes at the time) in the blood were capable of phagocytosis of pathogenic bacteria and foreign materials, and found that phagocytes in immunized animals were more active and had greater phagocytic activity than those in non-immunized animals, and subsequently proposed the theory of cell-mediated immunity, which was accepted by most people. However, in 1890, the German scientist Emil von Behring (a student of Robert Koch) and his assistant Shibasaburo Kitasato demonstrated that the serum from animals immunized with the diphtheria vaccine conferred stronger resistance upon recipient animals after being transferred, demonstrating for the first time that the factor determining resistance was present in the serum of the immunized animals. Based on this principle, he prepared diphtheria and tetanus toxin-positive sera and used them to successfully treat diphtheria and tetanus patients. These discoveries opened a new window to further unravel the immune mechanism. However, when Behring proposed the theory of humoral immunity, most people thought it was a fantasy and ridiculed it as "humors". "Humoral" was a pun at the time, standing for both the ancient Greek word "fluid" and "humor". Subsequently, many researchers found active components in serum that neutralized toxins, precipitated toxins, and agglutinated bacteria, naming the components in serum antitoxins, precipitins, and agglutinins, respectively, according to the role they played.

Robert Kock, a German physician with his own laboratory and a unique interest in the study of microorganisms, first discovered Bacillus anthracis as the causative agent of anthrax, subsequently found that a variety of epidemics were associated with the bacterium, proved that the causative agent of tuberculosis was Mycobacterium tuberculosis, and overturned the traditional notion that tuberculosis was a genetic disease. These findings marked the birth of medical microbiology. Thus, medical microbiology and immunology are considered to be twin brothers born at the same time. For his outstanding contributions to the study of immune mechanisms, Behring was awarded the Nobel Prize in Medicine in 1901, while the 1905 Nobel Prize in Medicine was awarded to his mentor, Robert Kock. Elie Michnikov was also awarded the Nobel Prize in Medicine in 1908 for his discovery of phagocytotic cells leading to cell-mediated immunity.

In the early twentieth century, however, the theories of humoral and cell-mediated immunity had been in opposition. The development of the cell-mediated immunity theory lagged relatively behind because of limitations in tissue cell culture techniques. While in 1939 Elvin Kabat immunized rabbits with OVA (ovalbumin) to demonstrate that the gamma-globulin (currently called immunoglobulin) in serum was responsible for immunity. As early as 1897, Paul Erlich called the active portion in the serum that neutralized the toxin as antibody(antikörper). Thus, it was clear that gamma-globulin in the serum was actually the antibody. In the 1940s, Merrill Chase transfused leukocytes from immunized guinea pigs into normal recipients, which gave the recipients immunity against tuberculosis. This experiment renewed hope for the theory of cell-mediated immunity, also called cellular immunity. In the 1950s, thanks to breakthroughs in cell culture technology, it was discovered that lymphocytes had both cellular and humoral immunity. Shortly afterward, in an experiment with Salmonella-infected chickens, Professor Bruce Glick of the University of Mississippi found that bursectomized chickens (with their bursa surgically removed) did not produce antibodies post-Salmonella infection, thus demonstrating that there were two main types of lymphocytes: T lymphocytes (T cells) derived from the thymus, which mediated cellular immunity, and B lymphocytes (B cells) derived from the bursa, which were involved in humoral immunity. These studies established that humoral and cellular immunities are two inseparable aspects of the function of the immune system. Thus, the contradiction between the theories of humoral and cellular immunities was completely resolved, and the two schools of thought shook hands and made peace.

1.2 The Formation of Immunological Theory and Its Development

In the early twentieth century, Jules Bordet of the Pasteur Institute demonstrated that animals could develop specific immunity to non-pathogenic heterologous substances by immunizing animals with red blood cells from other species of animals, which could then become immune against that red blood cell, and that this immunity could be passed on by serum transfer to homologous animals of different individuals. Kar Landsteiner showed that animals could be induced by injection with different organic compounds to produce specific antibodies against the different compounds, and found that antibodies could be specific against an infinite number of different kinds of antigens and that antibodies could distinguish between extremely subtle differences between antigenic molecules. Various speculations have been proposed to explain the antibody specificity. The representative theories are the selective theory and the instructional theory.

The earliest concept of the selective theory was proposed by Paul Ehrlich in 1900. He believed that cells in the blood express "side chain receptors" that bind and inactivate pathogens. Ehrlich believed that upon infection with a pathogen, the pathogen selects side chain receptors on the cell membrane and their binding is like "a key to a lock", where the pathogen selects the receptor and induces the cell to produce and release more side chain receptors with the same specificity. According to this hypothesis, the specificity of the side chain receptors exists prior to infection with the pathogen, and the antigen simply selects a specific receptor. The hypothesis is correct in many respects, except for one point, because the receptors mentioned exist in both the cell membrane bound and soluble forms, and the soluble receptors are not the released membrane receptors.

In the 1930s and 1940s, the selective theory was challenged by instructional theory. The instructional theory held that the antigen determined the specificity of the antibody and that the antigen served as a template, around which the antibody was formed by folding. Thus, the structure of antibody would have the configuration complementary to that antigen template. The instructional theory was originally proposed in 1930 by Friedrich Breinl together with Felix Haurowitz, and later it was further refined by Linus Pauling. By the 1960s, with the decryption of the structures of DNA, RNA, and proteins, it was discovered that the animals could produce antibodies against all compounds, and the instructional theory was completely abandoned and retired from history.

In the 1950s, the selective theory came back into favor as new research findings emerged. Niels Jerne, David Talmadge, and Macfarlane Burnet developed and refined the selective theory with super intelligence and insight, proposing the famous clonal selection theory. The core of this theory is that each lymphocyte expresses receptors that are specific for a distinct antigen, and the specificity of this receptor is determined before the lymphocyte is exposed to the antigen; the binding of antigen to the specific receptor activates the lymphocyte, and the activated lymphocyte proliferates into a clone of cells that have the same immunological specificity as the original (parental) lymphocyte. This theory has been further refined and developed, becoming one of the most fundamental core theories of modern immunology.

1.3 The Coming of the Era of Molecular Immunology

Since the 1950s, immunology has developed by leaps and bounds, both in theory and in practice, and has formed an independent and dynamic new discipline, deepening and improving the understanding of immunity. Animal immunity can be divided into innate and acquired immunities.

Innate immunity, the early defense ability of mammals to stop, inhibit, and eliminate pathogens, is the first line of defense against and destroying foreign pathogens. Innate immunity consists of four types of defensive barriers, including (1) anatomical barriers: also known as mechanical barriers, such as skin and mucous membranes; (2) physiological barriers: including temperature, low pH environments, and chemical mediators; (3) Phagocytic barriers: such as macrophages and neutrophils; and (4) inflammatory response barriers: such as antimicrobial active substances released by tissue damage and chemokines that attract phagocytes. Innate immunity is a physiological defense mechanism that animals are born with. Without innate immunity, animals simply cannot survive in their natural environment. Both lower and higher animals have innate immunity, but only higher animals have acquired immunity. Acquired immunity, also known as adaptive immunity, is the specific resistance generated in hosts in response to stimulation by antigens, and mainly consists of antibodies and specific T lymphocytes. Acquired immunity has four basic characteristics, including (1) antibody specificity, referring to the ability of the immune system to distinguish subtle differences between antigenic molecules, and antibodies can even distinguish the difference of one amino-acid molecule that exists in two molecules; (2) diversity, referring to the ability of the immune system to produce an endless number of antibody molecules with different specificities to recognize different molecular structures of antigens; (3) immune memory, referring to the immune system that, after primary immune response to antigen, has the ability to rapidly and powerfully respond to the same antigen again; (4) Recognition of self and non-self, under normal conditions, the immune system has tolerance to its own tissues and responds only to foreign (non-self) substances. If the tolerance breaks, the immune system will attack its own tissues, resulting in an autoimmune disease. Innate immunity and acquired immunity are not distinctly separate, and they are interrelated and mutually reinforcing to form the complete defense system of the organism. The deeper and finer mechanisms underlying these phenomena have always been the focus of immunological research.

The question that had puzzled immunologists for decades was why antibodies had such a vast diversity. Two theories had been proposed to explain this phenomenon, the germline theory and the somatic-variation theory. The germline theory maintained that the genome contains a large repertoire of immunoglobulin genes, which originated from germ cells such as eggs and sperm, and that a large repertoire of genes encoding immunoglobulins have been formed in the animal genome over a long period of evolutionary survival, i.e., all genes encoding antibodies were inherited from the parents and existed beforehand. In contrast, the somatic-variation theory maintained that the genome contains a relatively small number of immunoglobulin genes and that somatic cells generate a large number of specific antibodies through gene mutation and recombination. With the accumulation of data on the amino-acid sequence of immunoglobulins, a special phenomenon was discovered, namely, the constant region of immunoglobulins kept the amino-acid sequence constant, while the variable region varied greatly, and the diversity of immunoglobulins was mainly reflected in the variable region. The germline theory is difficult to explain why immunoglobulins have such a vast diversity in the variable region, which is composed of heavy and light chains, while the amino-acid sequence remains constant in the constant region. The somatic-variation theory is also difficult to explain why genes in the variable region mutate while the constant region remains unchanged. Even more puzzling, however, is that after amino-acid sequence analysis of the human myeloma protein, it was found that in some immunoglobulins, the variable region has the exact same amino-acid sequence, while the constant region has both γ and μ heavy chain types. A similar phenomenon was observed by Prof. Charles Todd of the California Institute of Technology when immunizing rabbits, where the sequence of the variable region that determines the specificity of the antibody remains the same for a particular antigen, but the constant region may have several heavy chains of α, γ or μ. It is entirely possible that antibodies of different isotypes (isotypes) may have the same sequence of the variable region, i.e., have the same antibody specificity.

In 1965, W. Dreyer and J. Bennett proposed the two-gene model hypothesis to explain the basic structure of immunoglobulins. The core of the model is that both the light and heavy chains are encoded by two separate genes, one gene encoding a variable region and the other a constant region. These two genes are joined together by some mechanism of recombination at the DNA level, forming a continuous DNA sequence that is transcribed and translated to form a light or heavy chain. Furthermore, there are tens of thousands of variable region genes in germ cells, while there is only one constant region gene. This recombination model explains why immunoglobulins exist with constant regions that are constant and variable regions that vary widely. However, the evidence supporting Dreyer and Bennett's experiments is insufficient, and although only one to two copies of the gene were found when the constant region genes of immunoglobulins were tested with nucleic acid probes, consistent with the predictions of the hypothesis, the results of these experiments have not yet led to the academic acceptance of the hypothesis because two genes encoding one protein violate the accepted principle of "one gene encoding one protein "and is unprecedented in any biological phenomenon. Science is often in the awkward position of not being able to understand a phenomenon, but new theories that can explain it cannot be proven because the experimental techniques and research methods fall behind, as was the case with immunoglobulin diversity research. Although the two-gene model hypothesis proposed by Dreyer and Bennett theoretically explains the relationship between immunoglobulin sequence and gene structure, experimental confirmation requires a breakthrough in experimental techniques in molecular biology.

In the field of molecular biology, the emergence of restriction endonucleases and DNA recombination technology provided the technical support to solve the long-debated issue of antibody diversity. In 1976, using this technology, Nobumichi Hozumi and Susumu Tonegawa of Basel Institute for Immunology, Switzerland, unraveled the mystery of antibody diversity, which brought immunology into the era of molecular immunology. Using isotope-labeled mRNA probe, they discovered that the genes encoding antibodies in the embryonic cell genome have multiple fragments, each was separated by other gene sequences, and that the variable region genes were separated from the constant region genes by a larger fragment containing an enzyme cleavage site, and that as embryonic cells differentiated into lymphocytes and reached the terminal plasma cell stage (antibody-producing myeloma cells), the variable region and constant region genes approach each other, and the genes inserted between them were eliminated, and both heavy and light chains follow this pattern. Further research by Susumu Tonegawa confirmed that the diversity of antibodies is mainly derived from gene recombination and somatic gene mutations. These findings confirmed the two-gene model theory proposed by Dreyer and Bennett that one gene encodes a variable region and the other a constant region. For his outstanding contributions to the study of antibody diversity, Susumu Tonegawa was awarded the Nobel Prize in Medicine in 1987.

1.4 Rapid Development of Molecular Immunology

Since the 1980s, techniques in molecular biology and molecular genetics have led to major breakthroughs in various aspects of immunology, and molecular immunology research has developed rapidly. With the introduction of the concepts of pathogen-associated molecular patterns (PAMPs) and pattern recognition molecules (PRMs) by Charles Janeway to the field of innate immunity research, especially in the late 1990s, Bruce Beutler discovered that the *LPS locus* in the mouse genome encodes the pattern recognition receptor TLR4, a receptor that recognizes bacterial LPS, opening up a new field of innate immune recognition of pathogens to generate an immune response. With the discovery of more pattern recognition receptors, innate immunity research is rapidly changing, and six major families of pattern recognition receptors (Toll-like receptors (TLRs), Nucleotide-binding oligomerization domain (NOD)-like receptors (NLRs), Retinoid acid-inducible gene I (RIG-I)-like receptors (RLRs), C-type lectin receptors (RLRs), Aabsent in melanoma 2(AIM2)-like receptors (ALRs), and cyclic GMP-AMP synthase (cGAS)) have been identified. The discovery of these receptors has unveiled the molecular mechanisms of recognition and activation of innate immune cells, and of generating immune responses. Bruce Butler was awarded the 2011 Nobel Prize in Medicine for his outstanding contributions to innate immunity. The discovery or re-recognition of new lymphocyte subpopulations, NKT cells, ILCs(innate lymphoid cells), Treg (regulatory T cells), Th17, and γδ T cells has brought molecular immunology research into a phase of rapid development and has become an important core component of life science and molecular medicine research.

Immune response and the underlying molecular mechanisms have always been at the forefront of immunological research. For many years, scientists have conducted in-depth studies on innate immune cells, antigen receptors of TB lymphocytes, antigen processing and presentation, immune recognition, immune cell activation, immune signal transduction, and immune tolerance. The application of monoclonal antibody technology and molecular biology techniques in the study of the structure and function of the cluster of differentiated (CD) antigens in TB lymphocytes and other immune cells has led to leaps and bounds in immune response research. More than 370 lymphocyte CD antigens have been identified and named so far, and their roles and mechanisms of action in the immune response are being investigated in depth.

Tumor-targeted immunotherapy has always caught public attention. In the 1990s, James P. Allison and Tasuku Honjo discovered an immune checkpoint (CTLA4/PD-1) that blocks the activation of T cells. Blockage of the immune checkpoint by monoclonal antibodies could restore the killing activity of effector T cells and promoted the efficacy of tumor therapy, opening up new pathways in immunotherapy for cancer, and James Allison and Honjo were jointly awarded the Nobel Prize in Medicine in 2018. In recent years, chimeric antigen receptor T-cell therapy (CAR-T) using CD19 as a target has flourished, and Carl June of the University of Pennsylvania School of Medicine successfully cured patients with acute B lymphoblastic leukemia, which set off a revolution in tumor immunotherapy and held a great promise for the cure of cancer. CAR-T technology, which uses specific tumor cell-associated antigens as therapeutic targets, is the most ingenious and successful application of molecular biology combined with immunology in the medical field and would have broad clinical applications. Similarly, CAR-NK, CAR-iPS, and so on technologies are being evaluated in clinical trials for tumor therapy.

Molecular immunological studies have shown that there is a complex regulatory network of immune responses within the immune system of animals and that the regulatory network between the neural, endocrine and immune systems plays a very important role in regulating the immune response of the organism. However, the main function of the immune system is to maintain its own stability (homeostasis) through the immune response, and the ultimate task is performed by macrophages, which maintain the balance of the body's internal environment and its own stability through phagocytosis and clearance of unwanted objects or materials (such as senescent or dead cells, foreign materials, antigen-antibody complexes, and so on.). However, each cell is a relatively independent individual, and the cell itself needs to maintain its own stability to ensure normal physiological functions. Autophagy is an important biological process by which cells remove abnormal or unwanted components (self-aggregated, abnormal or misfolding protein molecules, damaged or senescent organelles, pathogenic components, etc.) to maintain their own stability. Yoshinori Ohsumi won the 2016 Nobel Prize in Medicine for his discovery of the mechanisms for autophagy.

A tremendous progress has been made in antigen-antibody labeling techniques in the past 20 years, which generated a variety of efficient and sensitive detection techniques and methods, such as immunofluorescence, immunoenzymes, radioimmunoassay, immunoblotting techniques, immunoprecipitation techniques, chemiluminescence immunoassays, flow cytometry, immuno-PCR techniques, nano-antibody techniques, etc. These techniques and methods have been widely used in various research fields of life science and molecular medicine. In vaccine development, DNA recombinant technology and genetic engineering technology have created a brand new way for human and animal vaccine research, and various types of genetic engineering vaccines, such as genetically engineered subunit vaccines, genetically engineered recombinant live vector vaccines, gene deletion vaccines, DNA vaccines, etc., have been developed one after another. The development of new types of adjuvants has significantly improved the immunizing efficacy of vaccines. The development of protein synthesis technology has also opened up a new avenue for vaccine development, namely, the preparation of vaccines by synthetic peptides.

1.5 Characteristics of Veterinary Molecular Immunology and Its Applications

Immunology originated from the study of animal vaccines, but as research progressed, the study of human immunology using experimental animals (e.g., mice) as models far surpassed the level of immunological research on livestock. Although Veterinary immunology is primarily concerned with the immune control of infectious diseases in animals, many concepts and theories draw on human medical research findings to follow progress in immunology. Veterinary immunology research itself has characteristics and unique advantages in comparative immunological research, however, some phenomena are very difficult to explain by human medical immunological theories because of the differences in the composition of the immune system in different species of animals. With the progress of science and technology, Veterinary molecular immunology has also ushered in a period of rapid and vigorous development.

1.5.1 Prevention and Control of Animal Infectious Diseases by Vaccination

Prevention and control of animal infectious diseases is the main task of veterinary workers, and vaccination is one of the most important means of veterinary immunology to prevent and control animal infectious diseases. The eradication of some infectious diseases mainly relies on vaccines, such as the application of lapinized rinderpest vaccine to eradicate rinderpest in China in 1950s, and the prevention and control of infectious diseases by immunizing various types of animals. Thus, immunization with vaccines has been an important means of controlling infectious diseases in livestock and poultry. Traditional live attenuated vaccines and inactivated vaccines still play extremely important roles, and certain genetic engineering vaccines, such as porcine pseudorabies gene-deficient vaccine, have been applied to clinical practice. In recent years, reverse genetic technology has become an important tool for the efficient development of genetically engineered vaccines. However, to make full use of this technology to optimize or modify vaccines, the mechanisms of the pathogenic infection must be understood in advance, and only by elucidation of the mechanism of the pathogenic infection can the pathogen be modified to become an ideal vaccine candidate, especially those viruses that directly attack immune cells, such as African swine fever virus(ASFV), porcine reproductive and respiratory syndrome virus(PRRSV), and infectious bursal disease virus (IBDV). If the pathogenesis of the viral infection is not well understood, it is difficult to develop an ideal vaccine by traditional methods. In addition, passive immunization of diseased animal herds or avian flocks or individuals via injection with anti-serum or yolk antibodies has also achieved remarkable successes in the control or treatment of some infectious diseases, such as goose parvovirus infection, duck viral hepatitis, chicken infectious bursal disease, and some other viral diseases. Treatment of infectious diseases in animals by passive immunization early after infection can achieve better results. In addition, the new-born animals (or hatched chicks) can obtain natural passive immunization by acquiring maternally derived antibodies from colostrum (or egg yolk).

1.5.2 Diagnosis of Animal Infectious Diseases

The binding of Antibodies to the corresponding antigens is characterized by high sensitivity and specificity, and various serological and cellular immunoassay techniques have been established based on this feature. These techniques have been widely used in the diagnosis and monitoring of animal infectious and parasitic diseases. In addition, serological techniques allow serological typing and identification of newly isolated pathogens, as well as evaluation of clinical efficacy of vaccines.

1.5.3 Applications in Agriculture and Related Disciplines

Immunological technology is widely used in agriculture, food, and related disciplines due to its high sensitivity and specificity, such as the detection of harmful substances in food, trace active substances, pesticide, and veterinary drug residues, etc. It has greatly promoted the advances in agriculture, food, and related disciplines. In addition, the intersection of veterinary molecular immunology with multiple disciplines has led to the formation of many sub-disciplines and new research areas.

In conclusion, veterinary molecular immunology is flourishing and will certainly become one of the most important components of life sciences as research progresses.

Study Questions

1. When was the earliest time that immunization against smallpox appeared in China?
2. Which three vaccines were developed by Pasteur?
3. Where did the concept of antibodies come from?
4. What are the main contents of the theory of clonal selection?
5. How has antibody diversity been demonstrated? What are the implications of unraveling antibody diversity for the development of immunology?
6. Give examples of the main applications of veterinary immunology.

Chapter 2
The Immune System

Overview:
The mammalian immune system is a defensive system of the body that has evolved over time and includes immune organs, immune cells, and immune molecules. The various elements of the immune system work exquisitely together to provide for the normal physiological functions of the body by means of immune response.

2.1 Immune Organs

The immune organs of animals are composed of central immune organs, peripheral immune organs, and tertiary lymphoid tissues. The central immune organs, also known as primary immune organs, are the sites of lymphocyte development and maturation, including the thymus and bone marrow. The avian bursa of Fabricius is the primary site of B-lymphocyte maturation, serving as an important central immune organ. In addition, the liver, during fetal life, also functions as a central immune organ, but this function gradually fades after birth. Peripheral immune organs, also known as secondary immune organs, are the sites where mature lymphocytes reside, capture antigens, and produce immune responses, including the spleen, lymph nodes, and mucosa-associated lymphoid tissue. Tertiary lymphoid tissue refers to the diffusely distributed lymphocytes in tissues, usually residing under the skin, whose primary role is to participate in the inflammatory response and secrete cytokines and chemokines (Table 2.1; Fig. 2.1).

Hematopoiesis All cells in the blood are differentiated from hematopoietic stem cells (HSC). Stem cells can renew themselves by dividing and thus maintain a constant number. The formation and development of blood cells begins in the yolk sac during the first week of human embryonic development. In the embryo yolk sac, stem cells differentiate into primitive erythroid cells containing hemoglobin. By the

S. J. Zheng, *Veterinary Molecular Immunology*,
https://doi.org/10.1007/978-981-99-8929-4_2

Table 2.1 The immune organs and tissues, their distributions and functions

The immune organs and tissues	Distribution	Function
The central immune organs (also known as primary immune organs)	The thymus, bone marrow, bursa of Fabricius(avian)	The sites of lymphocyte development and maturation
Peripheral immune organs (also known as secondary immune organs)	The spleen, lymph nodes, and mucosa-associated lymphoid tissue)	The sites where mature lymphocytes reside, capture antigens, and produce immune responses
Tertiary lymphoid tissue	The diffusely distributed lymphocytes in tissues, usually residing under the skin	Participate in the inflammatory response and secrete cytokines and chemokines

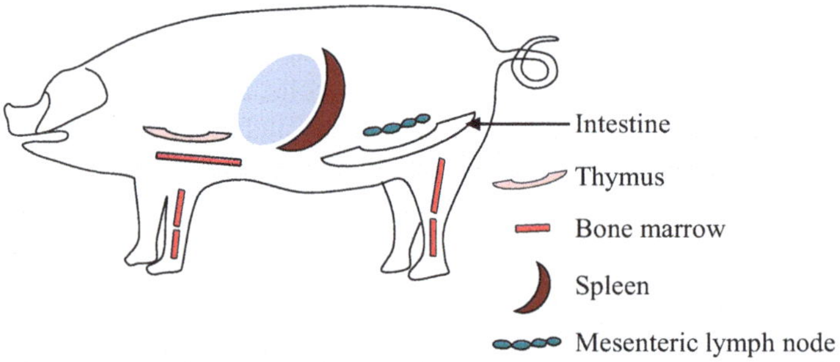

Fig. 2.1 Schematic diagram of the distribution of thymus, bone marrow, spleen, and mesenteric lymph nodes in pigs

third month of gestation, hematopoietic stem cells migrate from the yolk sac to the fetal liver and then to the spleen. During the third to seventh months of gestation, the liver and spleen are the major hematopoietic organs. As the hematopoietic stem cells differentiate, bone marrow gradually becomes the major hematopoietic organ. By the time of birth, the liver and spleen have largely ceased to produce blood.

Hematopoietic stem cells are pluripotent stem cells capable of differentiating in a variety of ways to produce red blood cells, granulocytes, monocytes, mast cells, lymphocytes, megakaryocytes, etc. The percentage of hematopoietic stem cells in the bone marrow is very small, usually 1 in 50,000. Because stem cells are self-renewing, their number remains essentially stable throughout the animal's life. However, when large amount of blood is required, hematopoietic stem cells proliferate rapidly and exhibit a tremendous capacity for proliferation and differentiation.

For a pluripotent stem cell, differentiation can occur early in hematopoiesis along one of the two differentiation pathways, i.e., either giving rise to lymphoid progenitor cells or myeloid progenitor cells (Fig. 2.2).

Hematopoietic stem cells (HSCs) divide and proliferate in the bone marrow, with half of the cells renewing themselves to maintain a constant number and the other half further proliferating and differentiating into multipotent progenitor cells, which

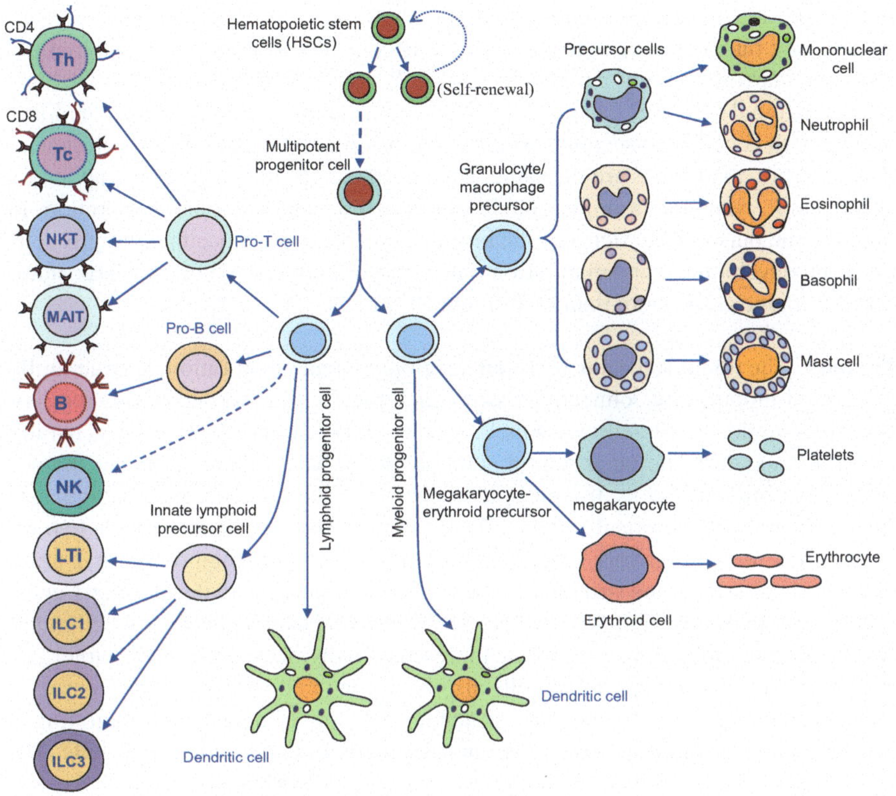

Fig. 2.2 Schematic diagram of formation and development of blood cells

further divide and proliferate into lymphoid progenitor cells and myeloid progenitor cells. The lymphoid progenitor cells divide and proliferate into pro-T cells, pro-B cells, NK cells, and innate lymphoid precursor cells, which further proliferate and differentiate into terminal cells. Myeloid progenitor cells further divide, proliferate and differentiate into granulocyte/macrophage precursor cells as well as megakaryocyte/erythroid precursor cells, which further proliferate/differentiate into different precursor cells as well as terminal cells (as shown in the figure). Conversely, dendritic cells can develop and differentiate from both lymphoid and myeloid progenitor cells.

Central Immune Organ The central immune organ, also known as the primary lymphoid organ, is the site of lymphocyte development and maturation, including the thymus and bone marrow. In birds, the bursa of Fabricius, the main site of B-lymphocyte development and maturation, serves as the central immune organ.

Bone Marrow or the bursa of Fabricius (bird) The bone marrow is the major hematopoietic organ of animals in the late embryonic and postnatal stage and is the site of origin of all blood cells, as well as the site of development, differentiation, and maturation of many immune cells. In the vast majority of animals, such as humans and rats, the bone marrow is where B cells are born, develop, and mature. The bone marrow contains cells important for B-cell development, such as the bone marrow stromal cells (see Chap. 5, Fig. 5.1). Bone marrow stromal cells are in direct contact with B cells and secrete various cytokines required for B-cell development. However, not all animals are the same, with some exceptions, such as the appendix of the rabbit, where is an important site for B-cell development and maturation. In contrast, the bursa of Fabricius of the bird is an important site accumulating B cells for B cell development and maturation. The bursa of Fabricius contains a large number of immune cells, more than 10 billion (10^{10}) cells.

Thymus The thymus, the site of T-cell development and maturation, is structurally divided into lobules by connective tissue. Each lobule is composed of two regions. The outer region is the cortex, which is dense with immature T cells, generally called thymocytes. The inner region is the medulla, where there are fewer thymocytes. In addition to T cells, thymic epithelial cells, dendritic cells, and macrophages are distributed in the cortical and medullary regions. These cells intermingle to form a three-dimensional network space that provides a developmental room for T cells. In the cortical area, some stromal cells are in direct contact with developing T cells, called nurse cells (see Fig. 4.1 in Chap. 4). Nurse cells usually have long cell membrane synapses and are surrounded by more than 50 thymocytes, forming larger cell clusters. Other thymic epithelial cells also have long cell synapses that cross over each other to form a network that can come into direct contact with migrating T cells and play an important role in the development and maturation of T cells. The thymus grows progressively with age, reaching its maximum size by puberty, and then gradually shrinks.

Peripheral Immune Organs Peripheral immune organs, also known as secondary immune organs, peripheral lymphoid organs, or secondary lymphoid organs, include lymph nodes, spleen, and various mucosal-associated lymphoid tissues (MALT), which is the site of antigen presentation and immune response. The gut-associated lymphoid tissue (GALT) also belongs to the peripheral lymphoid organs. Morphologically, lymph nodes can be roughly divided into three regions, namely the cortical, paracortical, and medullary regions. The outermost zone is the cortical zone, which contains lymphocytes (mainly B cells), macrophages, and follicular dendritic cells. Below the cortical zone is the paracortical zone, which is mainly populated by T cells, but also contains interdigitating dendritic cells. This area is also known as the thymus-dependent area because of the minimal number of cells in the paracortical area of the lymph nodes after thymic removal in mice. Correspondingly, the cortical area is also called the thymus-independent area. The innermost part of the lymph node is the medulla, with scattered lymphocytes, including antibody-secreting plasma cells.

When antigen carried by lymph fluid enters the lymph nodes, interdigitating dendritic cells and B cells in the paracortical area captures the antigen. The captured antigen is processed and presented by MHC class II molecules to helper T cells (Th), causing the activation of Th cells. Once Th and B cells are activated, the proliferating B cells form small follicles at the edge of the paracortical area. The B cells in the follicles differentiate into plasma cells, which secrete IgM and IgG. 4–6 days after antigen immunization, this plasma cell-containing lymphoid follicle reaches its maximum size.

Tertiary Lymphoid Tissue Tertiary lymphoid tissue is primarily the tissue with scattered distribution of lymphoid cells, which, in general, reside under the skin. The role of these lymphocytes is to participate in inflammatory response and secret cytokines and chemokines etc. The keratinocytes of the skin can secrete large amounts of cytokines causing local inflammatory responses. In addition, keratinocytes are induced to express MHC-II-like molecules, which play a role in antigen presentation. Intradermal Langerhans cells are also a kind of dendritic cells that travel to local lymph nodes after capturing antigens and differentiate into interdigitating dendritic cells that highly express MHC-II molecules and strongly activate Th cells. The majority of lymphocytes within the dermis are γδ T cells (expressing the antigen receptor TCR) and innate lymphocytes (not expressing TCR). These local lymphocytes act primarily against skin-invading microorganisms. The scattered CD4 and CD8 T cells in the subcutaneous layer are mainly already activated T cells (CTL) or memory cells.

2.2 Immune Cells

Immune cells refer to all cells involved in the immune response, including mainly lymphoid and myeloid cells. Lymphoid cells are the core cells of the immune system, not only including conventional lymphocytes (T and B cells) and unconventional lymphocytes (NKT, γδ T, B1, and MAIT cells), but also innate lymphocytes (NK, LTi, ILC1, ILC2, and ILC3 cells). T and B cells are mainly responsible for the acquired immune response and possess the attributes of diversity, specificity, memory, and self/non-self-recognition, which are the hallmarks of an acquired immune response. Myeloid cells are a population of non-lymphocytes differentiated from myeloid precursor cells, including monocytes, neutrophils or heterophil (in bird), eosinophils, basophils, megakaryocytes, and erythroid precursor cells in the blood. The main functions of these cells are to phagocytose and destroy microorganisms, present antigens, and secrete cytokines.

2.2.1 Lymphocytes

Lymphocytes make up 20–40% of the total number of white blood cells in the body. There are 10^{10}–10^{12} (average 10^{11}) lymphocytes in the human body. Lymphocytes carried within the blood or lymphatic fluid circulate in the body and move in and out of lymph nodes and local tissues to perform immune functions. Based on their functions and cell membrane composition, lymphocytes can be broadly classified as B cells, T cells, NK cells, NKT cells, γδ T, MALT, and ILCs. The surface antigens of lymphocytes are expressed during the process of cell differentiation and are thus called differentiation antigens. To identify and distinguish the different stages of lymphocyte development, specific antigens on the surface of the cell membrane can be detected by specific monoclonal antibodies. All monoclonal antibodies that bind to the same specific antigenic epitope are grouped together as a cluster of differentiation (CD) to define that antigen. The differentiation antigen of a lymphocyte is usually expressed as the cluster of differentiation of the antibody to which it binds and is therefore also referred to as the CD antigen. Each newly prepared monoclonal antibody is analyzed to determine if the antigen to which it binds is a known CD antigen. If it is not within a known CD antigen, the antigen to which the antibody binds should be named after the new CD. For example, CD3 is a molecule related to the T-cell receptor (TCR) complex, and anti-CD3 is a monoclonal antibody that binds to the CD3 antigen, sometimes written as αCD3. Currently, CD antigens have been named up to CD371. As immunology advances, the number of CD antigens will continuously increase.

B Cell The "B" in B cell is derived from the bursa of Fabricius. Avian B cells develop and mature in the bursa of Fabricius. For mammals, the bone marrow is the site of B-cell development and maturation, so it is appropriate to say that the "B" comes from the initials of the bone marrow. Mature B cells contain antigen receptors on their cell membranes; hence, the B-cell receptor is abbreviated for BCR for simplicity, which serves to recognize antigens (Fig. 2.3). In addition to the antigen-recognizing BCR molecules on the surface of B cells, there are molecules such as B220, MHC-II, CR1 (CD35), CR2 (CD21), FcγRII (CD32), B7-1 (CD80), B7-2 (CD86), and CD40. These molecules are associated with the function of B cells. Recent studies have revealed some differences in the development and function of B cells early in animal embryonic stage and after birth. Those B cells differentiated, developed, and matured from liver hematopoietic stem cells early during the embryonic stage are designated B1 cells. As the embryo develops, the bone marrow gradually replaces the liver in performing hematopoietic function, and B cells

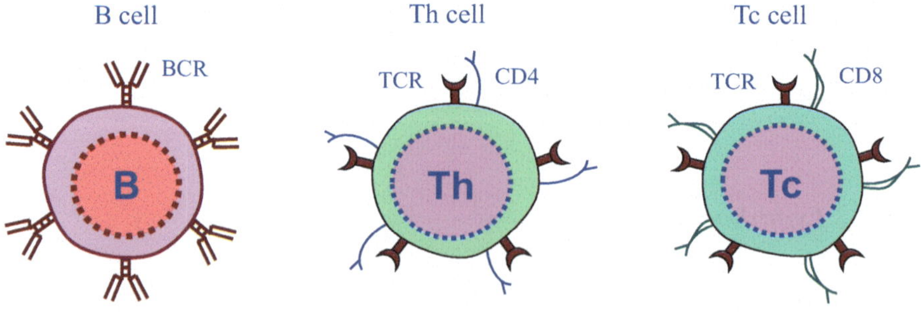

Fig. 2.3 B cells, T cells, and surface antigen receptors

differentiated, developed, and matured from bone marrow hematopoietic stem cells are called B2 cells. B1 cells do not require antigenic stimulation to produce natural antibodies to protect the fetus, and the main antibody types produced by B1 cells are IgM and IgG3, so B1 cells also belong to innate like lymphocytes, which are further divided into two subpopulations, B1a ($CD5^+$) and B1b ($CD5^-$), based on whether B1 cells express CD5. As an embryo develops and matures, liver hematopoiesis is replaced by bone marrow, and B1 cell development gradually ceases, but B1 cells continue to renew themselves in response to their own antigenic stimulation, retaining a certain number after birth and even throughout life, and acting primarily as antigen-presenting cells. B2 cells are conventional adaptive immune response B cells that require continuous differentiation and generation from bone marrow hematopoietic stem cells, which are continuously delivered to peripheral immune organs for accumulation, and participate in humoral immune response against specific antigens.

T Cells The "T" comes from the initial letter of thymus because T cells develop and mature in the thymus and have an antigen-binding receptor on their surface, called the T-cell receptor (TCR). TCR possesses the attribute of specificity, binding only to specific antigens. However, unlike the B-cell membrane receptor BCR, the TCR does not bind to free antigens and recognizes only antigens bound to its own MHC molecules. In other words, TCR of T cell recognizes only antigens presented by its own MHC. The TCR belongs to a polypeptide complex that includes CD3. Most T cells contain CD4 or CD8 membrane molecules (Fig. 2.3). In addition, mature T cells express molecules such as CD28 and CD45. T cells expressing CD4 recognize only antigens presented by MHC class II molecules, whereas CD8 T cells recognize only antigens presented by MHC class I. CD4 T cells are generally helper T cells (Th), whereas CD8 T cells are cytotoxic T cells (Tc). Th cells are activated by recognition of antigens presented by MHC-II class molecules. Activated Th cells begin to divide and proliferate, becoming effector cells. These activated Th cells secrete various cytokines that are directly involved in regulating the immune response. When cytokines change, the immune response changes as well. For example, the Th1 immune response produces cytokines that mainly promote inflammatory response, activate T cells and macrophages. Th2 immune response mainly activates B cells and promotes humoral immune response (producing antibodies). Th17 cells secrete IL-17 that promotes inflammatory response. Upon activation, Tc cells further differentiate and proliferate into effector cells called cytotoxic T lymphocytes (CTL). In contrast to Th cells, CTL cannot help B cells to secrete antibodies, but are able to kill cells infected by pathogens or abnormal cells that have undergone mutations.

NK Cells (Natural Killer Cells) NK cells are innate lymphocytes, large in size, containing granules, and without the typical surface markers of T or B cells. NK cells do not require antigen stimulation to kill target cells and are therefore called natural killer cells. Recent studies have shown that NK cells have a variety of roles, including antitumor, anti-virus, immunomodulation, and tissue growth promotion. Human NK cells make up 5–10% of the total number of lymphocytes in the peripheral blood. Although NK cells do not have antigen recognition receptors like TCR or BCR, they can recognize target cells in two ways: (a) NK cells can recognize target cells by recognizing reduced expression of MHC molecules on the surface of tumor cells or virus-infected cells or by expressing specific antigens; (b) NK cells recognize target cells by antibodies directed against tumor or virus-infected cells. According to the cell surface markers CD56 and CD16, NK cells are mainly divided into two subpopulations, namely: $CD56^{dim}CD16^{+}NK$ and $CD56^{bright}CD16^{-}NK$ cells. $CD56^{dim}CD16^{+}NK$ cells express CD16, which is the carboxy-terminal receptor of the Fc fragment (Fc receptor) of the IgG molecule. The process, by which NK cell with Fc receptor (CD16) that recognizes bound antibody on a target cell and subsequently causes lysis of the target cells, is called antibody-dependent cell-mediated cytotoxicity (ADCC), thus $CD56^{dim}CD16^{+}NK$ has antitumor and antiviral cytotoxic effects. In contrast, $CD56^{bright}CD16^{-}NK$ cells exert immunomodulatory effects and promote tissue growth (e.g., placental development during pregnancy) mainly through the secretion of cytokines (IFN-γ, TNF-α, GM-CSF, etc.) and chemokines (CCL3, CCL4, CCL5, etc.). In different local tissues (e.g., lung, uterus), NK cells may play completely different roles.

NK cells cause lysis of the target cells in two ways: (a) by releasing perforin and granzyme, and (b) by inducing apoptosis through membrane-bound receptors, i.e., by binding of FASL and TRAIL ligands on NK cell membranes to Fas and apoptotic receptors (DRs) on target cell membranes. The way in which NK cells distinguish between itself healthy cells and target cells (infected and tumor cells) that need to be killed has always been a focus of hot topic. Current findings suggest that whether NK cells exert killing effects is mainly related to activating receptor and inhibitory receptor-mediated signaling. These receptors include NKG2 family receptors, killer cell immunoglobulin-like receptors (KIRs), natural killer cytotoxic receptors (NKp30, NKp44, NKp46), cytokine receptors, TLRs, and adhesion molecules, which work together to regulate the killing function of NK cells. In recent years, with the successful application of CAR-T in oncologic therapy and the discovered side effects, CAR-NK technology has received great attention and is currently in the trial stage of cancer immunotherapy.

NKT Cells (Natural Killer T Cells) NKT cells are a new cell type discovered in recent years and are characterized by cells containing CD161 (CD161c also known as NK1.1) and the NKR-P1 surface marker as well as T-cell receptor (TCR). NKT cells differ from T cells in that their TCR does not bind to MHC molecules but to a molecule CD1, which is structurally similar to MHC. MHC primarily binds and presents protein-based antigens, whereas CD1 binds and presents glycolipid or lipid-based antigens. NKT cells specifically recognize APC-processed lipid antigen that binds to CD1 (e.g., cell wall components of certain gram-negative bacteria, endogenous lipids, etc.). Notably, α-GalCer, a glycolipid originally derived from a marine sponge, has a strong activating effect on NKT cells. α-GalCer is currently used as an important stimulant in the study of NKT cell function. Another difference between NKT cells and T cells, in addition to TCR-binding CD1 molecules, is that NKT cells have no immune memory and are therefore considered to be innate immune cells. NKT cells rapidly produce Th1 (IFN-γ) and Th2 (IL-4, IL-13) cytokines upon exposure to antigen, which regulate the immune response. Thus, NKT cells act as a bridge for the transition from innate to acquired immune response.

The most important feature of NKT cells is their ability to recognize lipid antigens presented by CD1 molecules. Compared with the TCR of conventional T cells, the TCR variable region of NKT cells is less diverse. The TCR α-chain of mouse NKT cells is more constant, mainly Vα14-Jα18, while the β-chain is mainly Vβ8.2, Vβ2, Vβ7. In human, the TCR α-chain of NKT cells is mainly Vα24-Jα18, while the β chain is mainly Vβ11. The distribution of NKT cells in vivo is relatively special, for example, the proportion of NKT cells in blood, bone marrow, thymus, spleen, and lymph nodes in mice is generally less than 1%, while in the liver they account for 20–40% of the total number of lymphocytes. So, what is the physiological significance of the enrichment of NKT cells in the liver? This question has not been satisfactorily addressed and needs to be studied in depth. However, if the Vα24/Vβ11 surface markers are used to define human NKT cells, NKT cells account for less than 1% of total lymphocytes in the human liver. It is still unclear whether this difference in NKT cells between mice and humans stems from species differences or whether the Vα24/Vα11 surface markers are inappropriate to define human NKT cells.

Based on advances in the study of NKT cells, Wingender and Kronenberg (2008) divided NKT cells into five subpopulations: (a) íNKT cells (í for invariant): this subpopulation of NKT cells expresses a constant Vα chain, Vα14í in mice and Vα24í in humans and is the classical NKT cell; (b) mNKT cells (m for mucosal): this subpopulation of NKT cells also expresses a constant Vα chain, Vα7.2í in mice and Vα19í in humans, and are associated with mucosal tissue; (c) vNKT cells (v for variant): this subpopulation of NKT cells can bind to CD1 molecules in complex with antigen, also called variant (variant) NKT cells; (d) χNKT or tgNKT cells: this subpopulation refers to those T cells that are not classified but are capable of expressing the NK cell receptor; (e) NK cell expressing receptor γδ TCR: this group of NKT cells is able to recognize and bind to antigens presented by CD1a, CD1b, and CD1c molecules. Obviously, according to this classification, using any one NKT cell subpopulation as a representative does not reflect the overall picture of NKT cells. Furthermore, it is still unclear how much the proportion of these NKT cell subpopulations varies between different species of animals or different individuals in humans. Although there is no uniform cell surface markers to represent all NKT cells, the most distinctive feature of NKT cells is their recognition of antigens presented by CD1 molecules, and therefore tetramers prepared as α-GalCer-CD1 are the primary means used to detect NKT cells.

Experimental evidence suggests that NKT cells are important immunomodulatory cells. Upon activation, NKT cells rapidly secrete a large number of cytokines, including both Th1 cytokines (IFN-γ) and Th2 cytokines (IL-4, IL-13) as well as Th17 cytokines (IL-17, IL-22), with all three types of cytokines sometimes produced simultaneously and sometimes with different preferences. NKT cells are divided into three subpopulations based on the cytokines produced and the site of residence, including: NKT1, which resides in the liver and spleen and expresses IFN-γ and IL-4; NKT2, which resides in the lung and expresses IL-4 and IL-13; and NKT17, which resides in the lymph nodes and expresses IL-17 and IL-22. NKT cells regulate the direction of the immune response by secreting these cytokines. NKT cells can be activated via the TCR binding to lipid antigens presented by CD1 molecules or can be directly activated in the presence of IL-12 and IL-18 without the TCR stimulation in in vitro assays. Since activated NKT cells in in vitro trials secrete large amounts of IFN-γ, which plays an important role in the antitumor process, it has given hope to clinical tumor treatment by activating NKT cells in tumor patients. To date, after several clinical trials, it has been found that the results obtained from clinical trials either by direct injection with NKT cell antigen (α-GalCer) or by injection with APC containing α-GalCer are quite different from what was expected, and the hope for treatment of tumors with activated NKT cells is difficult to reach. The results of these trials give good insight to biomedical practitioners that although the results obtained with animal models (e.g., NKT cell-deficient $V\alpha 14^{-/-}$ or $J\alpha 18^{-/-}$ mice) indicate that NKT cells play a very important role in anti-tumor immunity, it is difficult to simply reproduce the results obtained in experimental animals in the clinical treatment of tumor in patients. Thus, it is much more complicated to study on clinical immunotherapy in human than that of doing experiments merely with laboratory animals. In addition to their role in tumor immunity, NKT cells also play an important role in the autoimmune inflammation and in anti-defense response, especially in the early stages of pathogenic infection, where activated NKT cell regulates not only the innate immune response by secreting cytokines (see Fig. 3.1 in Chap. 3 for details) but also has important effects on adaptive immunity.

γδ T Lymphocyte γδ T cells are a type of T lymphocyte whose TCR complex is composed of two chains of γδ and CD3 molecules. Similar to NKT cells, γδ T cell TCRs recognize glycolipid antigens presented by CD1 molecules. γδ T cells are found in a small proportion of blood, lymphoid tissues and organs, accounting for less than 5% of spleen lymphocytes in mice and 1–10% of lymphocytes in human blood. γδ T cells accumulate mainly in the liver, skin, and mucosal epithelium (accumulating in the digestive tract, the genitourinary tract, lung, and other tissues). γδ T cells account for 10–20% of lymphocytes in mouse liver and 5–10% in human liver. γδ T cells develop and mature in the thymus, and naive γδ T cells develop into IL-17+ γδ T cells that produce IL-17. However, if the TCR binds to a CD1-presented antigen or the cell membrane CD70 receptor binds to a ligand during development, γδ T cells will develop into IFN-γ + γδ T cells that produce IFN-γ. Recent studies have found that the reason for a higher proportion of IL-17 + γδ T cells in the liver is that the cells are regulated by the intestinal flora, and that intestinal flora *E. coli* lipid antigens enter the liver through the hepatic portal vein and are presented by hepatocyte CD1 to activate IL-17 + γδ T cells, allowing them to proliferate and reach a dynamic balance. Similar to γδ T cells, there may be a similar reason for the higher proportion of NKT cells in the liver, but in-depth studies are needed to confirm this phenomenon. γδ T cells can be activated via TCR and non-TCR binding pathways (e.g., binding to PAMPs, TNF, superantigens, etc.) as well and are therefore considered to be innate immune cells. Since γδ T cells accumulate in the skin and mucosa, and upon activation they secrete cytokines such as IFN-γ, TNF-α, and IL-17 to regulate the immune response of monocyte-macrophages and neutrophils. In addition, this type of cells lysed pathogen-infected cells (e.g., Mycobacterium tuberculosis and Toxoplasma gondii-infected cells) by secreting perforin and granzyme. γδ T cells are the main type of cells early during anti-pathogenic infection. Furthermore, it was found that γδ−/− mice heal more slowly than wild-type (WT) mice after burns or trauma, and this result suggests that γδ T cells play an important role in burn recovery or trauma healing.

MAIT Cells (Mucosal-Associated Innate T Cells) The TCR of MAIT is composed of a constant alpha/beta chain. The alpha chain containing a combined Vα7.2-Jα33 fragment combines with the Vβ2 or Vβ13 of the beta chain to form the complete alpha/beta-TCR. Detection of $V\alpha7.2^+$ $CD161^+$ as markers can serve to identify MAIT cells, which, unlike other T cells, have a TCR that restrictively recognizes antigens presented by MR1 (MHC-I like related molecule), such as pathogenic antigens presented by MR1 that have structures similar to riboflavin (vitamin B2), folate (vitamin B9) metabolites, etc., which are commonly found in intestinal bacteria. The folate metabolite 6-formylpterin (6-FP) is an antigen recognized by the TCR of MAIT cells. IL-12 and IL-18 can act directly on MAIT cells to promote their activation. Superantigen-activated MAIT cells produce large amounts of IFN-γ, TNF-α, IL-10, and IL-17, which can lead to sepsis. Relatively little is known about the biological properties of MAIT cells.

It is important to clarify that lymphocytes include not only conventional lymphocytes (e.g., α/β-T cells and B2 cells, both traditionally referred to as T/B lymphocytes) and unconventional lymphocytes (e.g., NKT, γδ-T, B1, MAIT) but also innate lymphoid cells (ILCs). ILCs do not have antigen recognition receptors, but these cells play important roles in anti-pathogenic infection, antitumor, immune regulation, and tissue repair.

Innate lymphoid cells (ILC), a recently discovered large group of lymphoid cells, are characterized by having a lymphocyte morphology and residing in local tissues, but the cells do not have antigen recognition receptors. Based on cell development, transcriptional regulators expressed, and function, these cells are generally divided into five broad groups, including NK, LTi (lymphoid tissue-inducer), ILC1, ILC2, and ILC3. Classical NK cells were once classified as the ILC1 group due to their similarity to ILC1, and although NK cells have immunomodulatory effects, their prominent role is to kill target cells, so it is classified separately as a group. NK cells express the transcriptional regulator EOMES in addition to T-bet, and if the expression of EOMES is blocked during the developmental stage, it causes the cells to develop into ILC1 cells. LTi was once classified as an ILC3 group, but this cell is mainly associated with local lymphoid tissue formation and mainly expresses the transcriptional regulator RORγt (retinoic acid receptor-related orphan receptor-γt), cytokines IL-17 and TNF-α, and RANK and immunotoxins (lymphotoxin-α, –β, LT-α, LT-β) as well and is currently listed as a group. Unlike conventional lymphocytes, ILCs develop from innate lymphoid progenitor cells in the liver and intestine during embryonic development, colonize local tissues such as local skin, mucosa, and organs as the embryo develops and undergo self-renewal in local tissues after birth. They are the major cells in the early stage of local tissue to resist against pathogenic infection. ILC1, ILC2, and ILC3 resemble the conventional helper T cell (Th) differentiation subpopulations Th1, Th2, and Th17, respectively, in terms of transcriptional regulators and cytokine expression (see Chap. 9, Immunoregulation), e.g., ILC1 expresses the transcriptional regulator T-bet, which regulates the expression of IFN-γ and TNF-α, and plays an important role in combating intracellular pathogenic infections. If this group of cells is defective, it can lead to rheumatoid arthritis and inflammatory bowel disease (IBD). ILC2 expresses the transcriptional regulator GATA3, which regulates the expression of IL-4, IL-5, IL-9, IL-13, and amphiregulin (AR) in host response to helminthic or protozoal infections. Deficiency of ILC2 in host would lead to allergic reactions, such as asthma and allergic dermatitis. ILC3 expresses the transcriptional regulator RORγt, which regulates the expression of IL-17 and IL-22, and plays an important role in resistance to infection by extracellular microorganisms such as commensal bacteria or pathogens. If this group of cells is defective, it can lead to chronic inflammatory diseases such as inflammatory colitis and multiple sclerosis (MS) (see Fig. 2.4).

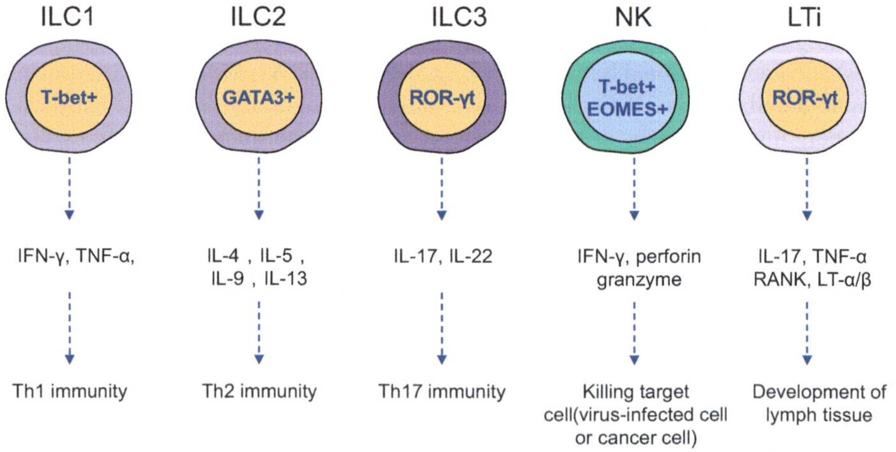

Fig. 2.4 Innate lymphoid cell populations and their functions. ILCs are divided into five groups (shown in the figure), each expressing specific transcriptional regulators and cytokines or specific molecules. With the exception of NK and LTi cells that have special functions, the other ILCs mainly play immunomodulatory roles

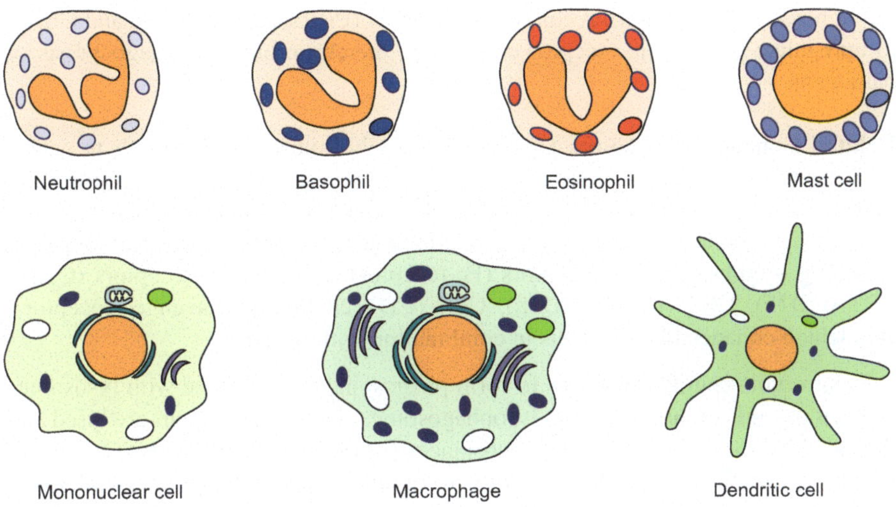

Fig. 2.5 Schematic diagram of myeloid cells

2.2.2 Myeloid Cells

Myeloid cells mainly include neutrophils, eosinophils, basophils, mast cells, monocytes/macrophages, dendritic cells, megakaryocytes, and erythrocytes (Fig. 2.5). Mammalian neutrophils and macrophages have a powerful ability to phagocytose, kill, and clear pathogens.

Neutrophil Neutrophils are the major cells among phagocytic granulocytes, produced by hematopoiesis in the bone marrow and released into the blood. Under normal conditions, there is a constant and large number of neutrophils in the blood, with approximately 50 billion neutrophils in adult body fluids, accounting for 60–70% of blood leukocytes, but in ruminants neutrophils account for only 20–30% of blood leukocytes. Neutrophils are generated quickly but have a short life span. Neutrophils in mouse blood have a life span of less than 18 hours, and in human blood they generally survive for 3–5 days. Neutrophils in human body fluids turnover approximately once a week and are mainly cleared by macrophages in the form of apoptosis to maintain homeostasis in the body. Neutrophils are very powerful in phagocytosis of pathogenic bacteria, and eliminate pathogenic bacteria by digestive enzymes (hydrolases, proteases, oxidases, peroxidases, lysozyme, fibrinases, etc.) and antimicrobial peptides in the phagolysosome. Neutrophils phagocytose pathogens through both opsonization and non-opsonization phagocytosis pathways. The opsonization phagocytosis pathway is the phagocytosis of pathogens through the binding of the opsonin–pathogen complex by the opsonin receptors on the cell surface. However, the non-opsonization phagocytosis pathway is the process by which phagocytes are activated by recognition of pathogen-associated molecular pattern (PAMPs, such as lipopolysaccharide, teichoic acid, glycoprotein, flagellin, etc.)

through cell surface pattern recognition receptors (PRRs, such as TLRs and CLRs) to upregulate phagocytic receptor expression, thereby promoting phagocytosis of pathogens.

Related Concept of Opsonin Opsonin is a soluble substance that binds to pathogens to promote phagocytosis, including antibodies, complement, fibronectin, mannose-binding lectin, milk fat globulin (also known as Lactadherin), etc. Opsonin receptor is a collective term for cell membrane receptors that bind to opsonin, including immunoglobulin receptors (Fc receptor), complement receptors (C1qR, CR1 (CD35), CR3 (CD11b/CD18), CR4 (CD11c/CD18)), scavenger receptors, etc., which enhance phagocytosis in combination with opsonin.

Neutrophils engulf pathogens to form phagosomes, which fuse with lysosomes and cytoplasmic granules to form lysophagosome, in which pathogens are lysed and degraded by various antimicrobial substances (ROS, AMP, superoxide (O_2^-, H_2O_2, HOCL, etc.), followed by clearance of neutrophils by macrophages to complete the whole process of phagocytosis, killing, and clearance of pathogens. However, birds do not have neutrophils. Instead, they have heterophils, which lack many degrading enzymes that are powerful in neutrophils, and thus birds are more susceptible to bacterial diseases than mammals. In recent years, it has been found that neutrophils can not only phagocytose and clear pathogens but also form neutrophil extracellular traps (NET) at the cost of their own lytic death, which capture and kill extracellular pathogens. By electron microscopy, NET formation is divided into three main stages: (1) cell morphology becomes flattened after neutrophil activation and the cytoplasm forms vacuoles; (2) the distinct structures of euchromatin and heterochromatin disappear, the nucleus lobes, and the inner and outer nuclear membranes produces gaps; (3) the nuclei enlarges and occupies most of the cytoplasmic space, and the nuclear membrane as well as the cytoplasmic granule membrane ruptures to release the contents participating in the formation of NET. Prior to formation of NET, cell membrane is intact, and in the final stage, the neutrophil dies, releases the NET, and exposes the cell death indicator such as phosphatidylserine (PS), which would be sensed by phosphatidylserine receptor (PSR) of macrophage for scavenging clearance. The composition of the NET is complex and subtle, mainly containing substances in the nucleus (DNA and histones), substances in the cytoplasmic granules (myeloperoxidase, histoproteinase, neutrophil tryptase, lactoferrin, penetrin-3, gelatinase, peptidoglycan-binding protein, antimicrobial peptides, etc.) and components of the cytoplasm (calprotectin, catalase, etc.). The death of neutrophil as a consequence of the formation of NET is called NETosis. NETosis is a form of cell death that can be broadly divided into suicidal NETosis and non-suicidal NETosis, also known as lytic NETosis and viable NETosis, respectively. The common NETosis is suicidal NETosis, which is mainly caused by the release of calcium ions into the cytoplasm of neutrophils due to the binding of pattern recognition receptors (PRRs) to the corresponding ligands (PAMPs) or the binding of other membrane receptors (Fc receptors, complement receptors, cytokine receptors, etc.) to the corresponding ligands. The release of intragranular material due to the high calcium ion concentrations in the cytoplasm causes nuclear membrane damage, ultimately leading to death. In contrast, non-suicidal NETosis, in which only part of the DNA and histones initially leave the nucleus as vesicles encapsulated in the nuclear membrane, with the nuclear membrane still relatively intact, can occur up to 30 minutes after LPS stimulation, but NETosis eventually occurs. In host anti-defense response, the main role of NET is to employ DNA and histone to adhere to and capture antigens, kill pathogens (bacteria or viruses) by antimicrobial peptides and lysozyme. Thus, NET plays an important role in the clearance of systemic infections, while it is also involved in septic shock process or causes tissue damage. Since NET contains various components within the cell, if not cleared in a timely manner, it will clog blood vessels and also cause autoimmune diseases, such as systemic lupus erythematosus (SLE). Reduction of DNase activity is closely related to the severity of animal vascular NET blockage and autoimmune diseases. Complements (e.g. C3, C1q etc.) are involved in the formation of NET and protect its degradation. After the NET is degraded by cytosolic DNase, it is subsequently cleared by macrophages. If mice lack DNase 1 and DNase 3, after neutrophil activation, the mice will die due to a large amount of NET that cannot be cleared and blocks the blood vessels.

Eosinophil Eosinophils are generated in the bone marrow, but migrate to the spleen to develop further. Eosinophils in the blood only survive for half an hour, but can survive for nearly 2 weeks after entering the tissues. Eosinophils also have the same function of phagocytosis and removal of pathogens. The phagocytes do not contain lysozyme, but the cytoplasmic granules contain large amounts of phosphatase and peroxidase, and these granules are red in eosin staining, which is distinctive. The phosphatases and peroxidases in the cytoplasmic granules of the eosinophils have a powerful ability to kill parasitic organisms. The most important feature of this cell, in addition to its ability to kill phagocytosed pathogens, is its ability to kill extracellular parasites. Upon infection with parasites, chemokines attract eosinophils to the site of infection, and eosinophils release granular substances, including cationic short peptides and lysophospholipase, which can effectively kill the parasite. Similar to neutrophils, eosinophils, in addition to phagocytosis, killing, and clearance of pathogens, produce EET (eosinophil extracellular trap) to capture and kill extracellular pathogens, resulting in EETosis. Similarly, EET causes tissue damage and is associated with respiratory diseases, autoimmune diseases, etc.

Basophil Basophils are the least abundant granulocytes and are found mainly in the blood, accounting for less than 1% of peripheral blood leukocytes. The life cycle of basophils is 2–3 days and the cells contain a large number of basophilic granules, which stain blue-purple. The cell membrane of basophils contains FcεRI, which is activated in the inflammatory response or upon binding to IgE-allergen complexes, leading to degranulation and release of Th2 cytokines (IL-4, IL-5, IL-13) as well as vasoactive granular substances, including histamine, 5-hydroxytryptamine, heparin, cathepsin, and tryptase, which have the effect of increasing vascular permeability. Basophils play a major role in local inflammation and allergic reactions, but basophil-induced allergic reactions are primarily chronic reaction.

Mast Cell Mast cells develop and differentiate from bone marrow hematopoietic stem cells and are now thought to share a common precursor cell (granulocyte/monocyte progenitor, GMP) with basophils. Mast cells were first discovered in 1878 by Paul Ehrlich. In the bone marrow, CD34+/CD117+ multifunctional precursor cells (pluripotent progenitor cells) develop and differentiate into immature mast cells that gradually develop into mature mast cells after migrating through the bloodstream to local tissues. Normal mast cells are 7–12 μM in diameter, round to ovoid in shape with a central round nucleus. The nucleus is often covered by a large number of cytoplasmic granules, and the granules stain blue. Mast cells often reside in local tissues (subcutaneous connective tissue, abdominal cavity) and can survive for a long time. The mast cells can be activated in two ways: (1) opsonin-mediated activation, mast cell membranes contain high-affinity FcεRI and complement receptors (C3aR, C5aR (CD88)), mast cell membranes bind to IgE-allergen complexes, causing degranulation and allergic reactions. Complement C3a or C5a-opsonized pathogen complexes can directly bind to mast cell complement receptors, inducing the activation of mast cells in favor of anti-defense response. When mast cells are activated, they release granule-associated active substances, including histamine, 5-hydroxytryptamine, heparin, histoproteinase, tryptase, and cathelicidin antimicrobial peptide. In addition, activated mast cells release Th2 cytokines (IL-4, IL-5, IL-13), which mainly regulate immune response. (2) Activation of mast cells by recognition of PAMPs via PRRs. Mast cells express a variety of PRRs including TLRs, NLRs, RLRs, CLRs, ALRs, cGAS, etc. (see Chap. 3 for details), and PRRs recognize pathogen-specific components to activate immune signaling pathways, which promote the expression of immune-related molecules, such as cytokines, chemokines, MHC-I/II, co-stimulatory molecules, etc. Mast cells express CD40L, which promotes B-cell activation for antibody production. In addition, mast cells have the function of phagocytosis and degradation of pathogens and can present antigens via MHC-I class molecules. Mast cells also form extracellular traps (ETs) that adhere to and kill pathogens with DNA, histones, and antimicrobial peptides, leading to METosis. Experimental results have demonstrated that mice deficient of mast cell are significantly less resistant to pathogenic infection and prone to death. Thus, mast cells play an important role in anti-defense response. Whether mast cells are anti-tumor or tumor-promoting is controversial because they have growth-promoting effects on some tumors but inhibitory effects on others. Further investigation will be needed to fully understand the biological functions of mast cells. Both mast cells and basophils can cause allergic reactions, but in contrast to basophils, the allergic reactions caused by mast cells are mainly acute processes.

Mononuclear Cell/Macrophage Mononuclear cell/macrophage is the collective name for mononuclear phagocyte in the blood. Monocytes/macrophages first originate in the yolk sac and liver progenitor during embryonic development, and some of them migrate to local tissues to develop into local macrophages for self-renewal, while others migrate to the bone marrow. After birth, monocytes that develop from bone marrow stem cells continuously enter local tissues to replenish the number of local macrophages. Mononuclear phagocytes are large and phagocytic, and their characteristic surface marker is the CD11b molecules on the surface of the cell membrane, but when they migrate to local tissues, they further differentiate into tissue-associated phagocytes, which increase more than fivefold in size and possesses different names, such as Kupffer cells in the liver, alveolar macrophage in the lung, macrophages in the spleen, histiocytes in connective tissue, and microglial cells in the brain. Monocytes/macrophages show a double positive markers of CD11b+/F4/80+ upon activation, and they are the core cells of the immune system that maintain homeostasis and are responsible for the clearance of pathogens, senescent or dead cells, antigen-antibody complexes, and various metabolic wastes, hence are named scavengers. Macrophages rely on scavenger receptors to perform their scavenging role. The scavenger receptor (SR) is a cell membrane receptor that binds to a variety of ligands and facilitates the clearance of foreign substances as well as their own abnormal substances. The underlying mechanism of SR-mediated clearance is to promote biological processes such as endocytosis, phagocytosis, adhesion, or signal transduction in scavenger cells, so that harmful substances are eventually eliminated. There are eight types of SR, named SR-A to SR-H. Different types of SR recognize and bind to different substances to be cleared, promoting the phagocytosis and clearance by macrophages. In addition to phagocytosis and clearance, macrophages also produce a large number of cytokines, regulating immune response. In recent years, it has been found that naïve macrophages differentiate into two different phenotypes of macrophages, called M1 and M2 macrophages, in response to different cytokines (IFN-γ or IL-4/13) that promote or suppress inflammatory responses, similar to the naive CD4 T cells (Th0) induced to differentiate into Th1/Th2 in response to IFN-γ/IL-4, producing different types of cytokines. M1 macrophages are differentiated under IFN-γ induction and are mainly characterized phenotypically by upregulated expression of MHC-II molecules, production of pro-inflammatory cytokines (e.g., TNF-α, IL-1, IL-6, IL-12, etc.) as well as i-NOS but downregulated expression of CD206 (mannose receptor). In contrast, M2 macrophages are differentiated under L-4 and IL-13 induction, and the phenotypes are characterized by upregulated expressions of CD206, transglutaminase and arginase, production of suppressive cytokines (mainly IL-10, TGF-β). IFN-γ induced activation of M1 macrophages belongs to classical activating process of macrophages, and the activated M1 macrophages produce TNF-α and IL-12. Both cytokines work together to activate NK and NKT cells, and the activated NK and NKT cells regulate the adaptive immune response by producing Th1 and Th2 cytokines (see Fig. 3.1 in Chap. 3 for details). Macrophages express all pattern recognition receptors (including TLRs, NLRs, RLRs, CLRs, ALRs, cGAS, etc.), recognize various pathogen-specific components, and initiate an appropriate immune response depending on the type of pathogen-associated molecular patterns (PAMPs) recognized by pattern recognition receptors (PRRs) (refer to the related sections in Chap. 3 for details). In addition, macrophages are professional antigen-presenting cells that present specific antigenic epitopes to CD4 and/or CD8 T lymphocytes via MHC molecules, activating the acquired immune response (refer to the related sections in Chap. 7 for details).

Dendritic Cell (DC) Dendritic cells are professional antigen-presenting cells and are widely distributed in the body, especially in tissues in contact with the outside, such as subcutaneous and mucosal tissues. DC is covered with long membrane extensions, resembling the dendrites of nerve cells, thus named dendritic cell. As DC has long membrane extensions that facilitate the quick movement of the cell to the nearby lymph node after capturing antigen, present antigens to lymphocytes, and elicit immune response, it is also called "sentinel cell." In addition to presentation of antigens, DCs also serve as scavengers, for example, CD103+ DCs can phagocytose apoptotic cells, produce TGF-β and IL-10, and promote immune tolerance. Dendritic cells express all pattern recognition molecules and secrete a variety of cytokines for immune regulation, including the direction of the immune response (Th1, Th2, Th17), antibody affinity maturation, and immune tolerance, depending on the nature of the recognized pathogen. Therefore, dendritic cells are also known as bridge cells for the transition from innate to adaptive immunity.

Platelets Platelets originate from cell-like vesicles formed by the lysis of megakaryocytes and do not contain a nucleus, but do contain mitochondrial DNA and RNA and some organelles. There are about 750 billion platelets in the body, and about 200 billion are produced from the bone marrow every day. After about 10 days of circulation of new platelets in the bloodstream, the cell membranes lose their sialic acid residues, while thrombopoietin (TPO) decreases, and the platelets show signs of senescence and are recognized by asialoglycoprotein receptor of hepatocytes (also known as Ashwell-Morell receptor, AMR), which activates the JAK2/STAT3 pathway, expresses TPO, and promotes the production of new platelets by megakaryocytes. AMR is essentially a CLR that recognizes the glycoproteins with N-acetylgalactosamine (GalNAc) residues. The mammalian AMR is composed of a heterodimer formed by two transmembrane glycoprotein molecules, hepatic lectin 1 (HL1) and 2 (HL2), primarily recognizing GalNAc, whereas the poultry genome contains only the gene encoding HL1, and avian AMR is a trimeric structure formed by HL1, mainly recognizing N-acetylglucosamine (GlNAc) residues. Senescent and dead platelets are eventually cleared by macrophages in the liver and spleen. Although platelets contain TLRs, the lack of a nucleus makes it difficult to activate complete immune signaling to generate a corresponding immune response, even after recognition of PAMPs. Platelet activation releases soluble mediators, such as CCL5, CXCL4, CD154 (also known as CD40L), which recruit neutrophils and monocytes to the site of inflammatory injury and engage granulocytes/phagocytes in the inflammatory response through CD40L-CD40 binding. Platelets are involved in NET adhesion to pathogens, promote phagocytosis and clearance of pathogens, promote B cell as well as dendritic cell activation through CD40L-CD40 binding, and regulate the adaptive immune response.

2.3 Immune Molecules

Immune molecule is a collective term for molecules involved in the immune response and generally refers to naturally soluble free molecules such as cytokines, chemokines, complement, and small molecular peptides against pathogenic microorganisms (including antimicrobial peptides and lysozyme). These immune molecules are involved in the immune response in different ways and play an important role in anti-defense response and immune regulation (see Sect. 3.1 in Chap. 3 for details).

2.4 Overview of the Basic Functions of the Immune System

The basic functions of the immune system include three aspects: (1) anti-defense (immunity against infection); (2) maintenance of homeostasis (removal of senescent and dead cells); and (3) immune surveillance (recognition and removal of cancer cells). Of these, maintenance of homeostasis is the central role of the immune system, and the important cells responsible for this role are macrophages. Macrophages maintain the normal physiological environment of the body by phagocytosis and clearance of senescent and dead cells and their remnants, foreign materials, antigen–antibody complexes, and so on. However, each cell is a relatively independent individual, and the cell itself needs to maintain its own homeostasis to ensure normal physiological functions. Autophagy is an important biological process by which cells remove abnormal components (self-produced abnormal molecules, damaged or senescent organelles, pathogenic components, etc.) to maintain their own homeostasis.

The organism maintains its homeostasis for normal physiological function at both macroscopic and microscopic levels. At the macroscopic level, an individual maintains a stable internal environment for normal activity of cells by macrophage cleaning of overall environment, while at microscopic level, cells depend on autophagy to maintain their normal physiological status. When a cell has a sudden condition (pathogenic infection, gene mutation, protein expression error, etc.), the cell, in addition to autophagy, transmits information outward by secreting cytokines and chemokines, recruiting relevant immune cells to generate an immune response and jointly deal with the local crisis, and eventually macrophages take the responsibility for phagocytosis and clearance of unwanted materials. The immune system functions by means of immune response, and for the convenience of explanation, the immune response process has been arbitrarily divided into innate and adaptive immune responses. All organisms, including single-cell organisms, plants as well as animals, have innate immunity, while only higher animals have developed adaptive immunity over long periods of evolution. Without innate immunity, organisms would not survive, whereas animals without adaptive immunity would experience recurrent infections.

2.4.1 Supplemental Reading: Cell Death

One of the core roles of the immune system is to maintain homeostasis in order to keep a normal physiological state through phagocytosis and clearance of senescent and dead cells, their remnants and foreign substances by macrophages. The main purpose of the immune response is to kill and remove pathogens, pathogen-infected cells, and tumor cells. Since killing is accompanied by cell death, the removal of dead cells and remnants is no doubt completed by the immune system. Cell death may be attributable to incidental forced death (non-programmed death) or to the programmed death of the cells themselves. Based on whether the cell membrane is lysed, cell death can be classified as non-lytic cell death (primarily apoptosis) and lytic cell death (including fatal necrosis, pyroptosis, and NETosis) (see Fig. 2.6).

2.4.1.1 Non-lytic Cell Death

Apoptosis is a non-lytic programmed cell death characterized by the overturn of the phosphatidylserine in the inner layer of the cell membrane to the outer layer, shrinking and rounded cell, condensed chromatin, fragmented DNA, and the formation of apoptotic body containing cell contents encapsulated in the cell membrane. Typically, apoptosis does not do any harm to the microenvironment, and apoptotic bodies are subsequently phagocytosed and cleared by phagocytes without causing a marked local inflammatory response. Three apoptotic pathways are recognized, namely the extrinsic pathway, intrinsic pathway, and perforin /granzyme pathway.

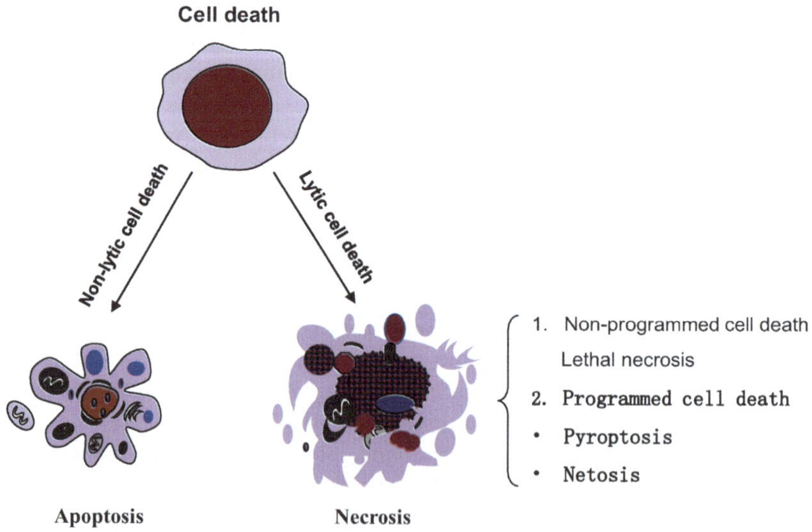

Fig. 2.6 Schematic diagram of different type cell death

(a) The extrinsic pathway, also known as the death receptor pathway, is the binding of a cell membrane death receptor (DR) to a corresponding ligand that causes apoptosis. The ligands that engage DRs to cause apoptosis are commonly members of the TNF superfamily, such as TNF-TNFR1, TRAIL(Apo2L)-DR4 (human) or DR5 (mouse) (see Chap. 10 for details), FasL (CD95L)-Fas, Tweak(Apol3L)-TRAMP, etc., generating apoptotic signals. Upon receipt of the signal by the death receptor, the intracellular domain is activated to recruit FADD, TRADD, and procaspase-8 or procaspase-10 to form the death-inducing signaling complex (DISC), which cleaves procaspase-8 to produce active caspase-8, which further cleaves the zymogen procaspase-3 to produce active caspase-3, caspase-3 cleaves ICAD to release CAD, which shears genomic DNA as well as Gelsolin (which is an actin-binding protein), ultimately leading to apoptosis.
(b) The intrinsic pathway, also known as the mitochondrial apoptotic pathway, is usually caused by increased permeability of the outer mitochondrial membrane due to intracellular apoptotic stimuli (radiation, toxin, hypoxia et al), which results in the release of cytochrome *c (Cyt-c). Cyt-c* binds to APAF-1 to recruit procaspase-9, forming an apoptosome, which cleaves procaspase-9 for activation. The activated caspase-9 further cleaves procaspase-3 to caspase-3, leading to apoptosis.
(c) Perforin/granzyme pathway, this is the third pathway by which cytotoxic T lymphocyte (CTL) and natural killer(NK) cell exert their apoptotic effects on tumor cells or virus-infected cells through secretion of perforin and/or granzyme. Perforin inserts into the cell membrane of target cells and forms pores for the subsequent entry of granzyme-contained vesicles to the target cell. The granzymes (such as A and B) are serine proteases, which, in addition to the activation of caspase-10 or induction of cytochrome-*c* release, can directly activate caspase-3, resulting in apoptosis.

The flip of phosphatidylserine (PS) from the inner membrane of the lipid bilayer of the cell membrane to the outer membrane is one of the main features of early apoptosis and is also the "eat me" signal from the cell. Macrophages (or dendritic cells) recognize the phosphatidylserine in the apoptotic cell membrane to engulf and remove the apoptotic cell or apoptotic body. Apoptosis is broadly divided into initial, developmental, and terminal stages (see Fig. 2.7).

2.4.1.2 Lytic Cell Death

Lytic cell death is characterized by rupture of the cell membrane, increased cellular osmotic pressure, swelling, and death by burst. Early on, this type of cell death was referred to as necrosis. In recent years, it has been discovered that necrosis can be subdivided into different types, including: lethal necrosis,

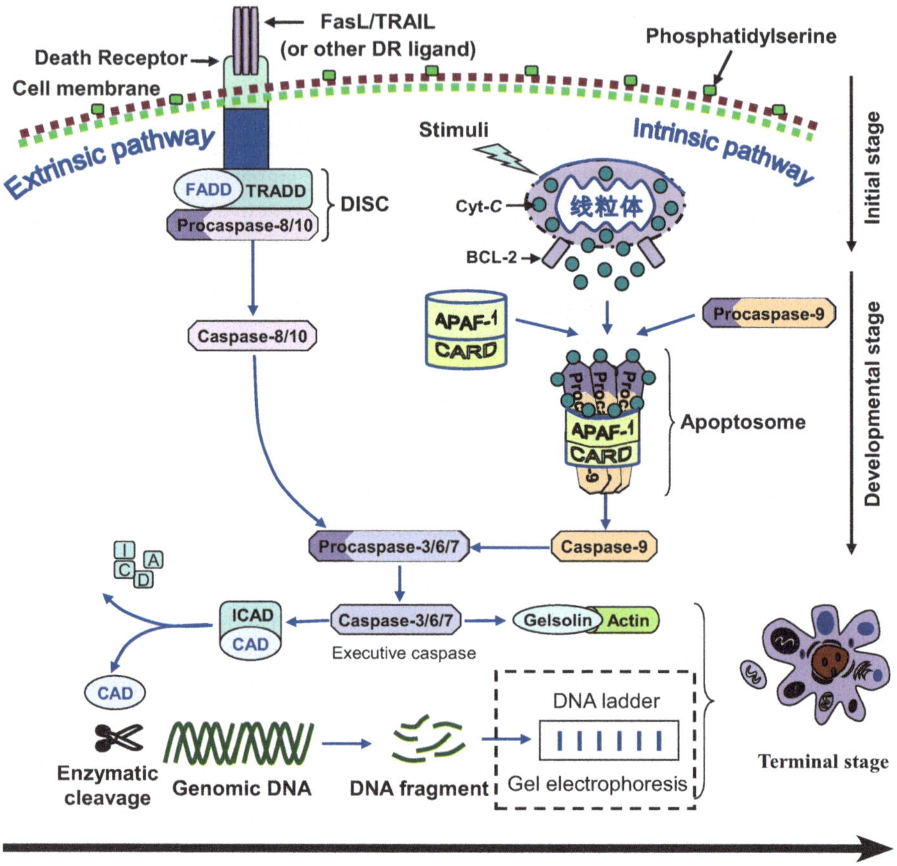
FasL/TRAIL
(or other DR ligand)
Death Receptor
Cell membrane
Phosphatidylserine
Extrinsic pathway
Intrinsic pathway
Stimuli
FADD
TRADD
Procaspase-8/10
DISC
Cyt-C
线粒体
BCL-2
Initial stage
APAF-1
CARD
Procaspase-9
Caspase-8/10
Apoptosome
Developmental stage
Procaspase-3/6/7
Caspase-9
ICAD
CAD
Caspase-3/6/7
Executive caspase
Gelsolin
Actin
CAD
DNA ladder
Enzymatic cleavage
Genomic DNA
DNA fragment
Gel electrophoresis
Terminal stage

Fig. 2.7 Schematic diagram of extrinsic and intrinsic pathways of apoptosis. Extrinsic pathway of apoptosis is also known as the death receptor pathway, and the intrinsic pathway known as the mitochondrial apoptotic pathway. The extrinsic pathway is mainly caused by the binding of TNF family members to cell membrane death receptors (DR), such as FasL-Fas, TRIAL-DR4/DR5, TNF/TNFR1, Tweak-TRAMP, etc. FasL forms a trimer that binds to Fas to activate the intracellular region of Fas to generate apoptotic signals, and the intracellular region is activated to recruit FADD, TRADD and procaspase-8 (or procaspase-10) form a death-inducing signaling complex (DISC), which cleaves procaspase-8 to produce active caspase-8. The active caspase-8 further cleaves zymogen procaspase-3/6/7 to produce active executive caspase-3/6/7, which performs irreversible death cleavage of the substrate, leading to apoptosis. For example, caspase-3 cleaves the substrate ICAD to release the nucleic acid endonuclease CAD, which shears genomic DNA, causing DNA breaks and chromatin fixation. In addition, caspase-3 cleaves Gelsolin causing actin dissociation, which allows cellular structural reorganization to form apoptotic bodies. Macrophages phagocytose and eliminate apoptotic cells by recognizing phosphatidylserine (PS) on the cell membrane. The intrinsic pathway is usually caused by the action of intracellular apoptotic stimulants (e.g., radiation, hypoxia, hypothermia, growth factor deficiency, viral proteins, intracellular mycotoxins, free radicals, etc.) on mitochondrial membrane protein BCL-2 family molecules, which affect the pro-apoptotic molecules (e.g., BAX, BAK) or anti-apoptotic molecules (e.g., BCL-2, BCL-XL) of BCL-2 family, resulting in increased permeability of the outer membrane mitochondria, promoting *Cyt-c,* release. *Cyt-c* binds to APAF-1 and recruits procaspase-9 through the CARD domain, forming an apoptosome that causes procaspase-9 cleavage and produces activated caspase-9, which further catalyzes procaspase-3/6/7 to produce caspase-3/6/7, leading to irreversible apoptosis. Apoptotic process is generally divided into the initial, developmental, and terminal stages. Related abbreviations and brief meaning: *APAF* apoptotic protease activating factor, *CAD* caspase-activated DNAse, *BAK* BCL2 antagonist killer 1, *BCL-2* B-cell lymphoma protein 2, *BCL-XL* BCL2-related protein, long isoform, *BAX* BCL2-associated X protein, *FasL* fatty acid synthase ligand, *CARD* caspase recruitment domain, *Caspase* cysteinyl aspartic acid-protease, *Cyt-c* cytochrome c, *DR4* death receptor-4, present in humans, *DR5* death receptor-5, present in rodents, *Fas* fatty acid synthase (receptor), *FADD* FAS-associated death domain, *ICAD* Inhibitor of CAD, *TRADD* TNF receptor-associated death domain, *TNF-α* tumor necrosis factor alpha, *TNFR1* tumor necrosis factor receptor 1, *TRAIL* TNF-related apoptosis-inducing ligand

pyroptosis, and NETosis, in which there is both programmed and non-programmed lytic cell death (Fig. 2.6).

Lethal necrosis, a non-programmed lytic cell death, is a consequence of direct destruction of cell membranes by exogenous stimulants (e.g., staphylococcal exfoliative toxins, or mechanical injury) that cause lysis of cells.

Pyroptosis, a programmed lytic cell death, mainly depends on the formation of plasma membrane pores by Gasdermins, and in most cases, it is caused by recognition of intracellular PAMPs by host cell PRRs to form inflammasome that plays an important role in anti-defense response (see Fig. 3.9 in Chap. 3).

NETosis, a programmed lytic cell death, is an extracellular trap produced by cells that recognize pathogens at the cost of death, serving to adhere (capture) and kill pathogens. NETosis is a new type of cell death only identified in recent years (see the Neutrophil section of this chapter for details).

Study Questions

1. What is the central immune organ? What is its role?
2. What are peripheral immune organs? What is their role?
3. Which cells are included in lymphocytes and which cells are included in ILCs?
4. What are myeloid cells? What cells do they include?
5. What are the characteristics of myeloid cells in birds?
6. What are the phagocytic cells? What are the non-phagocytic innate immune cells?
7. What is a scavenger receptor? What is its mechanism of action?
8. What is the biggest difference between apoptosis and necrosis?
9. How many apoptotic pathways are there? What caspases are involved in the activation of the different pathways?
10. What are DISC and apoptosome? and what are their roles?
11. What are pyroptosis and NETosis?

Chapter 3
Innate Immunity

Overview:
Innate immunity is the congenital resistance of organism to pathogenic infection, and is the first line of anti-defense of host to invaders, including natural physiological barriers (e.g., skin and mucosal barriers, blood-brain barrier, placental barrier, blood-testis barrier, etc.) and the innate immune response. This anti-defense mechanism exists from lower to higher organisms (including humans). The most direct killing and clearance of pathogenic microorganisms by innate immunity is achieved through the secretion of innate immune molecules and phagocytes in the organism. Innate lymphoid cells (ILCs) are important cells distributed at the skin and mucosal barriers to receive neural/endocrine signals to regulate immunity and play an important role in immune regulation, tissue repair, and maintenance of local lymphoid tissue stability. Among innate immune molecules, defensin plays a very important role. Defensins are a variety of small molecule polypeptides secreted by the mucous membranes of the respiratory, digestive, and genitourinary tracts, and by the epithelial cells of the skin. These small molecule polypeptides directly kill bacteria, fungi, or enveloped viruses by disrupting the membrane structure of pathogens. In addition, once the pathogen invades the organism, the complement in the body fluid is able to directly recognize the bacterial components, which activates the complement system and generates a cascade reaction to assemble a membrane attack complex that directly kills the pathogenic bacteria.

Phagocytes play a central role in the process of phagocytosis, killing and clearance of pathogens as well as maintaining their own stability. Phagocytes mainly include mononuclear macrophages, neutrophils, eosinophils, basophils, etc. These phagocytes distinguish foreign from self by recognition of the characteristic components of pathogenic microorganisms (pathogen-associated molecular patterns, PAMPs) through special receptors (pattern recognition receptors, PRRs), generating immune responses. Therefore, recognition of pathogenic microorganisms is the prerequisite and basis of immune responses. Non-phagocytic innate immune cells mainly

S. J. Zheng, *Veterinary Molecular Immunology*,
https://doi.org/10.1007/978-981-99-8929-4_3

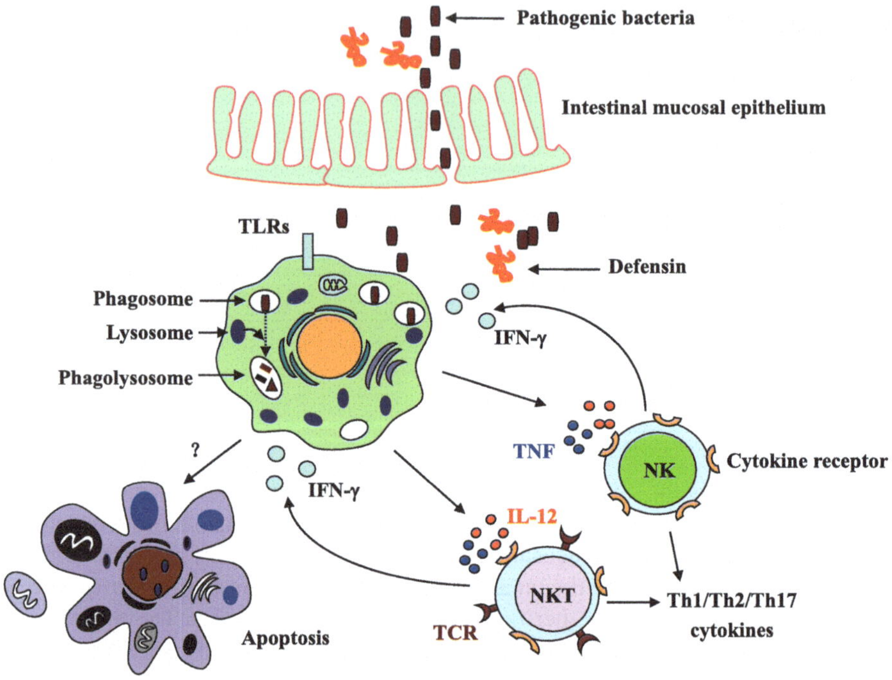

Fig. 3.1 Schematic diagram of the basic process of early innate immune response to pathogenic infection

include NK, NKT, γδ T cells, and so on, these cells not only have direct destructive effects on pathogen-infected cells but also further activate phagocytes by secreting cytokines to enhance their ability to phagocytose, kill, and clear pathogens. More importantly, NK/NKT cells play an important immunomodulatory role by secreting cytokines such as Th1, Th2, and Th17 cytokines, which regulate the inflammatory response as well as the type of adaptive immune response and move the immune response in the direction of cellular or humoral immune response. Thus, NK/NKT cells are a bridge for the transition from innate to adaptive immunity. Figure 3.1 illustrates the basic process of early innate immune response. This chapter primarily focuses on the innate immune molecules and cells discovered in recent years, mechanisms underlying defensin-induced destruction of pathogens, recognition of pathogens by phagocytes, and the molecular mechanisms of innate immune response.

Invasion of a host by a pathogenic microorganism first elicits an innate immune response in the host. Defensins, phagocytes, and NK/NKT cells play crucial roles in the killing and clearance of pathogens. Macrophages recognize and phagocytose invading pathogens to form phagosomes, which fuse with lysosomes to form phagolysosomes, where pathogens are degraded and killed by a variety of degradative enzymes. At the same time, macrophages secrete interleukin (IL)-12 and tumor necrosis factor (TNF). IL-12 and TNF activate NK and NKT cells, and the activated

NK/NKT cells secrete gamma interferon (IFN-γ), IFN-γ in turn further activates macrophages. The activated macrophages have a powerful ability of phagocytosis, killing, and clearance of pathogens, while activated NK/NKT cells regulate the immune response by secreting Th1, Th2 and Th17 cytokines. Some pathogens, such as *Listeria monocytogenes*, *Salmonella,* and some viruses, induce apoptosis in macrophages early after infection with excessive doses, and the physiological significance of this phenomenon remains to be investigated in depth.

3.1 Innate Immune Molecules

Innate immune molecules are small molecular peptides that are inherent in living organisms and are resistant to pathogenic infection. There are many types of these small-molecule peptides, and they are roughly divided into three major categories according to their mode of action: (1) small peptides that can directly destroy the outer membrane of bacteria and viruses, such as antimicrobial peptides (AMPs); (2) small peptides that can destroy pathogenic bacteria through enzymatic activity, such as lysozyme; (3) small peptides that can chelate metal ions like zinc, iron and so on, such as calprotectin and lactoferrin. These innate small molecules play a very important role in the process of innate immunity against infection.

3.1.1 Types and Effects of Antimicrobial Peptides

Antimicrobial peptides (AMPs), also known as host defense peptides (HDPs), are small molecular proteins (typically around 30aa) encoded by a single gene in the host that are not specific and have a broad-spectrum bactericidal or antivirus action as well as immunomodulatory effects. Antimicrobial peptides are ubiquitous in the biological realm, and both plants and animals or fungi improve their resistance by producing antimicrobial peptides. There are two main types of antimicrobial peptides: defensin and cathelicidin. These two antimicrobial peptides are usually expressed in the form of precursors and stored in the vesicles of cells (neutrophils in mammals, heterophils in birds, mucosal epithelial cells), and the antimicrobial peptide precursors are released and become mature and active antimicrobial peptides under the action of corresponding proteases after the host cells recognize PAMPs and are activated, exerting anti-defense and immunomodulatory effects.

3.1.1.1 Defensins

Defensins are small cationic polypeptides that contain six-disulfide-linked cysteine residues and are secreted by mucosal or skin epithelial cells and neutrophils, and have a broad-spectrum antimicrobial effect on pathogenic microorganisms. The structure of defensins is characterized by the presence of three conserved cysteine

Table 3.1 Types, structural characteristics, and functions of defensins

Type of defensin	Structural features (position of the cysteine disulfide bonds)	Animal species and main cells that express defensins	Functions	
			Anti-pathogen effect	Immunomodulatory effect
α-defensin	Cys^1 -Cys^6, Cys^2 -Cys^4, Cys^3 -Cys^5	Mammalian neutrophils, intestinal cells, and Paneth cells	Prevent virus from entering target cells by blocking viral binding to receptors; inhibit viral decapsidation, replication, and transport	Promotes dendritic cell maturation and upregulates the expression of CD83 (a maturation marker) as well as APC surface markers including CD80, CD86, CD40, HLA-DR, and the scavenger receptor CD91 (the ligand for this receptor is defensin)
β-defensin	Cys^1 -Cys^5, Cys^2 -Cys^4, Cys^3 -Cys^6	Vertebrate animal neutrophils (heterophils in birds),intestinal cells, Paneth cells, etc.	Lyses viral envelope; prevent virus from entering target cells by blocking viral binding to receptors; inhibits virus decapsidation, replication, and transport	Promotes dendritic cell maturation and upregulates the expression of CD83 (a maturation marker) as well as APC surface markers including CD80, CD86, CD40, HLA-DR, and the scavenger receptor CD91 (the ligand for this receptor is defensin)
θ-defensins	Cys^1 -Cys^6, Cys^2 -Cys^5, Cys^3 -Cys^4	Monkey and baboon neutrophils, intestinal cells, Paneth cells, etc.	Lyses viral envelope; prevent virus from entering target cells by blocking viral binding to receptors	?

disulfide bonds, an α helix, two β sheets, and multiple hydrophobic cationic amino acids. Based on the position of the cysteine disulfide bonds, defensins can be classified into α, β, and θ types. α defensin cysteine disulfide bonds are characterized by Cys^1-Cys^6, Cys^2-$Cys^{4,}$ and Cys^3-Cys^5; β-defensins, containing the cysteine disulfide bonds of Cys^1-Cys^5, Cys^2-Cys^4 and Cys^3-Cys^6, are found in all vertebrates and are mainly produced by mucosal epithelial cells, with slightly larger molecular mass than α-defensins. α- and β-defensins exist in various animals. In contrast, θ defensins are only found in non-human primates and have a circular structure with a cysteine disulfide bond characterized by Cys^1-Cys^6, Cys^2-Cys^5 and Cys^3-Cys^4. θ defensin is currently only found in monkeys and baboons, while the human genome contains only the pseudogene of this defense (see Table 3.1).

The distribution of each type of defensin varies, each type has different members, and has its own unique function. α-defensins are mainly produced by

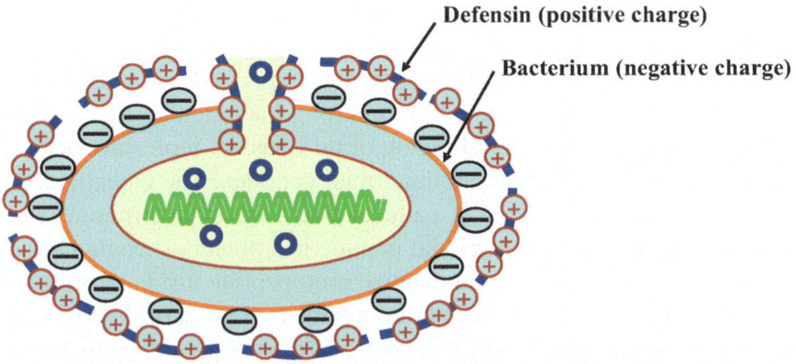

Fig. 3.2 Schematic diagram of the mechanism of defensin-mediated killing of pathogenic bacteria

mammalian neutrophils, mucosal goblet cells, paneth cells, etc., and are rarely expressed by other cells. β-Defensins were first discovered in 1991 in bovine epithelial cells. In terms of evolution, β-defensins appeared first. α- and θ-defensins are probably evolved from β-defensins, only β- defensins are found in avian species, and at least 25 subtypes of β-defensins have been found in avian species. The main function of defensins is to have a broad-spectrum antimicrobial effect on pathogenic microorganisms and modulate the immune response. The mechanisms of action of different defensins may vary, and the precise molecular mechanisms by which the various defensins act have not yet been fully elucidated. It is generally accepted that positively charged defensins are attracted to negatively charged bacterial cell membranes, and that defensins disrupt cell membrane permeability and cellular energy status by forming dimeric or multimeric ion channels in the bacterial outer membrane, leading to depolarization of the cell membrane, inhibition of respiration, and a decrease in cellular ATP content, ultimately leading to target cell death (Fig. 3.2). The antiviral effect of defensins is achieved by causing the virus to lose its biological activity by binding to viral outer membrane proteins, or by preventing the pathogen from binding to the target cell receptor to inhibit invasion. It is due to these specific mechanisms that defensins have two characteristics: (a) a broad antimicrobial spectrum; and (b) It is difficulty for pathogenic microorganisms to develop mutant resistant to defensins. Some defensins can stimulate the maturation of immune cells (dendritic cells), phosphorylate STAT, MAPK/p38, ERK1/2, etc., cause the expression of cytokines and chemokines, promote the degranulation of mast cells, and participate in the formation of granulocyte ET (extracellular trap) and ET capture and toxic killing of pathogens (For details, please refer to Chap. 2). As research progresses, a more comprehensive understanding of defensin function will be gained.

In general, pathogenic bacteria are negatively charged, and defensin molecules containing large amounts of cationic positive charges bind to the bacterium. The positively charged defensin and the negatively charged bacterial cell membrane attract each other, and the defensin forms transmembrane ion channels in the bacterial outer membrane by dimers or multimers, leading to the death of target cells.

3.1.1.2 Cathelicidin Antimicrobial Peptide

Cathelicidins, belonging to small short peptides expressed by myeloid precursor cells, neutrophils, and lymphocytes, have a strong bactericidal effect. Because of the high homology of the functional region of this class of molecules with cathelin drugs, they are called cathelicidin antimicrobial peptides. The cathelicidin antimicrobial peptide contains a feature of catheter-like domain in structure, and when expressed, the cathelicidin antimicrobial peptide is a precursor containing a signal peptide, which is stored mainly as an inactive protopeptide inside the cell. After the cell is stimulated, the signal peptide of protopeptide is cleaved off by serine protease to become a biologically active cathelicidin antimicrobial peptide, which is released outside the cell to perform bactericidal effects. According to the structural characteristics, cathelicidins can be divided into two types: alpha-helical peptides and hairpin peptides. Cathelicidins have antimicrobial and immunomodulatory effects, such as cathepsin LL-37, which promotes upregulated expression of DC CD86 and HLA-DR and proliferation of Th1 cells, and poultry cathepsin CATH-2, which promotes phagocytosis and antigen presentation by mononuclear phagocytes. Since different animal species differ greatly in the expression levels of Cathelicidins, whether such differences are related to species resistance to disease needs to be further investigated.

The activity of antimicrobial peptides is dose-dependent. The normal level of antimicrobial peptides in the fluids of human is 40 ng/mL, however under infectious conditions, their levels in body fluids exceed 1 mg/mL and can reach up to 170 mg/mL in cases of sepsis. In vitro studies, it was found that the activity of antimicrobial peptides is concentration-dependent, with low concentrations they would stimulate the secretion of cytokines by immune cells, and in medium concentrations, they are chemotactic, while at high concentrations, they can directly lyse tumor cells and bacteria. In addition, some antimicrobial peptides can directly bind to pathogen-associated molecular patterns (PAMPs), such as lipopolysaccharide (LPS) and cytosolic acid (LTA), to exert antimicrobial effects.

3.1.2 Bacteriolytic Enzyme

Bacteriolytic enzyme, also known as cell wall lysozyme, is a large class of enzymatic proteins widely present in nature, and is contained in a variety of tissues and secretory fluids in animals except for human, with a high content in egg white. It is a relatively stable basic protein with four disulfide bonds, consisting of 129 amino-acid residues with a relative molecular mass of 14,000. Its activity is relatively stable, and the bactericidal action of egg white lysozyme is mainly through cleaving the β(-1,4) glycosidic bond between the N-acetylneuraminic acid and N-acetylglucosamine of the peptidoglycan of the bacterial cell wall to lyse bacteria. Since the cell wall peptidoglycan content of Gram-positive (G+) and Gram-negative (G-) bacteria differs greatly, G+ bacteria cell walls are composed almost entirely of peptidoglycan and are therefore sensitive to lysozyme, while G- bacteria have low peptidoglycan contents and are in the inner wall layer and are poorly sensitive to lysozyme, so lysozyme is highly effective in killing Gram-positive bacteria and weakly effective against Gram-negative bacteria. Human and animal tears, saliva, nasal secretions, and breast milk have high levels of lysozyme. In phagocytes, lysozyme is mainly found in the lysosome. When phagocytes take in pathogenic bacteria, the phagosome (phagosome) fuses with the lysosome to form the phagolysosome (see Fig. 3.1), and the lysozyme then functions to degrade the pathogenic bacteria. Human lysozyme contains 130 amino-acid residues with four disulfide bonds, and its lytic activity is much higher than that of egg white lysozyme, which plays a very important role in the innate resistance to microbial infection in humans. Lysozyme is widely used in biomedicine and the food industry because of its strong destructive effect on cell walls. At present, transgenic cows generated in China using animal transgenic technology are able to secrete human lysozyme in their milk, laying the foundation for the extensive use of lysozyme.

3.2 Innate Immune Cells

The immune cells involved in the innate immune response are designated innate immune cells, including phagocytes (monocytes-macrophages, neutrophils, eosinophils, basophils, mast cells, nucleated erythrocytes (avian), etc.) and non-phagocytic innate immune cells (NKT, γδT, B1, MAIT, NK, LTi, ILC1, ILC2, ILC3, and other cells). All of these cells derived from hematopoietic stem cells, play an important role in the innate immune response and also regulate the direction of the immune response by secreting cytokines.

3.2.1 Phagocytes

Phagocytes is a general term for cells in the body that have phagocytic functions. All tissues in the body, especially mucosal tissues, have a large number of phagocytes. Mammalian phagocytes can be divided into two major categories with complementary functions: myeloid phagocytes and mononuclear phagocytes.

3.2.1.1 Myeloid Phagocytes

Myeloid phagocytes, containing a large amount of granule-like materials in the cytoplasm, are also called granulocytes. Granulocytes are divided into four subgroups: neutrophils (which are absent in birds and are replaced by heterophils), basophils, eosinophils, and mast cells, respectively. These granulocytes except for mast cells are characterized by lobulated and irregular nuclei and are collectively referred to as polymorphonuclear (PMN) cells. PMNs have a powerful ability to phagocytose and remove foreign pathogens, playing a very important role in the fight against pathogenic infections. Mast cells are mainly involved in allergic reaction and immunoregulation.

3.2.1.2 Mononuclear Phagocytes

Mononuclear phagocytes include the mononuclear phagocytes in the blood and also phagocytes in different tissues. These cells are characterized by a round, non-lobulated nucleus, and strong phagocytic capacity. Upon pathogenic infection, they can be rapidly recruited to the site of infection and phagocytose pathogens through pattern recognition receptors (PRRs) that recognize pathogen-associated molecular patterns (PAMP) such as carbohydrates, nucleic acids, and lipopolysaccharides of pathogens. After phagocytosis, phagosomes are fused with cytoplasmic lysosomes in the phagosome to become phagolysosomes, where pathogens are killed and cleared via degradation by various enzymes. In addition to the direct phagocytosis of pathogens, phagocytes can also phagocytose pathogen-infected cells, senescent and dead cells, cellular debris, and various residual wastes to maintain their physiological homeostasis at another level.

3.2.2 Non-phagocytic Innate Immune Cells

In the process of innate immune response, there are some cells that do not have phagocytic functions but have the killing effect (such as NK cells) as well as regulate the immune response by secreting cytokines. Thus, these cells belong to non-phagocytic innate immune cells, mainly including NKT, γδT, B1, MAIT, NK, LTi, ILC1, ILC2, ILC3, and other cells (see Chap. 2 for details). Among them, NKT, γδT, B1, and MAIT cells express antigen-specific receptors, except that B1 cells recognize antigens by BCR and mainly serve as antigen presenting cells, while NKT, γδT, and MAIT cells are activated by engagement of TCR with antigens presented by CD1 or MR1 (both CD1 and MR1 are molecules similar in structure and function to MHC-I class molecules), secret cytokines, and regulate the immune response. However, NK, LTi, ILC1, ILC2, and ILC3 cells belong to innate lymphoid cells (ILCs) and do not express receptors that recognize antigens (e.g., TCR or BCR), but mainly play an immunomodulatory role in immune response by recognizing corresponding ligands through cell membrane surface receptors. In addition, NK cells have a role in killing target cells, while LTi cells play an important role in local lymphoid tissue formation (see Chap. 2). In terms of immune regulation, NK cells express the transcriptional regulators T-bet and EOMES and secrete IFN-γ, perforin, and granzymes; LTi cells express the transcriptional regulator RORγt and secrete the cytokines IL-17 and TNF-α; ILC1 expresses T-bet and produces IFN-γ and TNF-α; ILC2 expresses the transcriptional regulatory factor GATA3, producing IL-4, IL-5, IL-9, IL-13, and bidirectional regulatory proteins; ILC3 expresses the transcriptional regulatory factor RORγt, producing IL-17 and IL-22 (see Fig. 2.4). ILC1, ILC2, and ILC3 are similar to Th1, Th2, and Th17 in terms of transcriptional regulators and cytokine expression profiles (see Chap. 9 Immunoregulation).

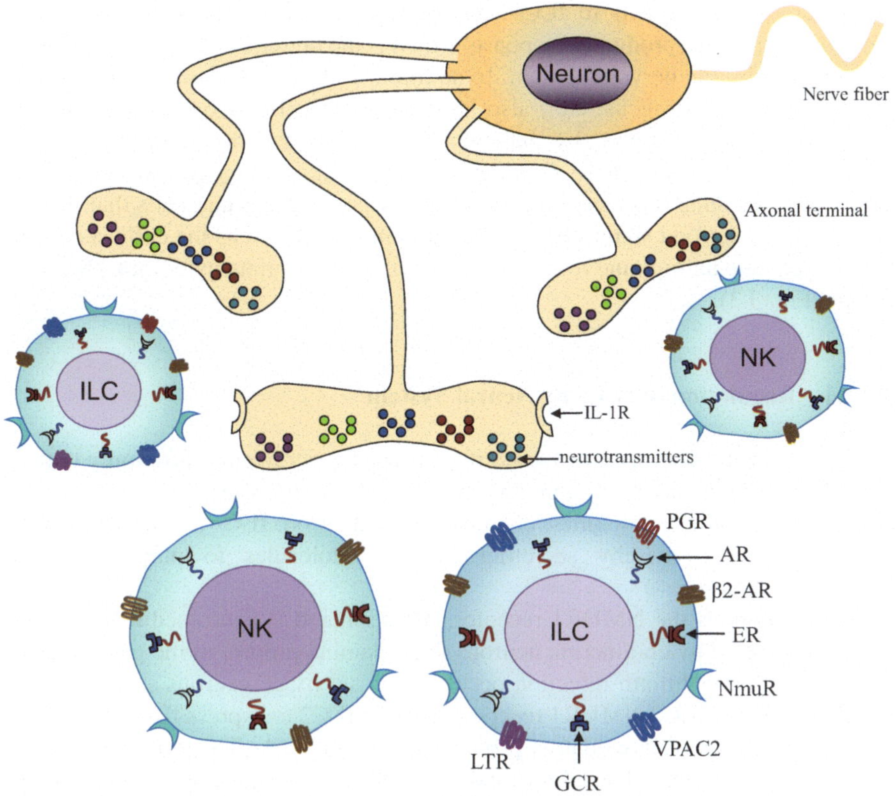

Fig. 3.3 Schematic diagram of the neural and endocrine systems regulating immune cells (ILCs and NK cells). The nervous system influences immune cells and regulates the immune response by secreting neurotransmitters. The immune system also influences the nervous system by secreting cytokines. Likewise, the endocrine system influences the immune response. ILC in the diagram mainly represents ILC2, but also stands for ILC1 and ILC3. *AR* androgen receptor, *β2-AR* β2-adrenergic receptor, *ER* estrogen receptor, *GCR* Glucocorticoid receptor, *LTR* leukotrienes receptor, *IL-1R* interleukin-1 receptor, *NmuR* neuromedin U receptor, *VPAC2* Vasoactive intestinal peptide receptor type 2

ILCs develop from innate lymphoid precursor cells in the liver and intestine during embryonic development, colonize local tissues such as local skin, mucosa, and various organs as the embryo develops, and self-renew in local tissues after birth, bearing local tissue characteristics. Except for NK cells, other ILCs are mainly distributed in local tissues, especially in the barrier tissues such as skin and mucosal surface, in close proximity to the neurons and microglia of the nerve endings of the autonomic nerve system (ANS), which includes the sympathetic nerve system (ANS) and the parasympathetic nerve system (PNS). Under normal and stable conditions, PNS regulates physiological functions by secreting mediators, such as food digestion, anergy conversion, and storage. The location of the ILCs determines the close relationship between the immune cells and the nervous system. The neuromediators produced by local neurons and the substances secreted by their target cells are transmitted as signals to ILCs and the ILCs recognize these ligand signals through receptors to produce a response, which is the prerequisite for the regulation of immunity by the nervous system. In addition, endocrine hormones (glucocorticoids, androgens/estrogens, etc.) also have regulatory effects on ILCs. Similarly, local tissue damage caused by pathogenic infections and the components (e.g., PAMPs) of pathogens can be regarded as danger signals recognized by ILCs to generate an immune response. Thus, the ILCs serve as important cells that receive neural/endocrine/food signals to regulate immunity and also as the earliest cells to locally generate an immune response by secreting appropriate cytokines. (See Fig. 2.4 and Fig. 3.3).

3.2.2.1 Regulation of ILCs by Neural System

ILCs colonize the skin and mucosal surface tissue barrier in close proximity to neurons, expressing a variety of signaling molecule receptors, which recognize neuromediators, hormones, cytokines, chemokines, and so on. ILCs are activated upon receiving signals, thereby secreting immune molecules to play a role in immunoregulation.

ILC2 expresses the NMUR1 receptor, whose ligand is neuromedin U (NMU), which is secreted by cholinergic neurons in the lamina propria of the gastrointestinal track. When helminth infection occurs, mucosal tissue neurons secrete NMU, which acts on the ILC2 NMUR1 receptor, causing ILC2 to express IL-4, IL-5, IL-9, IL-13, and bidirectional regulatory proteins that modulate the Th2 immune response and recruit eosinophilic, basophilic, and mast cells to participate in the antihelminth infection process. ILC2 also expresses the norepinephrine receptors β1-AR (adrenergic receptor) and β2-AR, and are regulated by epinephrine (EP) and norepinephrine (NE) that are mediators secreted by the sympathetic nervous system. NE acts on ILC2 cells via binding to β2-AR, suppressing the ILC2 immune response. Similarly, NE acts on NK cells by β2-AR, and suppresses IFN-γ, granzyme B, and perforin expression, thereby inhibiting NK cell cytotoxicity. In addition, ILC2 expresses VIP (vasoactive intestinal peptide) receptor type 2 (VPAC2). VIP is a neurotransmitter secreted by neurons in the lung and gastrointestinal tract that has the ability to promote Th2 and Treg (regulatory T cells, Treg See Chap. 9 Immunomodulation) differentiation. In local tissues (lung, gastrointestinal tract), VIP binds to the VIP receptor of ILC2 and promotes ILC2 expression of IL-5, while lung neuronal wound pain sensors have IL-5R, which recognizes IL-5, and the neurons produce VIP in response to IL-5, which creates a positive feedback reinforcement that, if uncontrolled, tends to produce large amounts of IL-5 and IL-13, causing local allergic reactions. In addition, the neuropeptide substance calcitonin gene-related protein (CGRP) secreted by neuroendocrine cells in the lung and gamma-aminobutyric acid (GABA) are associated with the regulation of ILC.

Substance P (SP) is a nociceptive substance released from neurons following stimulation of nociceptors. SP exerts a pro-inflammatory response (e.g., activation of eosinophils, neutrophils, mast cells, etc.) by binding to neurokinin receptors

(NK1R, NK2R, and NK3R), but SP inhibits NK cell degranulation and cytotoxicity.

Immune cells work closely with the nervous system in inflammation resolution and tissue repair. For example, in the inflammatory response, ILC3 secretes IL-1, which acts on the vagus nerve IL-1R, prompting the vagus nerve to secrete large amounts of acetylcholine that in turn acts on ILC3 to promote the secretion of tissue repair protective factor 1 (PCTR1). PCTR1 is an important anti-inflammatory response factor with the ability to recruit macrophages and enhance their functions of phagocytosis and clearance, suppressing local inflammatory responses and promoting local tissue repair.

In conclusion, the nervous system regulates immunity through ILCs, which plays a unique bridging role in the regulation of nervous and immune systems.

3.2.2.2 Regulation of ILCs by the Endocrine System

Glucocorticoids (GC) and sex hormones (androgens, estrogens) have a greater effect on ILCs, and GC has a broad inhibitory effect on immune cells, mainly by inhibiting NF-κB signaling pathway to suppress immune signaling, thereby inhibiting the production of pro-inflammatory cytokines such as IFN-γ, IL-6, IL-12, and TNF-α, while promoting the expression of immunosuppressive cytokines IL-10 and TGF-β. Similarly, GC has an inhibitory effect on NK and ILCs. It is currently found that uterine mucous ILC2 cells express estrogen receptor (ER), whereas pulmonary ILC2 cells do not express ER. It is unclear what role estrogen has in ILC2 function by promoting the proliferation of uterine ILC2. As ILC2 precursor cells express only the androgen receptor (AR) and not the ER, androgens affect ILC2 development, and the number of ILC2 in males is significantly less than that of females. It was demonstrated that testosterone acting on ILC2 cell AR inhibits the expression of IL-5 and IL-13, which explains the clinical phenomenon that the incidence of asthma and the number of blood ILC2 in female is significantly higher than that of males.

3.2.2.3 Nutrient Composition of Food and Regulation of ILC by Intestinal Commensal Bacteria

ILCs express a variety of food component receptors, such as the vitamin A (also known as Retinoic acid, RA) receptor and the vitamin D receptor, RA and the active vitamin D component VitD3, which mainly affect the immune response of ILCs and homeostasis. In addition, Lipids produced by food or intestinal commensal bacteria have a regulatory effect on the cellular immune response of ILCs and influence homeostasis. Arachidonic acid metabolites include a variety of prostaglandins (PGs) and leukotrienes (LTs) that act on the corresponding ILC membrane receptors and affect the migration, proliferation, and immune response of ILCs. It has been demonstrated that chronic deficiency of certain nutrients affects the immunity of the body.

3.3 Mechanisms of Recognition and Clearance of Pathogens or Self-Abnormal Substances by Innate Immune Cells

In recent years, the research in innate immunity has advanced rapidly, and particularly, the discovery of TLRs (toll-like receptors) has brought dramatic changes to the research field in innate immunity. It is now clear that host cells recognize pathogen-associated molecular patterns or damage-associated molecular patterns through pattern recognition molecules, distinguish non-self or self-abnormal substances, and initiate relevant immune responses to phagocytose, kill and remove foreign pathogens or self-abnormal substances.

3.3.1 Concepts, Types and Structures of Pattern Recognition Receptors and Pathogen-Associated Molecular Patterns

3.3.1.1 Concepts of Pattern Recognition Receptors and Pathogen-Associated Molecular Patterns

Pattern recognition receptors (PRRs), also known as pattern recognition molecules (PRMs), are specific molecules that are used by host cells to recognize pathogens. PRRs specifically recognize the characteristic molecular structures of pathogenic microorganisms. Target molecules specific to pathogenic microorganisms that can be recognized by innate immune cells are called pathogen-associated molecular patterns (PAMPs). PAMPs are evolutionarily conserved molecules shared by pathogenic microorganism, and are widely present in pathogens, such as endotoxin, peptidoglycan, flagellin, double-stranded RNA, non-methylated DNA, and mannose on yeast cell walls. Host cells recognize PAMPs through PRRs to initiate an immune response. Both the type and intensity of the immune response are related to the recognition of PAMPs by PRRs. Thus, specific recognition of PAMPs by PRRs is a prerequisite and basis for the innate immune response (Fig. 3.4). Six major families of PRRs have been identified, namely TLRs, NLRs, RLRs, CLRs, ALRs, and cGAS. These molecules are specifically recognized PAMPs and collaborate with each other to regulate the immune response and maintain the normal physiological state of the host.

Host cells specifically recognize PAMPs through PRRs to initiate an immune response and clear the pathogen.

3.3.1.2 Types and Structures of Pattern Recognition Receptors

TLRs

This family of receptors recognizes a wide range of pathogens, including bacteria, viruses, fungi, and protozoa.

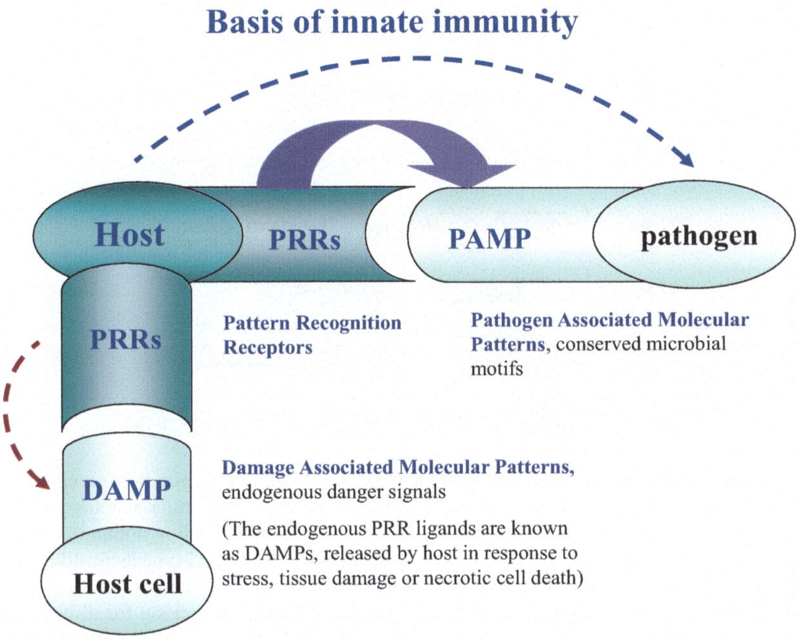

Fig. 3.4 Schematic diagram of the recognition of pathogen by innate immune cells

TLRs (Toll-like receptors), a large class of pattern recognition receptors, are membrane-bound proteins distributed on the surface of cell membranes and endosome membranes, with molecular structures very similar to that of interleukin-1 (IL-1). TLRs are structurally divided into extracellular, transmembrane, and intracellular regions. The extracellular region of TLRs containing leucine-rich repeats (LRR) is the site of recognition of PAMPs, and the intracellular region contains the TIR functional region that is responsible for the recruitment of adaptor proteins for signaling downstream upon TLR activation (Fig. 3.5). Twenty-six members of the TLR family (TLRs 1–26) have been identified, but the TLR members and number of TLRs vary in different animals.

There are 10 members of the human TLR family, including TLR1 to TLR10.

The mouse TLR family has 12 members, including TLR1 to TLR9, TLR11 to TLR13, and TLR10 is still unidentified.

The avian TLR family currently contains 10 members, such as TLR10 (TLR1A), TLR1 (TLR1B), TLR2A, TLR2B, TLR3, TLR4, TLR5, TLR7, TLR15, and TLR21.

The porcine TLR family is currently found to have eight members, including TLR1 to TLR6, TLR9, and TLR10. As research progresses, more members of the TLR family will possibly be identified.

Among the TLRs family, TLR3, TLR5, and TLR7 are evolutionarily conserved and expressed in all chordates, whereas TLR15 is mainly found in avian and reptilian species, and the recognized ligands are extracellular proteases.

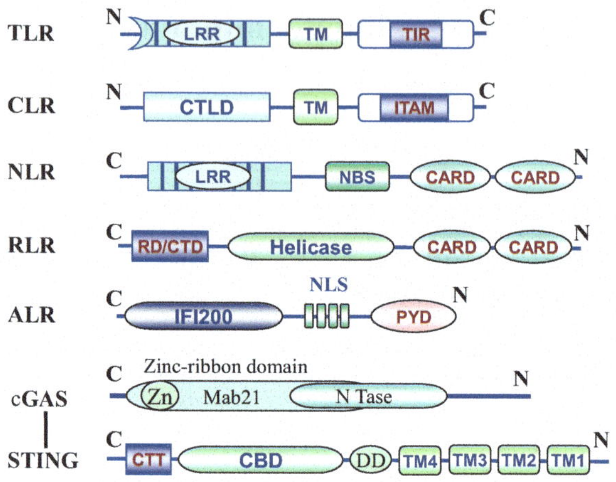

Fig. 3.5 Schematic diagram of the basic structure of the pattern recognition receptor (PRR). CARD (caspase activation and recruitment domain): a region that recruits adaptor proteins for signaling downstream after NLR or RLR activation; CBD (cGAMP-binding domain): a structural domain in which STING molecules bind to cyclic GMP-AMP; CTLD (C-type lectin-like domain): C-type plant lectin-like region that contains a structural domain for carbohydrate recognition domain (CRD); CTT: C-terminal tail; DD (dimerization domain): the region for dimerization of STING molecules; helicase domain: a region that recognizes viral nucleic acids; IFI200 (interferon (IFN)-inducible 200 aa repeats): interferon inducible 200 amino-acid repeat region that recognizes double-stranded DNA (dsDNA); ITAM (immunoreceptor tyrosine-based activation motif): an intercellular activation domain of transmembrane receptor that recruit adaptors for signal transduction; LRR (leucine-rich repeats): a region that recognizes pathogen-associated molecular patterns (PAMPs) of pathogenic microorganisms; Mab21 (C. elegans-Mab21): a protein of C. elegans; NBS (nucleotide-binding site): nucleic acid binding region where NLR monomers aggregate into trimers after activation; NLS (nuclear localization signal motif): nuclear localization signal structure, related to the localization of ALRs; N Tase: nucleotidyltransferase; PYD (Pyrin domain): this region is responsible for self-aggregation or interaction with downstream protein molecule; RD/CTD (repressor domain/C terminal domain): C-terminal repressor region, mainly inhibits the activation of RLRs; TIR domain [Toll/IL-1 receptor (TIR) domain]: intracellular functional region of TLRs, which recruits adaptor proteins to transmit signals downstream after TLR activation; TM (transmembrane domain): the region where receptor molecules are fixed on the membrane

Fish TLRs contain mammalian TLRs (TLR1, TLR2, TLR3, TLR4, TLR5, TLR7, TLR8, TLR9) and non-mammalian TLRs (TLR18, TLR19, TLR20, TLR21, TLR22, TLR23, TLR25, TLR26), but do not have TLR6, TLR10, TLR11, TLR12, TLR13, TLR15, and TLR24.

NLRs

This family of receptors primarily recognizes intracellular bacteria in the cytoplasm.

NLRs [nucleotide-binding oligomerization domain (NOD)-like receptors] are a class of pattern recognition receptors present in the cytoplasm and are responsible for recognizing pathogenic bacteria that enter the cytoplasm. NLRs are very conserved in species evolution and are not only found in animals but are also widely present in plants. The NLR basic structure has three functional regions, namely LRR, NBS (also known as NACHT), and CARD (or BIR, PYD) (Figs. 3.5 and 3.6). NLR is similar to TLR and relies on the LRR region to recognize PAMPs or DAMPs. NBS is the binding region where NLR monomers polymerize with each other to form trimers when NLR binds to PAMPs or DAMPs. When NLRs bind to PAMPs or DAMPs, the NLR structure is altered to form multimers that recruit ASC and Procaspase1/11 to form inflammasomes (inflammatory vesicles) to conduct signals downward (see Fig. 3.9). NLRs have a large number of family members (23 members in human and 34 in mouse) and are divided into four subgroups based on the N-terminal functional region, NLRA, NLRB, NLRC and NLRP, respectively (see Table 3.2).

NLRA, a subgroup that currently has only one member, CIITA, whose primary function is to promote the expression of MHC-II.

NLRB, a subgroup that currently has only one member of NAIP and whose primary function is to recognize type III secretion systems and flagellin.

NLRC, whose main members include NOD1-3, NLRC4, NLRC5, NLRX1, etc., whose main functions are recognition of type III and IV secretion systems and flagellin, activation of NF-kB, inflammatory vesicles, autophagy, and transduction of type-I interferon (IFN-α/β) signal.

NLRP, the main members include NLRP1-7, NLRP10, NLRP12, NLRP14, etc. whose main functions are activation of inflammasomes, NF-kB and type-I interferon (IFN-α/β) pathways as well as regulation of embryonic development.

RLRs

Members of this family primarily recognize viruses.

RLRs [Retinoic acid-inducible gene I (RIG-I)-like helicases receptors] are a class of pattern recognition receptors present in the cytoplasm and are responsible for the recognition of viral nucleic acids that appear in the cytoplasm. The basic structure of the RLR molecule has three functional regions, namely RD/CTD, Helicase, and CARD (Fig. 3.5). The Helicase domain (HD) is the site for viral nucleic acid recognition and is usually closed by the C-terminal repressor region (RD) with the N-terminal CARD overlay. Upon activation, the RD and CARD change the molecular conformation and expose the recognition region (HD) to recognize PAMPs (usually viral RNA) through ATP hydrolysis, dephosphorylation, and ubiquitination, and RLR activation signals through the CARD functional region down to the MAVS (mitochondrial antiviral signaling protein), which subsequently initiates the immune response via IRF3/7 as well as NF-κB signaling pathways. The main members of this family are RIG-I and MDA5 (melanoma differentiation-associated gene-5), but it is worth noting that chickens naturally lack RIG-I and rely mainly on MDA5 to recognize intracellular viral nucleic acids and initiate antiviral immune responses, while waterfowl (ducks and geese) express both RIG-I and MDA5. This is related to the high resistance of waterfowl to avian influenza because

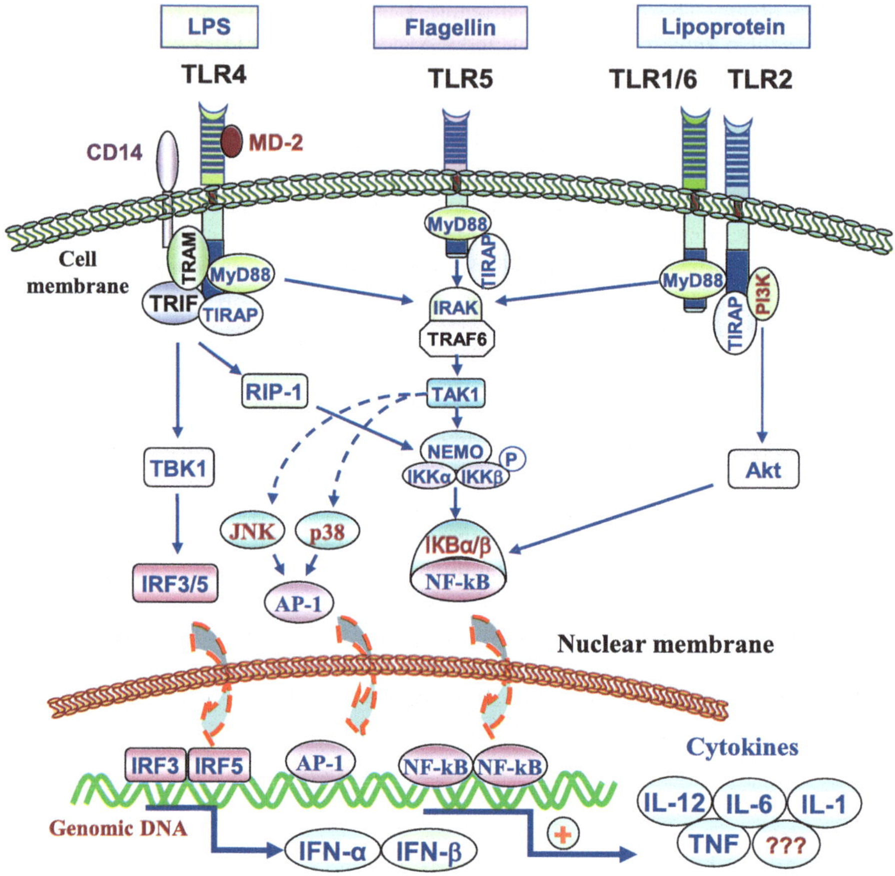

Fig. 3.6 Schematic diagram of the activation and signaling process of TLRs on the cell membrane. After TLRs bind to the corresponding ligands, the intracellular functional region TIR domain is activated, recruiting and activating MyD88, TIRAP, TRIF, TRAM, and other adaptor molecules. MyD88 is the most important adaptor molecule, and the signaling process via MyD88 requires interaction with TIR domain. Activated MyD88 recruits and phosphorylates IRAK, and activated IRAK further recruits and activates TRAF6, which activates TAK1.TAK1 activation eventually activates the NF-κB signaling pathway and also activates AP-1 via JNK and p38. After activation, NF-κB and AP-1 enter the nucleus and bind to the promoters of specific genes to initiate transcription and expression of downstream genes. In addition, recognition of ligands by TLR4 also recruits and activates TRIF, which activates TBK and ultimately activates the transcriptional regulator IRF. Upon activation, IRF enters the nucleus and binds to the promoters of type I-interferon genes to initiate transcription and expression of type I-interferon (IFN-α/β) genes. *MyD88* myeloid differentiation factor 88, *TIRAP* TIR domain-containing adapter protein, *TRIF* TIR domain-containing adapter-inducing interferon-β, *TRAM* TRIF-related adapter molecule, *IRAK* IL-1 receptor kinase, *TRAF* tumor necrosis factor receptor-associated factor, *TAK* TGF-β activated kinase 1, *NF-κB* nuclear factor kappa enhancer binding protein, *JNK* c-Jun-N-terminal kinase, *AP-1* activating protein-1, *IRF* interferon regulatory factor, *RIP* receptor interacting protein, *PI3K* phosphoinositol-3 kinase, *NEMO* NF-κB essential modulator, *IKK* inhibitor of NF-κB kinase, *IκB* inhibitor of NF-κB, *poly(I:C)* synthetic dsRNA based on viral nucleic acid structure dsRNA, *MAVS* mitochondrial antiviral signaling protein, *NBS* nucleotide-binding site; *CARD* caspase activation and recruitment domain

Table 3.2 Schematic diagram of the structure of the NLR subpopulation

NLR subpopulation	Member	Basic structure	Main function
NLRA	CIITA		Promote MHC-II expression
NLRB	NAIP		Sense type III secretion system and flagellin
NLRC	NOD1-3, NLRC4, NLRC5, NLRX1		Sense types III and Iv secretory system and flagellin, and activate NF-kB, inflammasome, autophagy, Type-I interferon pathway
NLRP	NLRP1-7, NLRP10, NLRP12, NLRP14		Activate inflammasome, NF-kB and Type-I interferon pathway, involved in fetal development

LRR (leucine-rich repeats): a leucine-rich repeat region that recognizes components of pathogens; NBS (nucleotide-binding site):a region for NLR monomers binding to aggregate into trimers; CARD (caspase activation and recruitment domain): cysteine aspartate enzyme activation recruitment structural domain that recruits adaptor proteins for downstream signaling after activation; AD (an acidic transactivation domain): acidic structural transformation domain; PYD (Pyrin domain): a region similar to CARD, associated with apoptosis, protein interactions during signal transduction; BIR (Baculovirus inhibitor-of-apoptosis repeat): Baculovirus apoptosis inhibitory region with a zinc-finger structural domain

RIG-I is the main pattern recognition molecule that recognizes the nucleic acid of highly pathogenic avian influenza virus to produce an antiviral effect. In addition, some members of the RLRs family lack RD/CTD at the C-terminus of their molecular structure, and the representative member LGP2 (laboratory of genetics and physiology 2), which functions similarly to RIG-I and MDA5, recognizes viral RNA to elicit an immune response.

CLRs

Members of this family primarily recognize fungi.

CLRs (C-type lectin receptors) are membrane-bound receptors with an extracellular fraction (CTLD) containing mannose-binding lectins that specifically recognize carbohydrates in fungal membranes, also known as the carbohydrate recognition domain (CRD) (Fig. 3.5). Such receptors are estimated to be over 1000 and can be divided into seven subgroups, most of which are associated with phagocytosis, antigen presentation, and maintenance of endogenous glycoprotein concentrations. Some members recognize specific components of fungal-like pathogens, transduce signals, and elicit immune responses. Dectin-1, Dectin-2, Dectin-3, Mincle, Mannose receptor (MR), and DC-SIGN are representative receptor molecules of this family, and these receptors recognize fungal PAMPs (mainly fungal glycolic anhydride, mannose, and polyacetylamino glucose (also known as chitosan)) through the extracellular region CTLD (C-type lectin-like domain), and the activated CLRs recruit the adaptor molecule tyrosine kinase Syk through the intracellular region ITAM (immunoreceptor tyrosine-based activation motif), which ultimately activates the NF-κB signaling pathway in host cell, leading to the expression of pro-inflammatory cytokines and chemokines. Some members of CLRs, such as MR and DC-SIGN, do not have a typical ITAM in the intracellular signaling region, but instead contain an ITAM-like domain (ITAM-like domain, also known as hem-ITAM domain). If CLRs are abnormal or absent, severe fungal infections can occur. Although CLRs are membrane receptors, they can be detected in serum or other body fluids.

ALRs

ALRs (Absent in melanoma 2 (AIM2)-like receptors) are a new class of pattern recognition receptors discovered in recent years (see Fig. 3.5). IFI200 is the region of ALR that recognizes DNA, mainly intracellular or nuclear pathogenic DNA or host own abnormal DNA. ALRs are self-inhibitory when at rest, and are activated upon recognition of dsDNA for signal transduction. Two well-studied ALRs are AIM2 and IFI16. AIM2 recognizes intracytoplasmic dsDNA (or synthetic dsDNA, e.g., poly(dA:dT)), whereas IFI16 recognizes both intracytoplasmic and nuclear abnormal DNA (pathogenic DNA or self-abnormal DNA). ALRs are not only involved in anti-defense responses but are also associated with autoimmune diseases (e.g., systemic lupus erythematosus, rheumatoid arthritis, multiple sclerosis, etc.). There are five members of the ALR family in humans and thirteen in mice.

cGAS

cGAS (cyclic GMP-AMP synthase), an important pattern recognition receptor that recognizes intracytoplasmic dsDNA, was discovered in 2013. cGAS molecules consist of 531 amino acids in human and 497 amino acids in porcine. cGAS has 160 to 330aa with a high homology with nucleotidyltransferase (NTase) sequence, hence named NTase structural domain, while the C-terminal 213 to 522aa portion is highly homologous to the nematode Mab21 (Male abnormal21) molecule, thereby called Mab21 structural domain. In the cGAS molecule, NTase partially overlaps with the Mab21 structural domain. cGAS is evolutionarily conserved, ranging from zebrafish to higher animals, and in particular, the NTase-Mab21 structural domain of cGAS is highly conserved among species and is the functional domain responsible for recognizing dsDNA. cGAS recognizes dsDNA (pathogenic dsDNA-PAMPs, or self aberrant DNA-DAMPs, such as DNA leaked from mitochondria or generated by genomic metabolic renewal) and changes its conformation, catalyzing the substrates adenosine triphosphate (ATP) and guanosine triphosphate (GTP) to produce cGAMP. cGAMP binds to and activates STING on the endoplasmic reticulum membrane, and the activated STING is transferred to the mitochondria to activate IRF3 and NF-κB signaling pathways to produce type-I or –III interferons (I- or III-IFN) as well as pro-inflammatory cytokines (IL-1, IL-6, IL-12, TNF, etc.) to exert anti-defense responses. Since cGAMP is a second messenger generated by cGAS after recognition of dsDNA and directly activates the STING signaling pathway, it is commonly referred to as the cGAS-STING signaling pathway. Although cGAS is genetically evolutionarily conserved among species, it is broad in recognizing dsDNA, unlike TLR9 and ALRs which are specific in recognizing dsDNA. cGAS not only recognizes pathogenic dsDNA to initiate immune responses and exert anti-defense effects, but also senses its own dsDNA and participates in antitumor, aging, and inflammatory processes.

3.3.1.3 Types of Pathogen-Associated Molecular Patterns Recognized by Pattern Recognition Receptors

The Main Types of PAMPs Recognized by TLRs

TLRs are membrane-bound molecules that primarily recognize pathogenic components outside the host cell and those in the endosomes.

TLR1 and TLR6 or TLR2 form a complex receptor that primarily recognizes microbial lipoproteins, glycoproteins, and other components outside the host cell.

TLR2 mainly recognizes lipoproteins and glycolipids outside the host cell, including yeast cell wall and peptidoglycan, and is usually expressed as a dimer, e.g., TLR2/2 recognizes lipoteichoic acid (LTA), TLR1/2 recognizes triacyl lipoprotein, TLR2/6 recognizes triacyl lipoprotein and peptidoglycan.

TLR3 primarily recognizes double-stranded ribonucleic acids (dsRNA) in endosomes.

TLR4 primarily recognizes lipopolysaccharides (LPS) in the cell walls of pathogenic bacteria outside the host cell.

TLR5 primarily recognizes the Flagellin of pathogenic bacteria.

TLR6 primarily recognizes microbial lipoproteins outside the host cell.

TLR7 primarily recognizes single-stranded nucleic acids (containing -UGUGU-sequences or Poly-U, or antiviral compounds) in endosomes.

TLR8 primarily recognizes ssRNA as well as RNA degradates.

TLR9 mainly recognizes microbial non-methylated CpGDNA and viral nucleic acids in endosomes.

TLR10 is currently found to be associated with the recognition of *Listeria monocytogenes (Lm)* and dsRNA. Since TLR10 recognizes dsRNA in endosomes mainly under low pH conditions, it competes with TLR3, which may be related to TLR10's inhibition of immune signal transduction. However, the ligands for TLR10 have not been fully determined (Table 3.3).

Table 3.3 Avian TLRs and the corresponding PAMPs

TLRs	Other names	PAMPs or ligand	pathogen
TLR1a	TLR1.1, TLR1/6/10/16	Diacetyl or triacetyl lipoproteins	Viruses
TLR1b	TLR1.2	Lipoproteins	Viruses
TLR2a	TLR2.1	Peptidoglycan (PG)	G+ bacteria
TLR2b	TLR2.2	Lipoproteins	Viruses
TLR3		Single-stranded RNA (ssRNA)	Virus
TLR4		Lipopolysaccharide (LPS)	G-bacteria
TLR5		Flagellin	G-bacteria
TLR7		Single-stranded RNA (ssRNA)	Virus
TLR15		Extracellular bacterial protease	Bacteria
TLR21		Non-methylated DNA (CPG DNA) or chromosomal DNA	Bacteria, viruses
TLR23*	?	?	?

*"?" It means that it is not clear at this time

The Main Types of PAMPs Recognized by NLRs

The NLRs have a large family of members that primarily recognize pathogenic components that appear in the cytoplasm. The representative members are NOD1 and NOD2. NOD1 primarily recognizes certain components of Gram-negative (G-) bacteria that enter the cytoplasm, such as gamma-glutamyl meso-diaminopimelic acid (Meso-DAP) in peptidoglycan (PGN), a structure that is primarily present in the PGN of G-bacteria, but small amounts are also present in the PGN of a few G+ bacteria (Listeria monocytogenes).NOD2 primarily recognizes certain components of Gram-positive (G+) bacteria that enter the cytoplasm, such as the muramyl dipeptide (MDP) in peptidoglycan (PGN), which is present in the PGN of all G- and G+ bacteria. In addition, NLRs also recognize proteins secreted by bacterial type III and IV secretion systems.

The Main Types of PAMPs Recognized by RLRs

RLRs are present in the cytoplasma and primarily recognize viral nucleic acids that enter the cell plasma.

The Main Types of PAMPs Recognized by CLRs

CLRs are receptors distributed on cell membranes that primarily recognize fungal β-glucans, glycoconjugates, and mannose, and act synergistically with TLRs.

The Main Types of PAMPs Recognized by ALRs

ALRs primarily recognize intracellular dsDNA. AIM2, a member of the ALRs family, is present in the cytoplasm and recognizes dsDNA in the cytoplasm, whereas IFI16 recognizes dsDNA in the cytoplasm as well as in the nucleus.

The Main Species of PAMPs Recognized by cGAS

The cGAS recognizes pathogenic dsDNA or its own abnormal dsDNA (e.g., DNA released from mitochondria or from genomic metabolic renewal).

3.3.2 Molecular Mechanisms by which Innate Immune Cell Recognizes Pathogenic PAMPs (or Self-DAMPs) to Generate an Immune Response

Innate immune cells play a key role in the killing and clearance of pathogens as well as in homeostasis. Immune cells recognize the characteristic components of pathogenic microorganisms, PAMPs (or self DAMPs), through PRRs to generate an immune response, which consists of two main aspects: first, recognition of pathogens and generation of immune signaling; second, phagocytosis, killing and clearance of invading pathogens.

3.3.2.1 Recognition of Pathogens by Immune Cells and the Process of Immune Signal Transduction

The Immune Response Signaling Process Induced by TLR Recognition of PAMPs

Since TLRs are membrane-bound molecules, some TLRs (TLR1, TLR2, TLR4, TLR5, and TLR6) are distributed on the surface of the cell membrane, and cells are activated for signal transduction through TLRs that recognize extracellular PAMPs (e.g., lipopolysaccharides, flagellins, lipoproteins, etc.). Other TLRs are distributed on the membrane structures of endosomes, including TLR3, TLR7, TLR8, TLR9, etc. These receptors specifically recognize PAMPs in endosomes, and are activated for signaling to initiate transcription and expression of downstream genes. Regardless of the distribution of TLRs, upon binding to the corresponding ligands, they are required to initiate transcription and expression of immune-related genes through transcriptional regulators such as NF-κB, AP-1, or IRF.

TLRs-mediated signaling pathways are mainly divided into two signaling pathways, MyD88 (myeloid differentiation factor 88) dependent and independent. MyD88-dependent signaling pathway controls the transcription and expression of related genes by activating NF-κB and AP-1 transcriptional regulators. After TLRs bind to the corresponding ligands, the intracellular functional region TIR domain is activated, recruiting and activating MyD88, TIRAP, TRAAM, and other adaptor molecules. MyD88 is the most important adaptor molecule, and the signaling process via MyD88 requires TIR-TIR interactions. MyD88 activation recruits and phosphorylates IRAK, activated IRAK recruits TRAF6, and the activated TRAF6 recruits and activates TAK1. Upon activation, TAK1 ultimately activates the NF-κB signaling pathway as well as AP-1 via JNK and p38. After activation, NF-κB and AP-1 enter the nucleus through the nuclear membrane gap and bind to the promoters of specific genes to initiate transcription and expression of downstream genes. NF-κB and AP-1 regulate the expression of a variety of immune-related genes, including pro-inflammatory response factors (IL-1, IL-6, IL-12, TNF, etc.), chemokines, immune costimulatory molecules, etc. The MyD88 independent signaling pathway is through the activation of TBK by the junction molecule TRIF, which ultimately activates the transcriptional regulator IRF, which upon activation enters the nucleus and binds to the type I interferon gene promoter to initiate transcription and expression of type I-interferon (IFN-α/β) genes (Fig. 3.6).TLR4, after recognizing LPS, can activate the IRF signaling pathway through TRIF adaptor molecules to initiate type I-interferon expression in addition to activating NF-κB and AP-1 signaling pathways through MyD88. In contrast, TLR2 recognizes the ligand and activates the NF-κB signaling pathway through PI3K adaptor molecules that activate Akt kinase (Fig. 3.6).

TLRs in endosomes primarily recognize viral nucleic acids and bacterial non-methylated DNA. After recognition of ligands, these TLRs also signal via two major pathways, MyD88-dependent and independent (Fig. 3.7). TLR3 signals primarily via the MyD88 independent signaling pathway, activating TBK1 and AKT via TRIF and PI3K, respectively, triggering IRF signaling pathway and initiating transcription and expression of downstream genes. Up to now, TLR3 has not been found to be associated with MyD88.

Immune Signaling Process Induced by NLR Recognition of PAMPs

As previously described, the NOD1 primarily recognizes peptidoglycan (PGN) of Gram-negative (G-) bacteria in the cytoplasm, while NOD2 recognizes PGN of Gram-positive (G+) bacteria. NOD binds to the corresponding ligands via LRR region, and the NBS regions of NOD monomers bind to each other to form trimers, followed by activation of the CARD functional region, recruitment and activation of RIP2, activation of TAK1 by RIP2 activation, and activation of NF-κB and AP-1 signaling pathways by TAK1 activation, initiating immune-related gene transcription and expression (Fig. 3.8).

In addition to recognition of bacterial peptidoglycans by NLRs (e.g., NOD1 and NOD2) to cause immune signaling and production of proinflammatory cytokines, certain members of the NLRC and NLRP subgroups, upon recognition of the corresponding PAMPs, recruit downstream molecules ASC and procaspase1 or

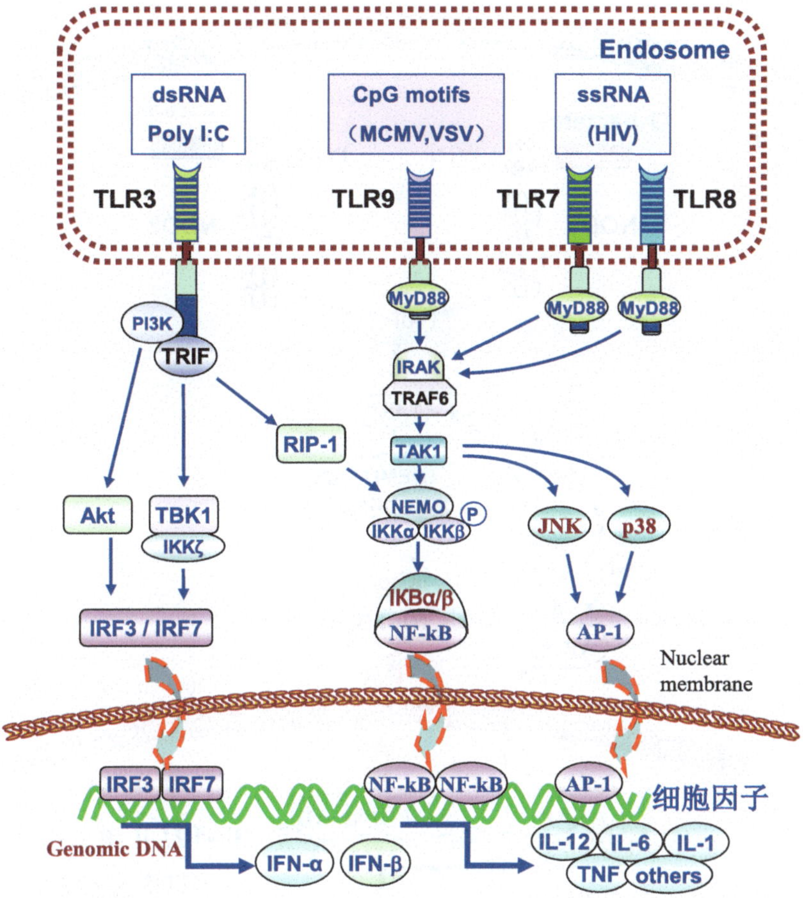

Fig. 3.7 Schematic diagram of the activation and signaling process of TLRs in endosomes. After TLRs recognize the corresponding ligands in endosomes, the TIR domain is activated, recruiting and activating MyD88, TRAF, PI3K, and other adaptor molecules. The signaling process via MyD88 requires TIR-TIR interactions. MyD88 activation recruits and phosphorylates IRAK, activated IRAK recruits and activates TRAF6, and TRAF6 activates TAK1. TAK1 activation eventually activates the NF-κB signaling pathway and also AP-1 via JNK and p38. The activated NF-κB and AP-1 enter the nucleus and bind to the promoters of specific genes to initiate transcription and expression of downstream genes. TLR3 recognizes ligands and recruits and activates TRIF, which activates TBK and Akt, and ultimately activates the transcriptional regulator IRF. The activated IRF enters the nucleus and binds to the type I-interferon gene promoter to initiate transcription and expression of the type I-interferon (IFN-α/β) gene. *MyD88* myeloid differentiation factor 88, *TIRAP* TIR domain containing adapter protein, *TRIF* TIR domain-containing adapter-inducing interferon-β, *TRAM* TRIF-related adapter molecule, *IRAK* IL-1 receptor kinase, *TRAF* tumor necrosis factor receptor-associated factor, *TAK* TGF-β activated kinase 1, *NF-κB* nuclear factor kappa enhancer binding protein, *JNK* c-Jun-N-terminal kinase, *AP-1* activating protein-1, *IRF* interferon regulatory factor, *RIP* receptor interacting protein, *PI3K* phosphoinositol-3 kinase, *NEMO* NF-κB essential modulator, *IKK* inhibitor of NF-κB kinase, *IκB* inhibitor of NF-κB, *poly(I:C)* synthetic dsRNA based on viral nucleic acid structure dsRNA, *MAVS* mitochondrial antiviral signaling protein, *NBS* nucleotide-binding site, *CARD* caspase activation and recruitment domain

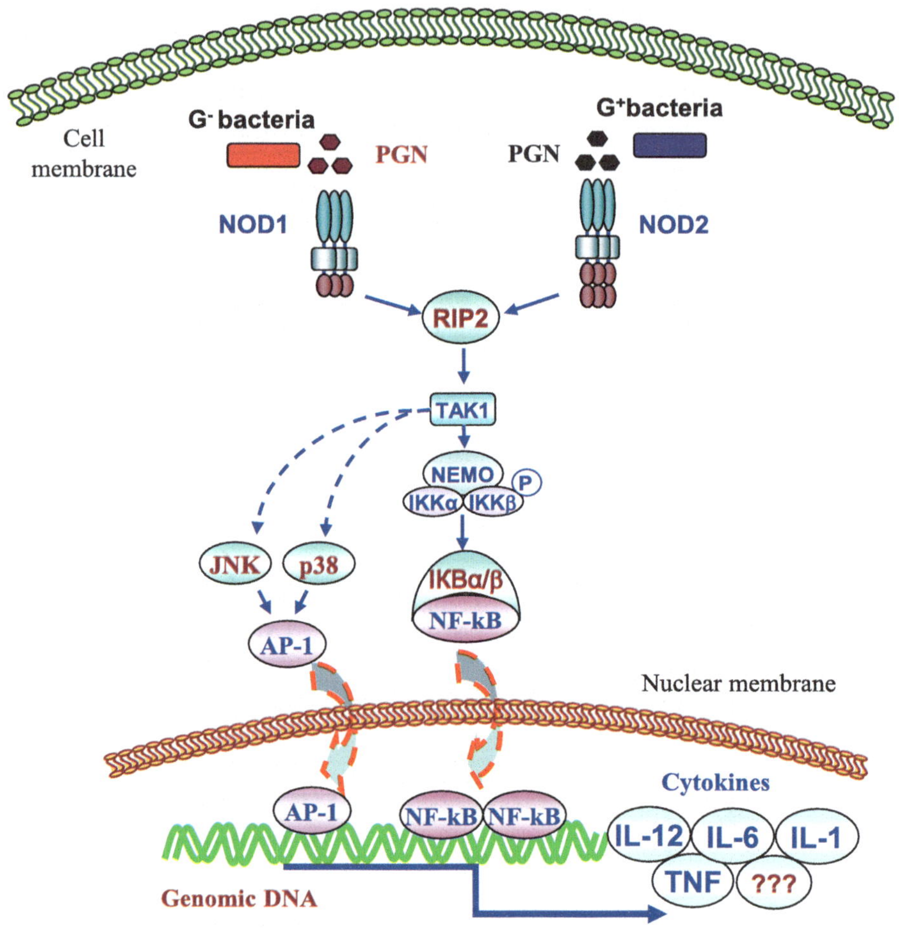

Fig. 3.8 Schematic diagram of the activation and signaling process of NLRs. NLRs are distributed in the cytoplasm and mainly recognize pathogenic components in the cytoplasm. NOD1 and NOD2 are representative members of the NLRC family. In normal cells without pathogenic infection, NOD exists as a monomer. When PAMPs are present in the cytoplasm of infected cell, NOD binds to the corresponding ligands via LRR region, and the NBS regions of NOD monomers bind to each other to form trimers, followed by activation of the CARD functional region, recruitment and activation of RIP2, activation of TAK1 by RIP2 activation, and activation of NF-κB and AP-1 signaling pathways by TAK1 activation, initiating immune-related gene transcription and expression. *NLRs* nucleotide-binding oligomerization domain (NOD)-like receptors, *PGN* peptidoglycan, *TAK* TGF-β-activated kinase, *NF-κB* nuclear factor kappa enhancer binding protein, *AP-1* activating protein-1, *RIP* receptor-interacting protein, *NEMO* NF-κB essential modulator, *IKK* inhibitor of NF-κB kinase, *IκB* inhibitor of NF-κB, *NBS* nucleotide-binding site, *CARD* caspase activation and recruitment domain

procaspase11 to assemble into inflammasome complexes, causing the release of pro-inflammatory cytokines (IL-1β and IL-18) and pyroptosis (Fig. 3.9).

Host cells recognize intracellular PAMPs or DAMPs through PRRs, which recruit ASCs, and Procaspase-1 (or Procaspase-11) to form inflammasomes, causing programmed lysis death, called pyroptosis. The inflammasome activation pathways are

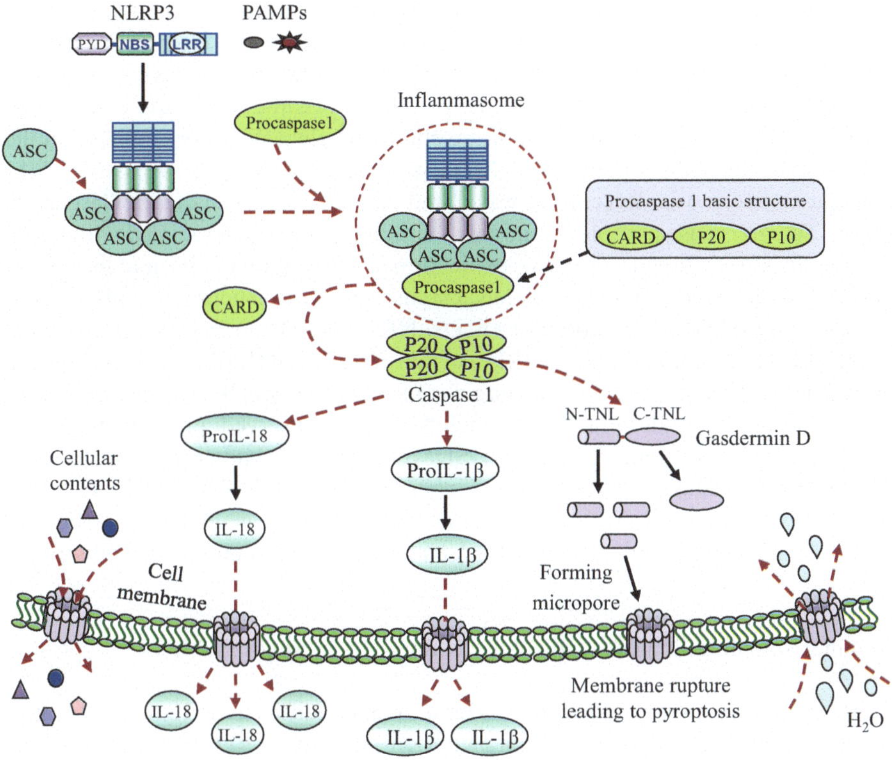

Fig. 3.9 Schematic diagram of the process by which NLRs recognize PAMPs to cause classical inflammasome formation and pyroptosis. NLRs (e.g., NLRP3), upon engagement with PAMPs, recruited ASCs and Procaspase1 (zymogen) to form an inflammasome complex, which has enzymatic self-cleaving activity and cleaves Procaspase1, releasing the CARD domain and forming the enzymatically active dimeric structure 2X (P20-P10), i.e., Caspase1. Caspase1 cleaves the aspartate target sites in the ProIL-1β and ProIL-18 molecules to produce mature IL-1β and IL-18; meanwhile, Caspase1 cleaves Gasdermin D to produce two fragments, the N-terminal (N-TNL) and C-terminal (C-TNL). The N-TNL fragment is inserted into the cell membrane to form micropores that assist in the release of IL-1β and IL-18 through the micropores to the outside of the cell, but the micropores also allow the influx of extracellular water, a increase in cellular osmotic pressure, and an overflow of cellular contents, leading to pyroptosis. *CARD* caspase activation and recruitment domain, *ASC* apoptosis-associated speck-like protein, *NLRP3* NLR with pyrin domain-containing Protein 3, *ProIL-1β/18* Pro-interleukin 1β/18

divided into classical and non-classical pathways based on the type of inflammasome activation caspases: 1) the classical inflammasome pathway, characterized by activation of Caspase-1 and production of IL-1β and IL-18; 2) the non-classical inflammasome pathway, characterized by activation of Caspase-11 and production of IL-1α and IL-18. In both pathways, the activated caspase-1 or caspase-11 cleaves the substrate Gasdermin D (GSDMD), releasing the N-terminal structural domain (N-TNL) and the C-terminal structural domain (C-TNL). N-TNL binds to the inner cell membrane to form micropores in the cell membrane that, in addition to assisting in the release of IL-1α/β as well as IL-18 from the cell, the cell membrane micropores allow for extracellular water influx, increased cellular osmotic pressure, and overflow of cell contents, causing pyroptosis. Since the cleaved N-TNL of GSDMD has a high affinity to Phospholipid and Cardiolipin, N-TNL can also be inserted into the membrane of organelles (e.g., endoplasmic reticulum, Golgi apparatus, lysosomes, etc.), resulting in the swelling and cleavage of organelles and the overflow of organelle contents. GSDMD is a substrate protein of caspase-1/4/5/11 that causes pyroptosis. Recently it was found that caspase-3 can cleave Gasdermin E (GSDME) to cause pyroptosis and that familial deafness is associated with mutations in the GSDME gene. GSDME expression is low in tumor cells, and increasing GSDME expression in tumor cells has a significant tumor suppressive effect, suggesting that GSDME has antitumor effects. NK cells, when releasing granzyme B to target cells, can directly cleave D270 of substrate GSDME (the same cutting site as cleaved by Caspase-3), inducing target cells (e.g., tumor cells) to undergo pyroptosis. (Of note, Gasdermin family members are conserved proteins including GSDMA, B, C, D, E, and DFNB59. GSDMD is the best-known membrane pore-forming protein causing pyroptosis, but other members of GSDM family were also found to be involved in pyroptosis. Interestingly, mice are genetically deficient of GSDMB, while up to now only GSDMA and E have been found in chickens.)

Immune Signaling Process Induced by RLRs Recognition of PAMPs

RLRs are present in the cytoplasm. Under normal conditions, the Helicase recognition region of RLRs is covered by its own RD/CTD and CARD structural domains. Upon viral invasion, the conformational change of RLRs exposes the Helicase functional region, which recognizes viral nucleic acids in the cytoplasm. After recognizing viral nucleic acids, the RLRs activate the CARD region of MAVS on the mitochondria through the CARD region, and the activated MAVS-CARD region activates the NF-κB and IRF signaling pathways to initiate the expressions of antiviral immune response genes. RIG-I and MDA-5 are representative members of the RLR family, while avian species are naturally deficient in RIG-I, thereby MDA-5 playing a major role in chicken response against viral infection. PolyI:C (a synthetic dsRNA that mimics viral nucleic acids) specifically binds to MDA-5 rather than RIG-1. RLRs recognize the naked phosphate group at 5′ terminus of viral nucleic acids. Since the phosphate group at the end of the host RNA in the cytoplasm is covered, i.e., capped (capping), it is not recognized by RLRs, which provides the material basis for RLRs to distinguish viral RNA from host RNA (Fig. 3.10).

Immune Signaling Process Induced by CLRs Recognition of PAMPs

CLRs are distributed on the cell membrane and recognize mainly extracellular fungal β-glucans, and act synergistically with TLRs. upon recognition of the corresponding ligands, CLRs recruit and activate Syk kinase through the intracellular functional region ITAM. Syk kinase activates Syk, and Syk activates CARD9. The

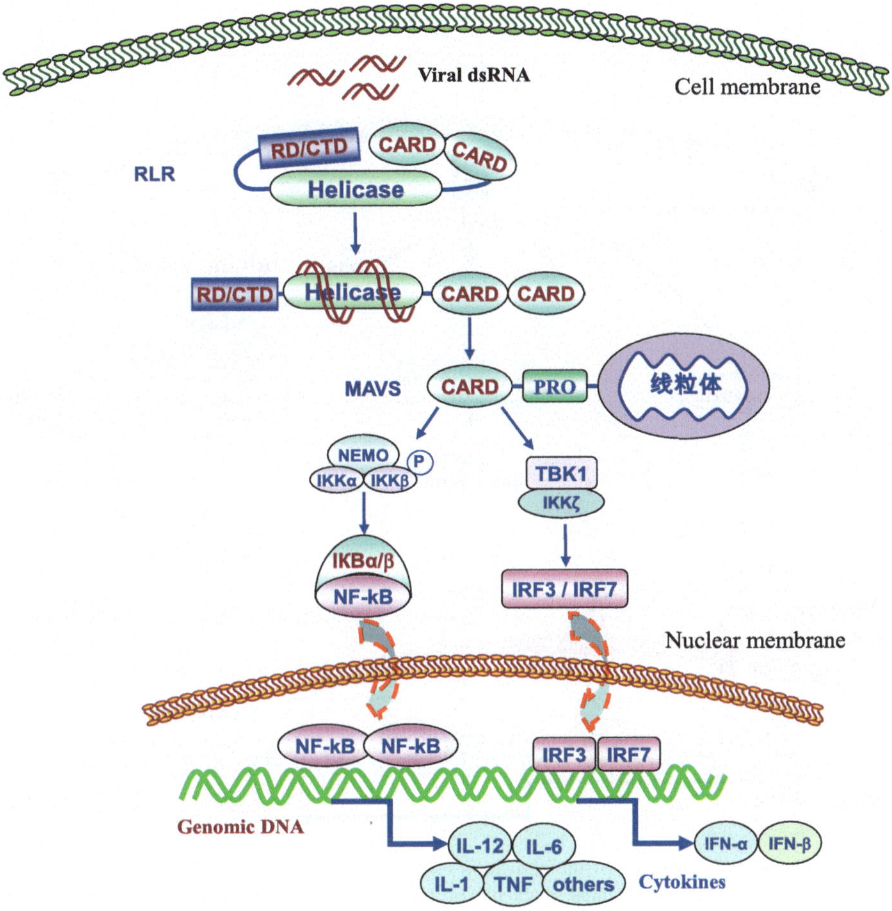

Fig. 3.10 Schematic diagram of the activation and signaling process of RLRs. Upon viral invasion, The conformational changes of RLRs expose the Helicase functional region, which recognizes viral nucleic acids in the cytoplasm. RLRs bind to the corresponding ligands and, through the CARD region, activate the CARD region of MAVS on mitochondria. The activated MAVS-CARD region activates the NF-κB and IRF signaling pathways, initiating the expression of immune-associated genes. *RLRs* retinoic acid-inducible gene I (RIG-I) like helicases receptors, *NF-κB* nuclear factor kappa enhancer binding protein, *IRF* interferon regulatory factor, *NEMO* NF-κB essential modulator, *IKK* inhibitor of NF-κB kinase, *IκB* Inhibitor of NF-κB, *MAVS* mitochondrial antiviral signaling protein, *CARD* caspase activation and recruitment domain, *TRAF* tumor necrosis factor receptor-associated factor, *TANK* TRAF-associated NF-κB activating factor, *TBK* TANK-binding kinase

activated CARD9 activates MALT1 and BCL-10, and MALT1/BCL-10 activates NF-κB and AP-1 signaling pathways. In addition, Syk activation triggers the Calcineurin-dependent NFAT signaling pathway, which initiates the expression of immune-related genes. In recent years, it has been found that fungal infection-induced Syk activation leads to the production of Reactive oxygen species (ROS),

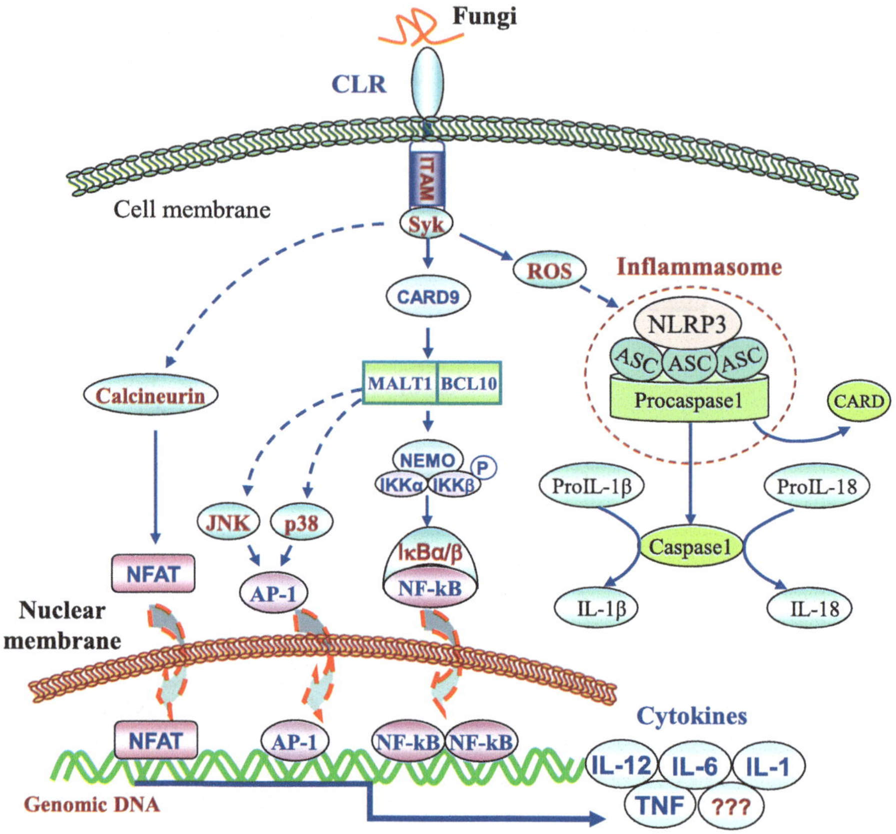

Fig. 3.11 Schematic diagram of the activation and signaling process of CLRs. *AP-1* activating protein-1, *ASC* apoptosis-associated speck-like protein, *CARD* caspase activation, and recruitment domain, *IKK* inhibitor of NF-κB kinase, *IκB* inhibitor of NF-κB, *JNK* c-Jun-N-terminal kinase, *IRF* interferon regulatory factor, *NEMO* NF-κB essential modulator, *NFAT* nuclear factor of activated T cells, *NF-κB* nuclear factor kappa enhancer binding protein, *IRF* interferon regulatory factor, *NLRP3* NLR with pyrin domain-containing Protein 3, *ROS* reactive oxygen species

K^+ efflux, and lysosomal histone protease release, which activates NLRP3, recruits ASC as well as Procaspase-1 to form inflammasome that produces Caspase-1. Caspase-1 cleaves ProIL-1β as well as ProIL-18, yielding mature IL-1β and IL-18 that participates in the inflammatory response (Fig. 3.11).

Immune Signaling Process Induced by ALR Recognition of PAMPs

ALRs are a new family of pattern recognition molecules discovered recently, which mainly recognize intracellular or nuclear pathogenic DNA or self-abnormal DNA. dsDNA recognition region (e.g., IFI200) is in a self-closed state when ALRs are at rest. When ALRs recognize dsDNA, signals will be generated and transduced to activate the STING signaling pathway on the endoplasmic reticulum to initiate

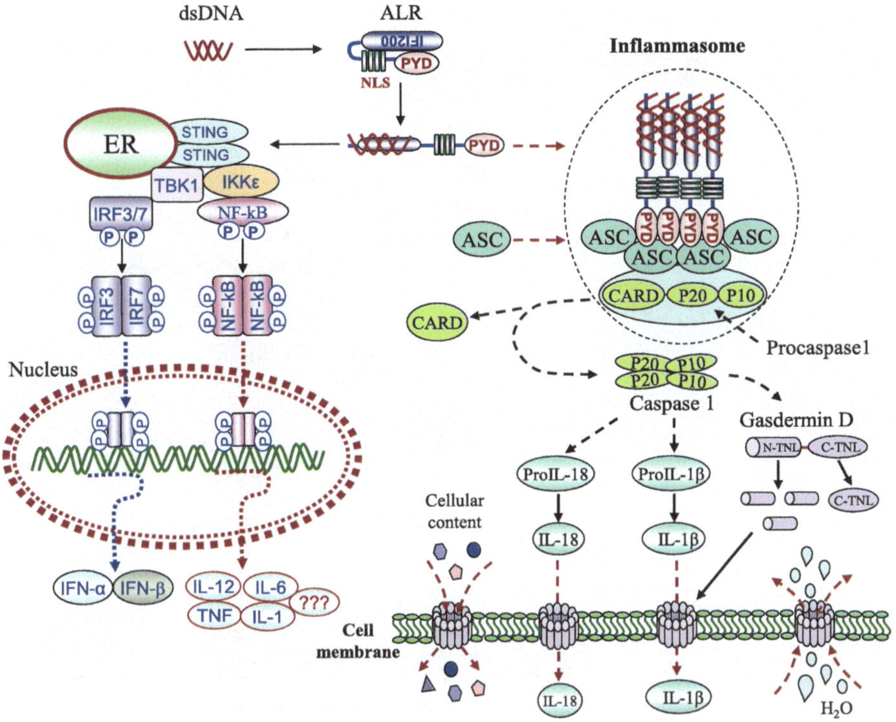

Fig. 3.12 Schematic diagram of the activation and signaling process of ALRs. *ASC* apoptosis-associated speck-like protein, *IFI200* interferon (IFN)-inducible 200 aa repeats, *NLS* nuclear localization signal motif, *PYD* Pyrin domain, *STING* stimulator of interferon genes, *NF-κB* nuclear factor kappa enhancer binding protein, *IRF* interferon regulatory factor, *IKK* inhibitor of NF-κB kinase, *IκB* Inhibitor of NF-κB, *CARD* caspase activation and recruitment domain, *TRAF* tumor necrosis factor receptor-associated factor, *TANK* TRAF family member-associated NF-κB activator, *TBK* TANK-binding kinase

I-type interferon expression via TBK1 activation of IRF3/7 or pro-inflammatory cytokine expression via activation of the NF-κB signaling pathway (see Fig. 3.12). In addition, ALRs recognize dsDNA and recruit ASC as well as Procaspase-1 to form inflammasomes and elicit an inflammatory response. Two types of ALRs are well studied, AIM2 and IFI16. AIM2 recognizes dsDNA in the cytoplasm, whereas IFI16 recognizes abnormal DNA both in the cytoplasm and in the nucleus. ALRs are not only involved in the anti-defense response but also associated with autoimmune diseases.

Immune Signaling Process Induced by cGAS Recognition of PAMPs

cGAS (cyclic GMP-AMP synthase), a pattern recognition receptor that recognizes intracytoplasmic dsDNA, was discovered in 2013. The receptor contains the NTase-Mab21 functional domain and is responsible for recognition and binding to dsDNA. cGAS recognizes both pathogenic dsDNAs and self-aberrant DNAs. cGAS

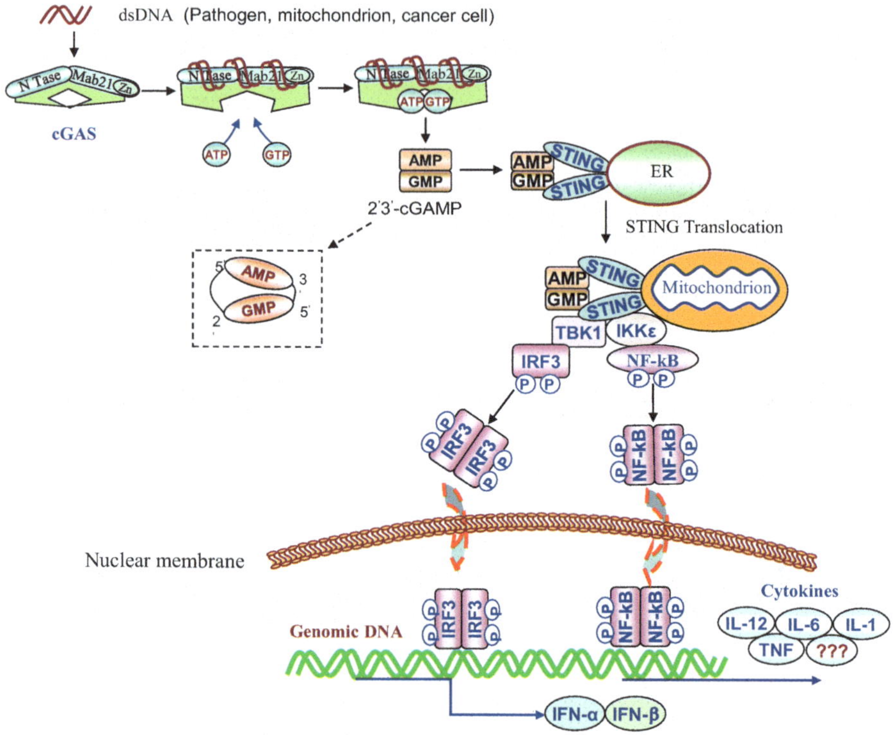

Fig. 3.13 Schematic diagram of the activation and signaling process of cGAS. *AMP* adenosine monophosphate, *ATP* adenosine triphosphate, *CARD* caspase activation and recruitment domain, *cGAS* cyclic GMP-AMP synthase, *cGAMP* cyclic GMP-AMP, *GMP* guanosine monophosphate, *GTP* guanosine triphosphate, *IKK* inhibitor of NF-κB kinase, *IRF* interferon regulatory factor, *IKK* inhibitor of NF-κB kinase, *IκB* Inhibitor of NF-κB, *Mab21* Male abnormal 21(a male nematode abnormal molecule 21), *NTase* Nucleotidyltransferase, *NF-κB* nuclear factor kappa enhancer binding protein, *STING* stimulator of interferon genes, *TBK* TANK-binding kinase, *TANK* TRAF-associated NF-κB activating factor, *TRAF* tumor necrosis factor receptor-associated factor

undergoes a conformational change upon recognition of dsDNA and catalyzes the substrates adenosine triphosphate (ATP) and guanosine triphosphate (GTP) to produce the second messenger cGAMP, which binds to and activates STING on the endoplasmic reticulum. The activated STING is translocated to the mitochondria to activate IRF3 and NF-κB signaling pathways, regulating the expression of interferon (I- or III-IFN) and pro-inflammatory cytokines (IL-1, IL-6, IL-12, TNF, etc.) to exert anti-defense or pro-inflammatory responses. Since cGAMP is a second messenger generated by cGAS after recognition of dsDNA, it directly activates the STING signaling pathway and is therefore commonly referred to as the cGAS-STING signaling pathway. cGAS not only recognizes pathogenic dsDNA to initiate the immune response and exerts anti-infective effects, but also recognizes self dsDNA and participates in antitumor, aging, and inflammatory processes (Fig. 3.13).

3.3.2.2 Phagocytosis, Killing and Clearance of Pathogens by Phagocytes

The animal consists of different tissues and organs, and the basic components of tissues and organs are cells. The core function of the immune system is to maintain homeostasis and guarantee normal physiological activities. The primary task of maintaining homeostasis is to maintain a sound internal environment at two major levels: (1) macroscopically, through phagocytosis, foreign substances (pathogens) and self-waste products (senescent and dead cells or debris) are phagocytosed and removed to maintain a healthy and stable internal environment; (2) microscopically, through autophagy, abnormal substances in cells (pathogenic components, misfolding or aggregated proteins, etc.) to maintain the intracellular environment and ensure normal physiological activities of cells. Thus, the intracellular environment is maintained by autophagy, while the extracellular environment requires phagocytosis and clearance by macrophages to maintain stability, and the substances removed by phagocytosis include pathogens, dead cells, tissue debris, immune complexes, metabolic wastes, etc.

Phagocytosis is the basic ability of cells to acquire substances from the outside, and all cells have this ability. Phagocytes are divided into professional phagocytes and non-professional phagocytes: professional phagocytes are myeloid cells with prominent phagocytic capacity, mainly including neutrophils, monocytes/macrophages, etc., which are responsible for the removal of pathogens, dead cells or cellular debris, antigen-antibody complexes, and various metabolites and degradants, while non-professional phagocytes have a weak phagocytic capacity and mainly include endothelial cells, fibroblasts, dendritic cells, and so on.

During pathogenic infection, neutrophils are the first to reach the site of infection and have a powerful phagocytic and killing effect on pathogens, phagocytosing more than 50 bacteria in a few minutes. Neutrophils phagocytose pathogens through two pathways: opsonophagocytosis and non-opsonophagocytosis. The opsonophagocytosis pathway depends on transmembrane receptors FcR (FcgRII (CD32), FcgRIII(CD16), FcgRI(CD64), FcaR(IgA receptor)) and complement receptors CR(C1qR,CR1(CD35), CR3(CD11b/CD18 or MAC-1), CR4(CD11c/CD18)) mediated-phagocytosis of pathogens by phagocytes, and these receptors enhance phagocytosis by neutrophils. The non-opsonophagocytic pathway relies on the recognition of PAMPs (e.g., lipopolysaccharide, cytosolic acid, glycoprotein, flagellin, etc.) by cell surface pattern recognition receptors (PRRs) to phagocytose pathogens. Neutrophils phagocytose pathogens to form phagosomes, which fuse with lysosomes as well as cytoplasmic granules, and pathogens are killed by a variety of antimicrobial substances (ROS, AMP, superoxide (O_2^-, H_2O_2, HOCL, etc.)). After phagocytosis and killing of pathogens, neutrophils then undergo apoptosis and are cleared by macrophages, and thus the entire process of pathogen killing and clearance by phagocytes has been completed. However, unlike mammals, avian species do not have neutrophils, which are replaced by heterophils. Heterophils lack many of the digestive enzymes that neutrophils are powerful at, making avians more susceptible to bacterial infection. Although neutrophils play a major role in phagocytosis and the killing of pathogens, macrophages are ultimately needed for the clearance

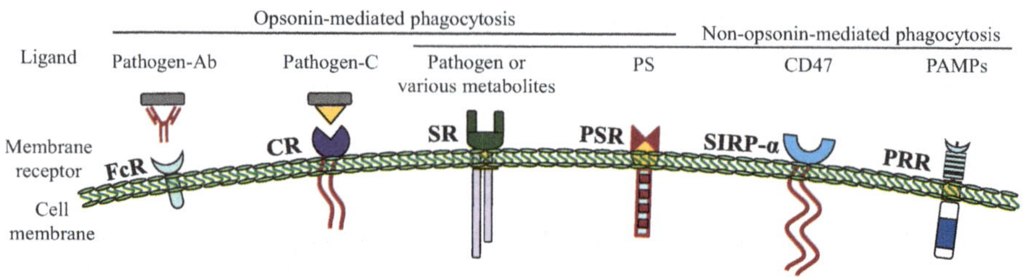

Fig. 3.14 Schematic diagram of the receptors by which phagocyte exerts phagocytosis. *Ab* antibody, *C* Complement, *CR* Complement receptor, phagocytic CR mainly includes C1qR, CR1(CD35), CR3 (CD11b/CD18 or MAC-1), CR4 (CD11c/CD18), and so on; FcR: Immunoglobulin Fc receptor, the FcR of phagocytes mainly includes FcgRI (CD64), FcgRII (CD32), FcgRIII(CD16), FcaR, and so on; *SR* Scavenger receptor, mammalian SR mainly includes SR-A~SR-B, SR-D~SR-L family, each family containing multiple receptors; *PS* phosphatidylserine, an early marker for apoptosis; *PSR* phosphatidylserine receptors, mainly including BAI-1 (Brain-specific angiogenesis inhibitor-1) and stabilin-2; *SIRP-α* Signal regulatory protein-α, an α-signal regulatory protein that recognizes and binds to the ligand CD47; *PRR* Pattern recognition receptor

of pathogens to end up. During pathogenic infection, regardless of how the immune response proceeds, phagocytosis and clearance are ultimately required to end the process. It is like a war that ultimately relies on macrophage phagocytosis and clearance to end the war and clean up the battlefield. The following is a brief description of the basic process of phagocytosis, killing and clearance of pathogens by macrophages, as well as the process of autophagy by cells to maintain the stability of the internal environment, using pathogenic infections as an example.

Basic Processes of Phagocytosis and Removal of Foreign Materials by Macrophages

Macrophages have a very prominent phagocytic and scavenging capacity and are responsible for the removal of pathogens, senescent and dead cells or debris, antigen-antibody complexes, and various metabolites or degradates, thus named Scavenger. The macrophage phagocytosis is mainly in the form of receptor-mediated phagocytosis and direct phagocytosis. The receptor-mediated phagocytosis is caused by cell membrane receptors binding to the corresponding ligands. The receptors associated with phagocytosis include six major categories: complement receptors (CR), Fc receptors (FcR), PRRs, scavenger receptors (SR), phosphatidylserine receptors (PSR), and α-signal-regulated proteins (SIRPα). Each class of receptors, except SIRPα, contains multiple receptors. These membrane receptors recognize ligand-induced phagocytosis, and phagocytosis is classified as either opsonophagocytosis or non-opsonophagocytosis, and some receptors can cause both types of phagocytosis (see Fig. 3.14).

Macrophages exert phagocytosis through various membrane receptors in vivo, however, macrophages cultured in vitro can phagocytose glass or plastic beads with smooth surfaces, suggesting that macrophages do not rely solely on membrane receptors to phagocytose foreign materials. No doubt, membrane receptors binding to corresponding ligands can greatly facilitate phagocytosis. The membrane receptors and their roles associated with macrophage phagocytosis are as follows.

FcR: Immunoglobulin Fc receptor. FcRs on macrophage membranes mainly include FcgRI (CD64), FcgRII (CD32), FcgRIII (CD16), FcaR, etc. Macrophages capture the Ag-IgG complex via the membrane receptor FcgR binding to IgG-Fc, and capture the Ag-IgA complex via the membrane receptor FcaR binding to IgA-Fc, followed by phagocytosis of the Ag-Ab complex by macrophages, a process known as antibody-mediated opsonophagocytosis. Opsonophagocytosis is the process of phagocytosis induced by recognition of opsonin by membrane receptors. Opsonins are soluble molecules that bind to microbials, marking them for phagocytosis, for example, antibodies and complement components are important and commonly seen opsonins that can induce efficient phagocytosis.

CR: Complement receptor, the CRs of phagocytes are mainly C1qR, CR1 (CD35), CR3 (CD11b/CD18 or MAC-1), CR4(CD11c/CD18), and so on. The binding of IgG to Ag exposes the complement binding site at the heavy chain CH2 domain, which binds to complement C1q, and activates the classical complement pathway (see Chap. 11 Complement System), producing a membrane-attacking complex killing pathogen; meanwhile, macrophages phagocytose the Ag-Ab complex via C1qR binding to C1q or exert opsonophagocytosis via membrane CR1 and CR3 binding to C3b.

SR: Scavenger receptor, there are various types of SR, but not all receptors mediating phagocytosis are SR. Currently, it is considered that SR must have three characteristics: (1) acting as cell surface receptor, (2) having multiple ligands, and (3) involved in phagocytosis and clearance of foreign or self-abnormal substances. In addition, SR ultimately removes degradants or harmful substances through mechanisms such as endocytosis, phagocytosis, adhesion, and signal transduction. Mammalian scavenger receptors mainly include 11 large families such as SR-A, SR-B, SR-D–SR-L, each containing multiple receptors (different receptors are responsible for scavenging different substances), while SR-C is only found in Drosophila. For example, macrophages bind to and scavenge fungi, heat shock proteins (HSP), and low-density lipoproteins (LDL) via SR-F1 (also known as SCARF-1). Macrophages play two roles through SR: opsonophagocytosis or non-opsonophagocytosis. In addition to removing pathogens, they mainly remove harmful substances, degraded substances, and conformationally altered waste products produced by their own metabolism, which are crucial for maintaining self-stability and ensuring normal physiological activities of the body.

PSR: phosphatidylserine receptor, a collective term for a variety of membrane receptors that bind phosphatidylserine (PS). BAI-1 (brain-specific angiogenesis inhibitor-1) and Stabilin-2 are commonly seen PSR. Under normal conditions, PS is located in the inner membrane of the lipid bilayer of the cell membrane, and if PS flips over to the outer membrane, it indicates that the cell is undergoing early apoptosis. Therefore, PS flipping over to the outer membrane is one of the main features of apoptosis and is also a "eat me" signal sent by the cell. When the outer membrane of apoptotic cells exposes PS, the apoptotic cells can be phagocytosed directly by macrophages via PSR binding to PS, or taken in by opsonophagocytosis through the binding of PSR to PS-associated soluble molecules (Protein S, Gas6, C1q, etc.).

SIRP-α: signal regulatory protein-α, a transmembrane receptor expressed by macrophages that binds to ligand CD47 to produce a signal to inhibit myosin-II-mediated phagocytosis. CD47 is a cell membrane receptor, a self-recognition marker (Self-marker) of the immune system. CD47 is a cell membrane receptor that binds to macrophage SIRP-α to inhibit its phagocytosis. For example, normal erythrocytes express CD47 to avoid phagocytosis and clearance by macrophages, but at the end of their life span (approximately 120 days), CD47 expression declines significantly in senescent erythrocytes, which is eventually phagocytosed and cleared by macrophages. The same is true for other cells, which are phagocytosed and cleared by macrophages as foreign materials if the cell membrane is deficient in CD47. Apoptotic cells have decreased CD47 expression, which facilitates the clearance of these cells by macrophages.

PRRs; pattern recognition receptors recognize PAMPs. Macrophages recognize PAMPs through PRRs to generate immune signals and upregulate phagocytic receptor expression, which promotes the phagocytosis of pathogens by macrophages.

The process of phagocytosis and clearance by macrophages is divided into three main phases: an initiation phase of phagocytosis, a phagosome formation phase, and a phagosome maturation phase (see Fig. 3.15). The basic process of phagocytosis and clearance of pathogens by macrophages, taken as an example, is as follows.

Initiation of phagocytosis: macrophages recognize pathogens through the direct binding of membrane PRRs to PAMPs, which may directly initiate phagocytosis or indirectly trigger phagocytosis by promoting the expression of non-opsonic receptors that bind to pathogens. In addition, macrophages can capture pathogens through membrane receptor binding to opsonins that bind to pathogen, which promotes phagocytosis. The signal transduction for phagocytosis is initiated by the engagement of membrane receptors with corresponding ligand, causing actin cytoskeletal changes to form a pseudopod that protrude and surround the receptor–pathogen site and finally encapsulate the pathogen.

Phagosome formation: Fusion of cell membrane pseudopods surrounding the pathogen allows phagocytosis of the pathogen, forming the early phagosome.

Phagosome maturation: the process by which the early phagosomes fuse with late endosomes to form late phagosomes and further fuse with lysosomes to form phagolysosomes is called phagosome maturation. During phagosome maturation, the pH of the phagosomal environment decreases, and the pH within the phagolysosome drops to about 5~ 5.5, which facilitates the degradation of pathogens.

Basic Processes of Autophagy

The main function of the immune system is to maintain homeostasis through the immune response, and the ultimate task is performed by macrophages, which maintain the balance of the internal environment and homeostasis through phagocytosis

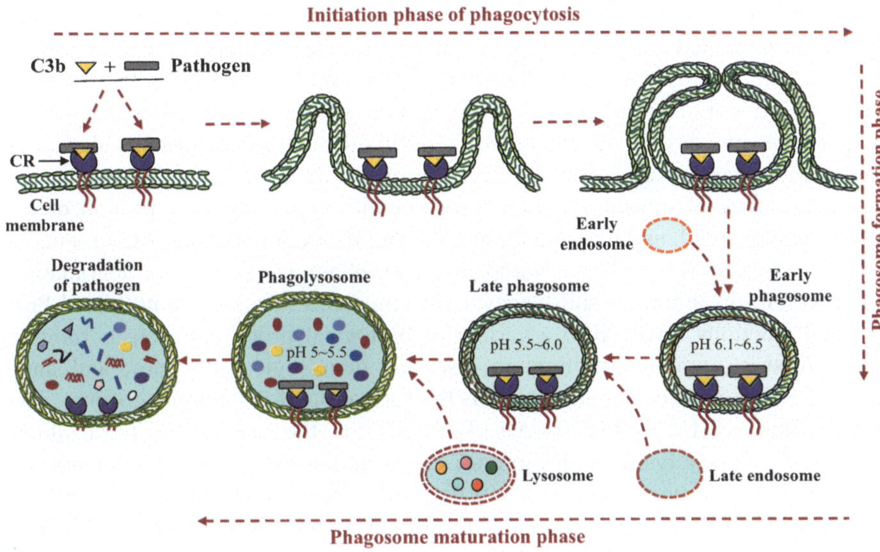

Fig. 3.15 Schematic diagram of the process by which macrophage phagocytose and clear pathogens. Complement (e.g., C3b) binds to pathogens (bacteria) to form complexes to exert opsonin effects, macrophages bind to complement through complement receptors (CR) to capture pathogens, causing changes in the actin cytoskeleton, cell membrane protrude to form pseudopods, pseudopods surround pathogens causing membrane fusion triggering endocytosis. The newly formed phagosomes detach from the cell membrane and fuse with early endosomes to form early phagosomes. The early phagosomes carry the markers of membrane proteins such as Rab5, EEA1, and hvPS34, which acidify the internal environment (pH 6.1~ 5.5) by accumulating hydrogen ions through the membrane protein V-ATPase. Early phagosomes further fuse with late endosomes to form late phagosomes, which carry the markers of membrane proteins such as Rab7 and LAMPs, and the internal environment is further acidified by V-ATPase (pH 5.5~ 6.0). Late phagosomes further fuse with lysosomes to form phagolysosomes, a process called phagosomal maturation. The phagolysosomal internal environment is further acidified (pH 5~ 5.5) by the action of V-ATPase, and pathogens are degraded and killed by various degradative enzymes (such as cathepsins, proteases, lysozymes, and lipases, hydrolase, NADPH oxidase, etc.). *C* complement, *CR* complement receptor, *EEA1* early endosome antigen 1, *hvPS34* human vacuolar protein-sorting 34, *LAMPs* lysosomal-associated membrane proteins, *Rab* is a GTPase (GTPase) that It is a small G protein with a molecular weight of 20~30kD, with more than 70 types, widely localized to membrane-structured organelles and involved in protein vesicle formation, transport, and fusion, *NADPH* Nicotinamide adenine dinucleotide phosphate, also known as reductive coenzyme II, is a hydrogen donor for a variety of anabolic reactions in the body

and clearance (e.g., senescent and dead cells, foreign materials, antigen-antibody complexes, and so on.). However, cells are the basic units of tissue and organ structure, and the stability of internal environment of the cells themselves needs to be maintained to ensure normal physiological activities. Autophagy is a biological process by which cells take cytosolic abnormal substances (misfolded molecules, cohesions, damaged or senescent organelles, etc.) into the autophagosome for recycling through lysosomal degradation, which is an evolutionarily conserved lysosomal degradation pathway. If the material taken into the autophagosome for degradation is the component of invading pathogens, this process is designated xenophagy.

There are three types of autophagy: Macroautophagy, Microautophagy, and Chaperone-mediated autophagy (CMA), all of which essentially remove abnormal substances through lysosomal degradation to maintain the stability of the intracellular environment and safeguard normal physiological activities. Microautophagy refers to the process by which lysosomes directly engulf abnormal substances through membrane invagination for degradation, while CMA refers to the process by which abnormal substances enter lysosomes through chaperone protein molecules to be degraded, and macroautophagy is the most common type of autophagy, which is generally referred to as autophagy. Autophagy is a process of degradation and recycling of abnormal substances, consisting of a series of complicated biochemical reactions involved in a variety of biological processes such as tumors, diabetes/obesity, aging, autoimmune diseases, neurological diseases, and anti-defense response. A core of autophagy is the formation of the autophagosome. The ULK1 complex (ULK1-FIP200-Atg13-Atg101) is the sensor for intracellular amino-acid (mainly lysine and arginine) levels and stress stimuli. Under normal physiological conditions, activated mTORC interacts with and inhibits the activation of the ULK1 complex, but under nutritional deprivation or stress, AMPK is activated, and AMPK inhibits mTORC activity and causes activation of the ULK1 complex, initiating the autophagic process (see Fig. 3.16).

In general, the initiation of autophagy is located at the membrane edge of organelles such as mitochondria, endoplasmic reticulum, or endophagosomes. The activated ULK1 complex directly recruits and activates Vps34/BECLIN1, and further recruits VPS15/Atg14L to form the Vps34-BECLIN1-VPS15-Atg14L complex, which catalyzes and allows the addition of phosphatidylinositide-3-phosphate to lipid-like molecules at the autophagy initiation site, forming the core of the autophagosome initiation site (nucleation). The Vps34-BECLIN1-VPS15-Atg14L complex is regulated by the VMP1-AMBRA1 complex or other regulatory molecules. The entire process of autophagy is broadly divided into five phases: initiation, formation of double-membrane nucleation, formation and extension of phagophore, engulfment of unwanted materials and formation of autophagosome, formation of autolysosome and degradation of engulfed cargo (see Fig. 3.16), as follows.

Initiation: In response to certain factors (e.g., amino-acid starvation or low levels of energy or other stress), phosphoadenosine-activated protein kinase (AMPK) is activated. AMPK inhibits mTORC, and activates the ULK1 complex (ULK1-FIP200-Atg13-Atg101), initiating the autophagic process.

Formation of double-membrane nucleation: the ULK1 complex activates phosphatidylinositide 3-kinase (Vps34) as well as BECLIN1, which recruits VPS15 and Atg14L to form the type III PI3K complex (Vps34-BECLIN1-VPS15-Atg14L) complex, which recruits molecules such as Atg2, Atg9, and WIPI, and adds phosphatidylinositol-3-phosphate to assemble a membrane core as a starting point for the formation of autophagosomes. The catalytic role of the PI3K complex is regulated by the VMP1-AMBRA1 regulatory complex.

Formation and extension of phagophore: after membrane core formation, the lipid membrane grows in two ways: (1) catalyzed by the Vps34-BECLIN1-

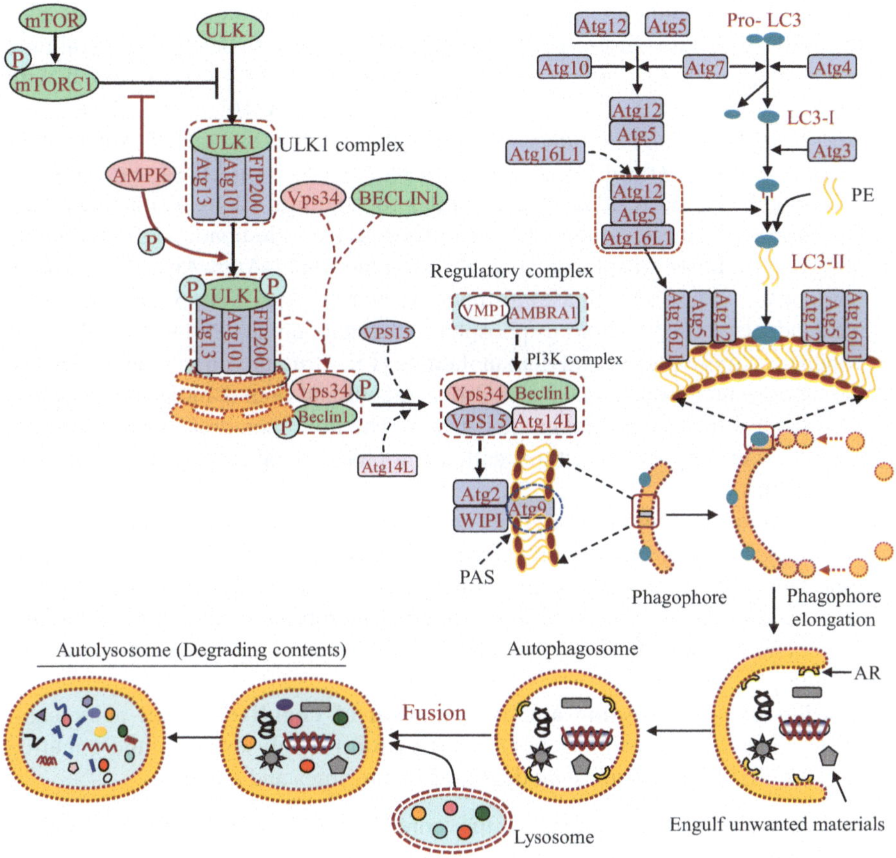

Fig. 3.16 Schematic diagram of the process of autophagy. Abbreviation and annotation where necessary, *AMPK* Adenosine monophosphate-activated protein kinase, a sensor for intracellular nutrient (mainly lysine and arginine) levels, energy metabolism and stress stimuli, activates AMPK if nutrient or energy decrease below threshold, *Beclin1* a protein with a molecular weight of 60 kD, this protein interacts with BCL2 to inhibit Sinbis virus replication, thus designated Beclin1, and in autophagy Beclin1 interacts with Vps34 to further form a PI3K complex with Vps15/Atg14L. *mTOR* mammalian targets of rapamycin, *mTORC* mammalian targets of rapamycin complex, *Atgs* autophagy related genes, *AR* autophagy receptor, mainly binds to ubiquitin, galactose lectin, etc. on the surface of the molecule, *FIP200* FAK-family interacting protein of 200 kDa (molecular weight), *LC3* microtubule-associated protein 1 light chain 3, *PE* phosphatidylethanolamine, *PAS* Phagophore assembly site, the core site of phagophore formation, *ULK1* Unc-51 like autophagy activating kinase 1, autophagy-activating kinase 1 similar to Unc-51, forming a complex with FIP200, Atg13 and Atg101, *Vps15/34* vacuolar protein sorting molecule 15/34, belonging to the type 3 phosphatidyl inositol 3 kinase catalytic subunit, forming PI3K complex with Beclin1 as well as Atg14L, regulated by regulatory complex (VMP1-AMBRA1), *WIPI* WD-repeat *protein* interacting with phosphoinositides; *PI3K* phosphatidylinositol 3-kinase

VPS15-Atg14L complex, the membrane core (i.e., the initiation site) keeps adding phosphatidylinositol-3-phosphate-lipid molecules to keep the membrane extension growth; (2) catalyzed by Atg7 and Atg10, Atg12 forms a complex with Atg5 (Atg12-Atg5), which further recruits Atg16L1 to form the Atg12-Atg5-Atg16L1 complex, while the microtubule-associated protein light chain 3 precursor (Pro-LC3) is catalyzed by Atg4 to form LC3-I, exposing the C-terminal glycine residue, Atg7 (similar to ubiquitin El ligase) activates the C-terminal glycine residue of LC3-I to form a disulfide bond, Atg3 (similar to ubiquitin E2 ligase) displaces Atg7, allowing the Atg12-Atg5-Atg16L1 complex (similar to ubiquitin E3 ligase) to function, and the Atg12-Atg5-Atg16L1 complex induces LC3-I to bind to phosphatidylethanolamine (PE) to form LC3-II, which contains a phosphatidylethanolamine (PE) structure directly embedded in the vesicle membrane, making LC-II a membrane-bound molecule (whereas LC3-I is a soluble molecule in the cytosol). The lipid membrane grows continuously to form cup-like structures called autophagosomal precursors or autophagic vesicles.

Engulfment of unwanted materials and formation of autophagosome: The autophagic vesicle engulfs unwanted materials (e.g., misfolded proteins, nonfunctional cohesions, pathogenic components, senescent or damaged organelles, etc.) via autophagic receptors in the internal membrane, and as the membrane grows and extends, the ends meet and fuse to form a complete vesicle structure, the autophagosome.

Formation of autolysosome and degradation of engulfed cargo: autophagosomes fuse with lysosomal membranes to form autolysosomes, and engulfed cargo in the autolysosomes are degraded by enzymes and recycled.

Maintenance of homeostasis is the core function of the immune system. Intracellular environment stability is maintained through the process of autophagy (removal of pathogenic components, misfolded proteins/cohesions, damaged or senescent organelles, etc.) at a microscopic level, whereas in vivo extracellular environment, homeostasis needs to be maintained through phagocytosis (removal of pathogens, dead cells and debris, immune complexes, metabolic wastes, etc.) by macrophages at a macroscopic level. Thus, the immune system plays a role in maintaining homeostasis and safeguarding the normal physiological activities of the organism at both microscopic and the macroscopic levels.

3.3.3 NF-κB Signaling Pathway

The innate immune response is a physiological defense response of the organism against infection by foreign pathogenic microorganisms, and this defense mechanism exists in all lives from lower to higher organisms, including humans. The innate immune response begins with the recognition of specific components of pathogenic microorganisms (pathogen-associated molecular patterns, PAMPs) through pattern recognition receptors (PRRs). As mentioned above, there are six major families of PRRs, including TLRs, NLRs, RLRs, CLRs, ALRs, and cGAS-STING. These receptors recognize the corresponding pathogenic components and initiate different levels of immune responses against different pathogens through specific signaling pathways, reflecting physiological response processes that are both independent and complementary to each other. Although different PRRs mediate different immune responses, all can initiate transcription and expression of immune-related genes through activation of NF-κB signaling pathways. It is thus clear that the NF-κB signaling pathway plays a very important role in the immune response.

3.3.3.1 Basic Structure of NF-κB, IκB, and IKK Family Members

NF-κB/Rel is the most important transcriptional regulator, regulating the transcription and expression of a variety of genes, and is closely related to physiological processes such as cell activation and proliferation, immune response, and inflammatory response. The family has five members, RelA, RelB, C-Rel, p100/p52 (NF-κB2), and p100/p50 (NF-κB1) (Fig. 3.17). These members all share the same RHD (Rel homology domain) functional region. NF-κB is tightly controlled by the upstream molecule IκB (inhibitor of NF-κB). The IκB family also has five members, IκBα, IκBβ, IκBε, IκBγ, and BCL3. With the exception of BCL3, the other IκB

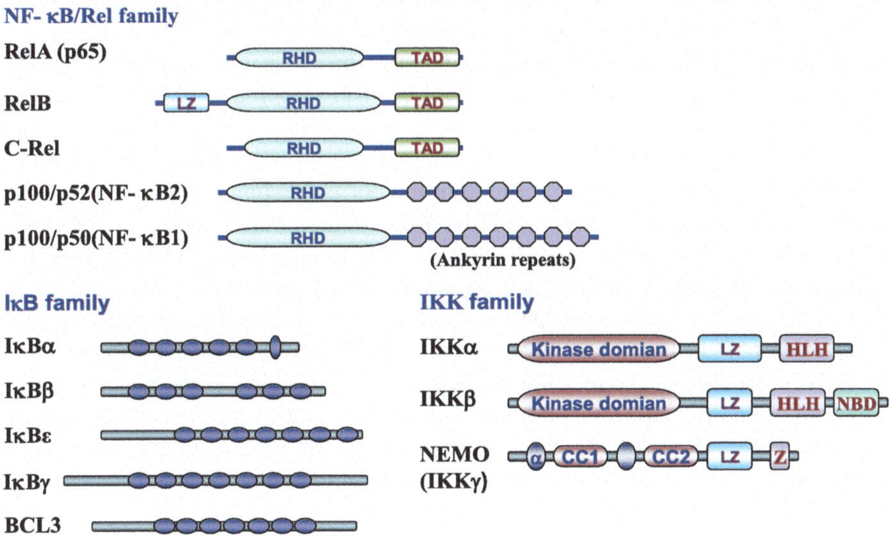

Fig. 3.17 Schematic diagram of the basic structures of NF-κB, IκB, and IKK family members. *LZ* leucine zipper domain, *HLH* helix-loop-helix domain, *NBD* NEMO-binding domain, *RHD* Rel homology domain, *TAD* transactivation domain, *NF-κB* nuclear factor kappa enhancer binding protein, *NEMO* NF-κB essential modulator, *IKK* inhibitor of NF-κB kinase, *IκB* inhibitor of NF-κB

family members are responsible for tightly controlling the activity of the transcriptional regulator NF-κB. In turn, IκB is controlled by the upstream molecule IKK (IκB kinase), and the IKK family has three members, IKKα, IKKβ, and IKKγ (also known as NEMO).IKK receives upstream signals that activate IκB and unrestricted IκB from NF-κB, thus NF-κB moves from the cytoplasm to the nucleus, binds to specific promoters, and regulates the transcription and expression of target genes.

3.3.3.2 NF-κB Signaling Process

NF-κB signaling is the best-understood molecular signaling process, which is divided into a classical pathway (Canonical pathway) and an alternative pathway (Non-canonical pathway).

1. **Classical pathway of NF-κB signaling:** In the classical pathway of NF-κB signaling, releasing NF-κB from the restraint of IκB is the key to activate the NF-κB signaling pathway. When cell receptors (TNFR1/2, TCR, BCR, PRRs) bind to the corresponding ligands, the intracellular activation domain of the receptor initiates recruitment, and activation of adaptor molecules to transduce signals downstream to activate TAK1, and the activated TAK1 activates IKKα and IKKβ. IKKα/IKKβ are the kinases of IκB. The activated IKKα/IKKβ phosphorylates IκB. In the presence of SCFbTrCP-type E3 ligase, the phosphorylated IκB is ubiquitinated and degraded by proteasomes, allowing p50/RelA to be released and translocate to the nucleus, where it binds to the promoter of a specific gene and initiates transcription and expression of the related genes (Fig. 3.18).
2. **Alternative pathway of NF-κB signaling:** In the alternative pathway of NF-κB signaling, degradation of NF-κB p105 and p100 to p50 and p52 is critical for activation of this signaling pathway. The formation of p52/RelB heterodimers is a hallmark of the NF-κB signaling alternative pathway. When cell receptors (BAFF-R, LTβ-R, CD40R, etc.) bind to the corresponding ligands, they recruit and activate the adaptor molecule NIK. The activated NIK activates IKKα, causing IKKα to bind p105 or p100, which is ubiquitinated by SCFbTrCP-type E3 ligase and degraded via the 26s proteasome. The degradation of p100 produces p52, which can form a p52/p52 homodimer or a p52/RelB heterodimer (the resulting p50 usually interacts with p65 to form a p50/RelA heterodimer). p52/RelB heterodimer can enter the nucleus directly and bind to the promoter of a specific gene (promoter) to initiate transcription and expression of the related genes (Fig. 3.18), while the p52/p52 homodimer can form a complex with Bcl3 that regulates the transcription and expression of the related genes.

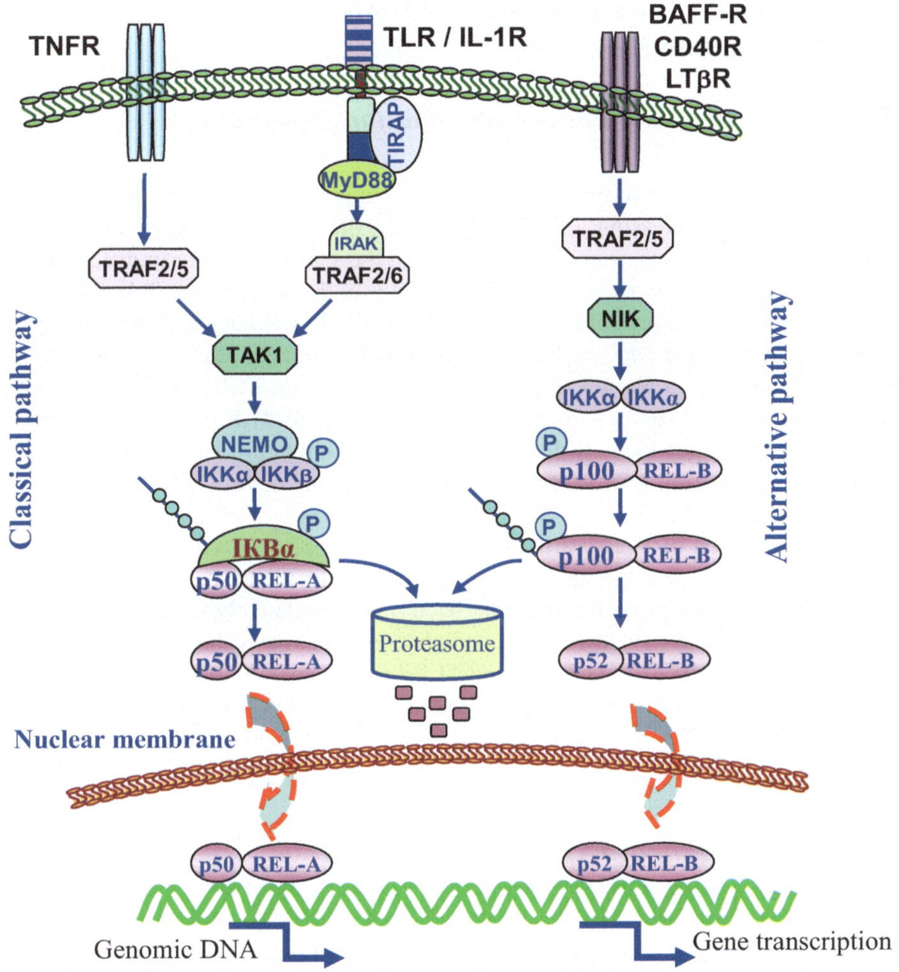

Fig. 3.18 Schematic diagram of the NF-κB signaling process. Classical pathway or Canonical pathway: cell receptors (TNFR1/2, TCR, BCR, or PRRs) bind to the corresponding ligands, and subsequently recruit and activate the adaptor molecules to signal downstream to activate TAK1, which activates IKKα and IKKβ. IKKα/IKKβ are the kinase of IκB. The activated IKKα/IKKβ phosphorylates IκB. In the presence of SCFbTrCP-type E3 ligase, the phosphorylated IκB is ubiquitinated and degraded by proteasomes, allowing p50/RelA heterodimer to be released and translocate to the nucleus, where it binds to the promoter of a specific gene and initiates transcription and expression of the target genes. Alternative pathway or non-Canonical pathway: cell receptors (BAFF-R, LTβ-R, CD40R, etc.) bind to the corresponding ligands, recruiting and activating the adaptor molecule NIK. The activated NIK activates IKKα, causing IKKα to bind p105 or p100, which is ubiquitinated by SCFbTrCP-type E3 ligase and degraded via the 26s proteasome. The degradation of p100 produces p52, which can form a p52/p52 homodimer or a p52/RelB heterodimer. p52/RelB heterodimer can enter the nucleus directly and bind to the promoter of a specific gene (promoter) to initiate transcription and expression of the related genes. *TAK* TGF-β activated kinase, *NF-κB* nuclear factor kappa enhancer binding protein, *NEMO* NF-κB essential modulator, *IKK* inhibitor of NF-κB kinase, *IκB* inhibitor of NF-κB, *NIK* NF-κB-inducing kinase

3.4 Expression and Regulation of Pattern Recognition Receptors

3.4.1 Expression of Pattern Recognition Receptors

Pattern recognition receptors (PRRs) are differentially expressed in different cells, with phagocytes expressing all PRRs. However, the expression of PRRs can also vary in different species or in different states in the same species or in different tissues of the same species or even in different states of the same tissue. Take TLRs for example, mast cells express TLR2, TLR4, TLR6, and TLR8, but not TLR5. Different subpopulations of dendritic cells express different TLRs, e.g., MDC (myeloid dendritic cells) in human blood express TLR1, TLR2, TLR4, TLR5, and TLR8. PDC (plasmacytoid dendritic cell) expresses only TLR7 and TLR9. Cells at different developmental stages express different TLRs, e.g., immature DCs express TLR1, TLR2, TLR4, and TLR5. TLR3 is expressed only in mature DCs. Each tissue expresses at least one TLR.

The distribution of TLRs varied in different regions of the same tissue, such as low expression of TLR4 at the end of the intestinal villi to avoid excessive cell response to LPS produced by intestinal bacteria, and relatively high expression of TLR5 in the crypt to recognize pathogenic bacteria that enter the crypt (Fig. 3.19). Renal epithelial cells express TLR2 and TLR4 to recognize G-bacteria and prevent septic nephritis, and ocular corneal epithelial cells and vascular endothelial cells express TLR4.

3.4.2 Regulation of Expression of Pattern Recognition Receptors

3.4.2.1 Expression of TLRs Is Regulated by Cytokines

Colony-stimulating factor (CSF) decreases TLR9 expression in macrophages and inhibits CpG DNA-induced inflammatory cytokines. Macrophage movement inhibitory factor (MIF) decreases TLR4 expression. IFN-γ enhances

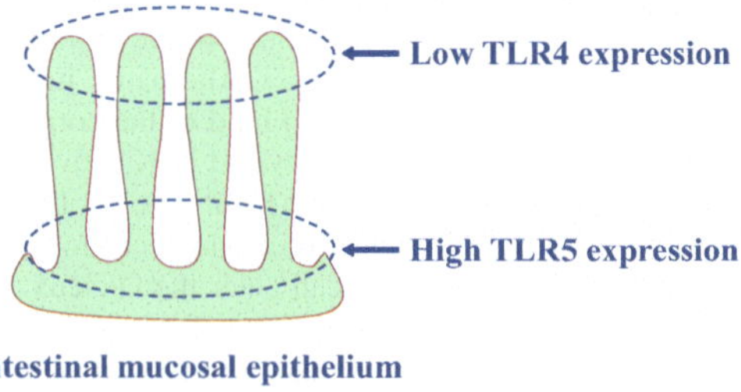

Fig. 3.19 Schematic illustration of the differential expressions of TLR in the intestine. TLR4 expression was low in the distal end of intestinal villi, and TLR5 expression was high in the crypt

monocyte-macrophage TLR4 expression. Inflammatory cytokines such as IL-2, IL-15, IL-1β, IFN-γ, and TNF-α induce TLR2 expression. The neuron short peptide VIP (vasoactive intestinal peptide) decreases the expression of TLRs by inhibiting PU-1 and NF-κB.

3.4.2.2 Expression of TLRs Is Regulated by Pathogenic Infections and Pathogenic Components

Avian mycoplasma infection enhances TLR2 expression and suppresses TLR4 expression. *E. coli* infection enhances TLR2 expression in γδ T cells. LPS enhances TLR2 expression in macrophages and adipocytes, and decreases TLR4/MD2 complex expression. Viral infection enhances macrophage TLR1, TLR2, TLR3, and TLR7 expression.

3.4.2.3 Collaboration between Pattern Recognition Receptors

TLRs and NLRs are both independent of each other and working with each other, e.g., the production of active IL-1β requires the joint participation of TLRs and NLRs. The binding of TLRs to the corresponding ligands activates the NF-κB signaling pathway and initiates the expression of IL-1β precursors, which must be cleaved by NLRs-mediated Caspase-1 to become IL-1β to be biologically active. Thus, collaboration between PRRs plays an important role in antidefense response.

3.5 Applications of Innate Immune Molecules

3.5.1 Application of Antimicrobial Peptides

Antimicrobial peptides (AMPs) as broad-spectrum bactericidal or antiviral innate immune molecules have a wide range of prospects. An in-depth research on them as antibiotic substitutes has been conducted. From the results of the past decade or so, although AMPs have some antibacterial or antiviral biological activity, the antibacterial effect is still inferior to that of antibiotics. In order to develop AMPs into antibiotic substitutes for clinical use in animals, further screening, optimization, and improvement will be definitely required. However, AMPs can be used as immunoprophylactic, adjuvant, or immunomodulator. For example, injection of poultry cathepsin DCATH-2 (D-amino-acid analog of chicken cathelicidin-2) into the allantoic cavity of 18-day-old chicken embryos resulted in the protection of the respiratory and digestive mucosa of chicks and reduced bacterial infections after birth. The use of bovine cathelicidin (indolicidin, 13aa) as an antigenic adjuvant may induce the immune response toward the humoral immune response. In experimental clinical studies for the treatment of human diseases, exciting results (more than ten-fold reduction in bacterial load compared to controls) have been achieved with human neutrophil defensin-1 (HNP-1) in the treatment of tuberculosis as well as E. coli infections. No clinical application has been reported in immune modulation so far.

In summary, antimicrobial peptides have three main applications: 1) as bactericidal agents; 2) as prophylactic immune agents; and 3) as adjuvants.

3.5.2 Applications of Pattern Recognition Molecules

Stimulation of TLRs by bacterial components elicits APC-dependent and non-dependent signaling transduction that generates innate and acquired immunity. Recent studies have shown that TLR signaling can move T cells toward a Th1 or Th2 immune response. Activated TLRs can also directly or indirectly affect the function of regulatory T cells. Therefore, TLRs are required in both immune activation and immune regulation. They have been applied in three areas: (a) regulation of immune response; (b) prevention and treatment of infectious diseases; and (c) drug development: mimics or antagonists of TLR ligands are used as immune enhancers, such as CpGDNA.

Study Questions

1. What is the basic process by which antigenic infection induces an innate immune response?
2. What are the innate immune molecules?
3. What is defensin? What are its characteristics? What is the mechanism by which defensins kill pathogenic bacteria?
4. What is the mechanism by which lysozyme kills pathogenic bacteria?
5. What are the effects of nervous and endocrine systems and food on the regulation of ILCs?
6. What are pathogen-associated molecular patterns (PAMPs)?
7. What are pattern recognition receptors (PRRs)? What are the major families of PRRs? What are the main types of PAMPs recognized by each PRRs family? What is their role in the immune response?
8. What are inflammasomes? What are the pathways of inflammasome activation? and what are their characteristics?
9. What are autophagy and heterophagy? What are the similarities and differences between phagocytosis and autophagy?
10. What is the NF-κB signaling pathway?

Chapter 4
T-Cell Development, Maturation, and Activation

Overview:
In the thymus, 10^7 thymocytes are produced daily, but the vast majority of cells undergo apoptosis during positive and negative selections, and only 1%~3% of thymocytes eventually develop and mature to become $CD4^+$ or $CD8^+$ T cells. Those T cells that cannot bind to their own MHC during development are induced to apoptosis, a process called positive selection. Thus, thymocytes that survive positive selection are able to bind to their own MHC, which determines the MHC-restricted nature of T cells, i.e., T cells only recognize cells that express their own MHC. However, those thymocytes that bind too strongly to their own MHC or MHC+Ag are likewise induced to apoptosis, a process called negative selection, because cells that bind too strongly to their own MHC or MHC+Ag will be activated when they enter peripheral blood and tissues, attacking their own tissues and causing autoimmune diseases. T cell development and maturation are central to immunology and are the focus of this chapter.

4.1 Thymus

4.1.1 Basic Structure of the Thymus and Thymocytes

The thymus is a central immune organ, and the murine thymus originates from the third pharyngeal pouch in the early stage of embryogenesis. The starting structure contains the endodermal-derived epithelial rudiment surrounded by neural crest-derived mesenchyme. Thymic development is regulated by the transcriptional regulator Foxn1 (Forkhead box N1), a molecule that plays a key role in the proliferation and differentiation of thymic epithelial cell (TEC) and in the formation of a three-dimensional TEC network characteristic of the thymic parenchyma. All T cells develop and mature in the thymus. In mammals, the thymus is located in the thorax above the heart and is divided into two lobes. However, the avian thymus is located

S. J. Zheng, *Veterinary Molecular Immunology*,
https://doi.org/10.1007/978-981-99-8929-4_4

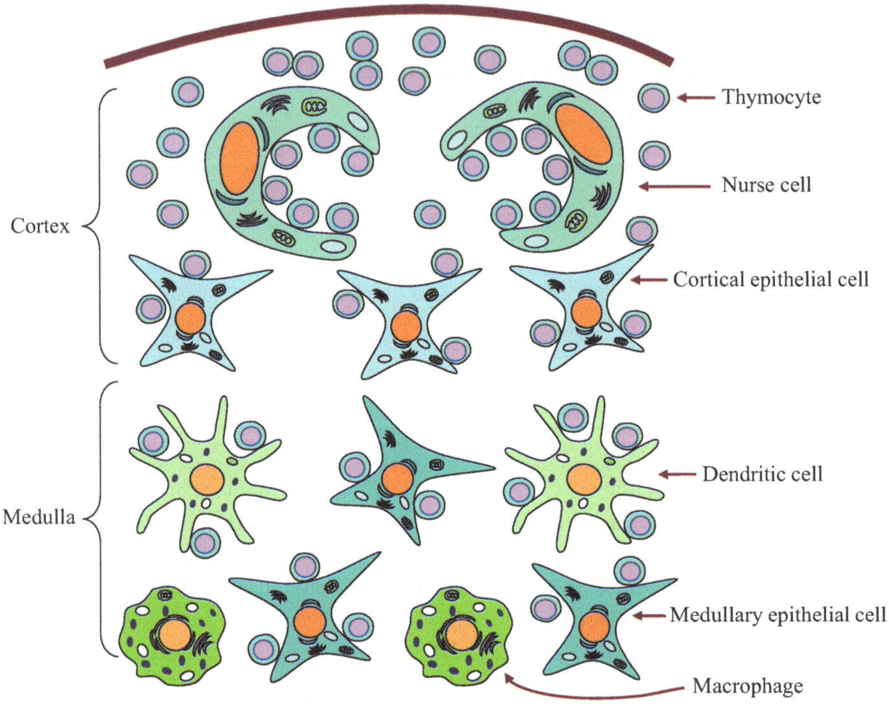

Fig. 4.1 Schematic diagram of the major cell types in the thymus

on both sides of the cervical trachea and is divided into multiple lobes (three to six lobes). According to the anatomy, the thymic organ consists of the cortex and medulla. The immature T cells in the thymus (i.e., before they mature and leave the thymus) are called thymocytes, and the cortical area contains thymocytes in addition to nurse cells and cortical epithelial cells. The medullary zone contains mainly thymocytes, dendritic cells, medullary epithelial cells, and macrophages (Fig. 4.1).

The epithelial cells, dendritic cells, and macrophages in the thymus form an interlocking three-dimensional structural network with which thymocytes come into direct contact to complete the development and maturation process. In the cortical region, there is a special type of epithelial cell, called the nurse cell, which is in direct contact with several thymocytes and assists in the development and maturation of thymocytes.

4.1.2 The Thymus Is the Site of T-Cell Development and Maturation

All T cells develop and mature in the thymus, and an important function of the thymus is to select T-cell populations against pathogenic infection and tumors. Unlike the bone marrow, however, the thymus does not contain hematopoietic stem cells per se, the thymocytes develop from lymphoid progenitors that migrate to the thymus (see Fig. 2.2 in Chap. 2). Early in embryonic development, hematopoietic stem cells produced by the liver migrate to the thymus and develop in the thymic microenvironment to mainly form γδ T cells, but as the embryo matures, the number of generated γδ T cells decreases. The bone marrow gradually replaces the liver in the production of hematopoietic stem cells from mid- to late term of embryonic development. The lymphoid progenitors derived from the hematopoietic stem cells leave the bone marrow, circulate in the blood and lymphatic systems, and enter the thymus, where they develop to form αβT cells. After the birth of the animal, αβ T cells increase significantly, while γδ T cells markedly decrease. This chapter focuses on the development, maturation, and activation of αβT cells in animals after birth.

In the thymus, 10^7 thymocytes are produced daily, but the vast majority of cells undergo apoptosis via positive and negative selection, with only 1–3% of thymocytes eventually developing into mature $CD4^+$ or $CD8^+$T cells. Lymphoid progenitors are differentiated from bone marrow hematopoietic stem cells and migrate through the bloodstream to the thymus (Fig. 4.2). During thymocyte development, different fragments of the TCR gene (V-D-J) undergo random combinations, rearrangements, and even mutations, giving rise to thymocytes with TCRs of different specificity and different affinity, but for the same cell, the TCRs are identical. Those thymocytes that cannot bind to their own MHC are induced to apoptosis during development, a process called positive selection. Thus, thymocytes that survive

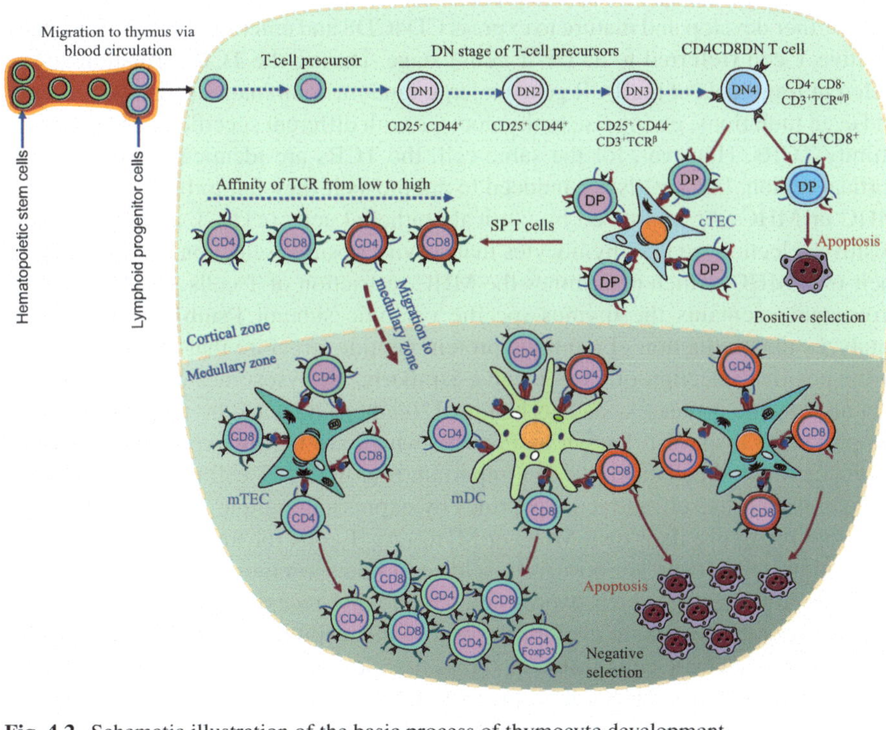

Fig. 4.2 Schematic illustration of the basic process of thymocyte development

positive selection are able to bind to their own MHC, which determines the MHC-restricted nature of T cells, i.e., T cells only recognize cells that express their own MHC. However, those thymocytes that bind too strongly to their own MHC or MHC + Ag are likewise induced to apoptosis, a process called negative selection, because cells that bind too strongly to their own MHC or MHC + Ag are activated when they reach peripheral tissues and attack their own tissues, causing autoimmune disease. Therefore, negative selection determines T-cell tolerance to self-tissue. Positive and negative selections work exquisitely and skillfully together, like a school that is always subjecting its trainees to rigorous selection and brutal elimination, producing an elite who is able to defend against invaders and remains loyal to his country as well! Its exact molecular mechanism is not yet clear, and the precise process of thymocyte development and maturation embodies the profundity of immunology and the mysteries of life!

Bone marrow hematopoietic stem cells proliferate and differentiate into lymphoid progenitor cells, which enter the thymic cortical zone through blood circulation and develop and differentiate into T-cell precursor cells in the cortical zone. At this stage, thymocytes do not express complete TCR, nor CD4CD8 ($CD4^-$ $CD8^-$), and belong to the double negative (DN) stage of T-cell precursors. At this stage, there are three developmental stages of DN1-3 according to the expression of CD25 and CD44 (as shown in the figure), and T-cell precursors at stage DN3 express TCR-β chains, with CD25 positive and CD44 negative. T-cell precursors further develop into the DN4 stage, i.e., the CD4CD8 DN T-cell (DN T-cell for short) stage, where DN T cells express the complete TCR complex ($CD3^+$αβTCR), and DN T cells further develop and mature to express CD4CD8 and enter the CD4CD8 double-positive T cell (referred to as DP T cells) stage. During the TCR formation stage, different fragments of the TCR gene undergo random combinations, rearrangements and even mutations, giving rise to thymocytes with different specificity and different affinity TCRs. However, for the same cell, the TCRs are identical. In the thymic cortical region, DP T cells are induced to apoptosis if they fail to bind to their own MHC or MHC+Ag expressed by cortical epithelial cells (cTEC), a process called positive selection, so that thymocytes that survive positive selection are able to bind their own MHC, which determines the MHC restriction of T cells. cTEC's thymoproteasome contains the thymus-specific catalytic subunit Psmb11 (β5t), which catalyzes the production of a unique protein peptide structure that plays a key role in the positive selection of CD8 T cells. Similarly, the lysosomes of cTEC contain thymus-specific proteases, such as histone L (cathepsin L) and thymus-specific serine protease (Tssp, also known as Prss16), which are highly expressed in thymic cortical epithelial cells and play an important role in CD4T cell positive selection. cTEC thus selects self-antigenic peptides by expressing unique self-antigenic peptides that recognize their own MHC or MHC+Ag T cells for further developmental maturation. DP T cells are positively selected to develop into CD4 or CD8 single-positive (SP) T cells, which express the chemokine receptor CCR7 and begin to migrate from the cortical to medullary zone. In the medulla, myeloid epithelial cells (mTEC) and dendritic cells (mDC) express high levels of CCL19 and CCL21, two ligands that bind to CCR7, which promote CD4 and CD8 T-cell migration in the medulla. mTEC and mDC are key APCs for T-cell development, expressing large amounts of tissue-restricted antigens (TRAs), presenting antigens via MHC-I and MHC-II molecules, and those SP T cells that bind too strongly to their own MHC or MHC+Ag are induced to apoptosis, a process called negative selection. Because T cells that bind too strongly to their own MHC or MHC + Ag, once move to peripheral tissues, will be activated to attack self-tissues and cause autoimmune disease. T cells that survive negative selection do not attack self-tissues, i.e., they are tolerant to self-tissues, which is also known as central immune tolerance. Meanwhile, a small number of CD4 SP T cells are differentiated into $FOXP3^{+}$CD4T cells after negative selection, called regulatory T cells (Treg). Treg mainly inhibit peripheral self-reactive T-cell activation and are responsible for peripheral immune tolerance. Medullary epithelial cells and dendritic cells play a key role in negative selection.

4.2 Pioneering Experimental Findings in T-Cell Development and Maturation

As described in the first section of this chapter, thymocytes undergo a positive selection process during maturation, in which those thymocytes that recognize self-MHC are able to develop and mature, while those thymocytes that do not recognize self-MHC are induced to apoptosis. Thus, the positive selection process ensures that all mature T cells recognize antigens presented by self-MHC and not those presented by other MHCs, i.e., MHC restriction of T cells. This discovery was made by Australian scientists Doherty and Zinkernagel in the mid-1970s (1974) from Cytotoxicity assay (Fig. 4.3).

Isotope 51Cr-labeled mouse peritoneal macrophages were used as target cells and mixed with splenocytes from LCMV-inoculated (immunized) mice in culture for cytotoxicity assay (Doherty and Zinkernagel, JEM 1975).

In brief, Doherty and Zinkernagel obtained peritoneal mononuclear macrophages from two strains of mice of different genetic backgrounds ($H\text{-}2^{k}$ or $H\text{-}2^{d}$) and labeled LCMV-infected or normal cells with isotope 51Cr as target cells. Mice of the CBA/H strain (genotype $H\text{-}2^{k}$) were inoculated with LCMV, and 1 week after inoculation, the spleens of the mice were removed and splenocytes were isolated, which contained a large number of specific cytotoxic T lymphocytes (CTL). When splenocytes from the immunized $H\text{-}2^{k}$ mice were mixed with normal target cells of the same strain ($H\text{-}2^{k}$) in culture, no cytotoxic activity of CTL were observed occurred (Fig. 4.3A). However, when $H\text{-}2^{k}$ CTL was mixed with LCMV-infected target cells of the same strain ($H\text{-}2^{k}$) of mice in culture, CTL killed and lysed the target cells (Fig. 4.3B). If the same CTL Cytotoxicity assay was performed with LCMV-infected target cells from another strain of mice (BALB/c, genotype $H\text{-}2^{d}$), CTL did not kill heterologous target cells, i.e., $H\text{-}2^{d}$ target cells were not lysed (Fig. 4.3C). This landmark finding reveals that the cytotoxic activity of CTL is not only strictly specific to killing target cells, but also follows MHC restriction (Fig. 4.4), opening a window for further exploration of the mechanisms of cellular immunity. For this

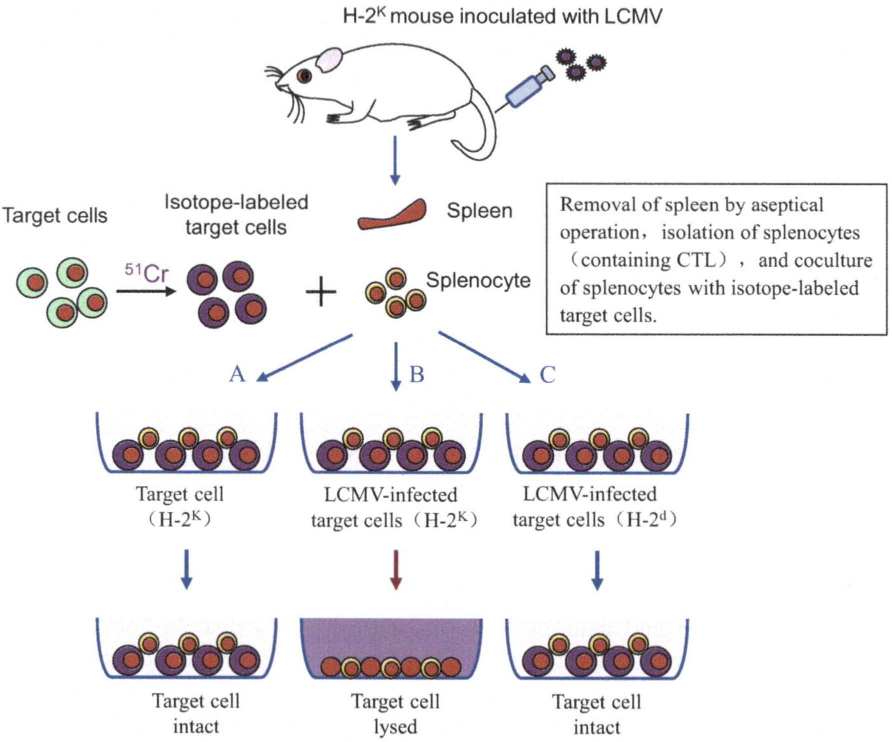

Fig. 4.3 Schematic diagram of a brief procedure for a cytotoxicity assay

far-reaching discovery, Doherty and Zinkernagel were awarded the 1996 Nobel Prize in physiology and Medicine.

From the above, it is clear that mature T cells recognize antigens presented by self-MHC or homozygous genetic background MHC through TCRs and kill target cells. During thymocyte development, different fragments of the TCR gene (V-D-J) undergo random combinations, rearrangements, and even mutations, giving rise to TCRs with different specificities or different affinities for the same specificity. Theoretically, the animal should contain T cells that recognize various antigens presented by MHC. In addition to these T cells that recognize self MHC, are there really no T cells in the body that recognize other (non-self) MHCs? To answer this question, Zinkernagel and Doherty crossed the CBA/H (H-2^k) and C57BL (H-2^b) strains of mice as parents to obtain the F1 generation (H-$2^{K/b}$), further generated chimeric mice, and found that only those thymocytes that recognized self MHC could develop to maturity, whereas those thymocytes that recognized non-self MHC could not develop into mature T cells (Fig. 4.5).

In the experiment with chimeric mice as mentioned above, bone marrow hematopoietic stem cells from F1 generation mice (H-$2^{K/b}$) generated lymphoid progenitor cells, which migrated to the thymus through the bloodstream and proliferated

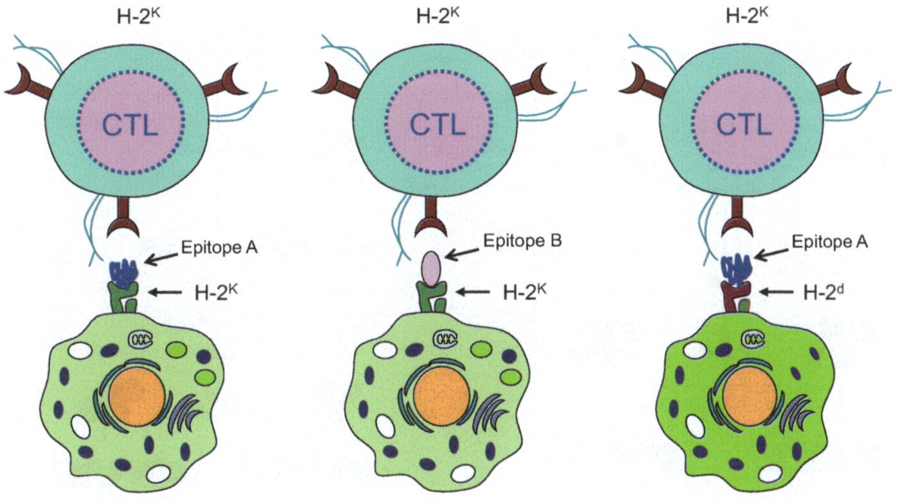

Fig. 4.4 Schematic diagram of the requirement of self MHC and antigen in CTL-mediated killing of target cells. The specific CTL of H-2^k mouse recognizes only antigens (epitope A) presented by antigen-presenting cells (APCs) of the same genetic background, but not other antigens (epitope B), reflecting the specificity of CTL for killing target cells. However, if an APC of a different genetic background (H-2^d) presents an antigen specific for that CTL, the CTL does not recognize the antigen (A epitope) presented by the H-2^d of the heterozygous genetic background and fails to kill the target cell

and differentiated in the thymic cortical region into the thymocytes with TCRs recognizing a variety of target cells with different genotypes including H-$2^{k/b}$, H-2^k and H-2^b. In the H2^b thymus, only the thymocytes that TCRs recognized H-2^b were selected to continuously develop and mature, while thymocytes recognizing H-2^k are induced to apoptosis, so that in the spleen of chimeric mice only CTL recognizing H-2^b and no mature T cells recognizing H-2^k are present, which is the reason why only H-2^b target cells are killed by mature splenocytes selected by the H-2^b thymus after immunization. The results of this experiment suggest that the MHC plays an important role in the development and maturation of thymocytes in the thymus (Fig. 4.6). This process determines the MHC restriction of T cells.

4.3 T-Cell Receptors

4.3.1 Discovery of the T-Cell Receptor

According to experimental observations, the key to T-cell cytotoxicity of target cells is that the T-cell receptor (TCR) specifically recognizes antigens presented by the MHC of target cells. It was therefore hypothesized that since the TCR specifically

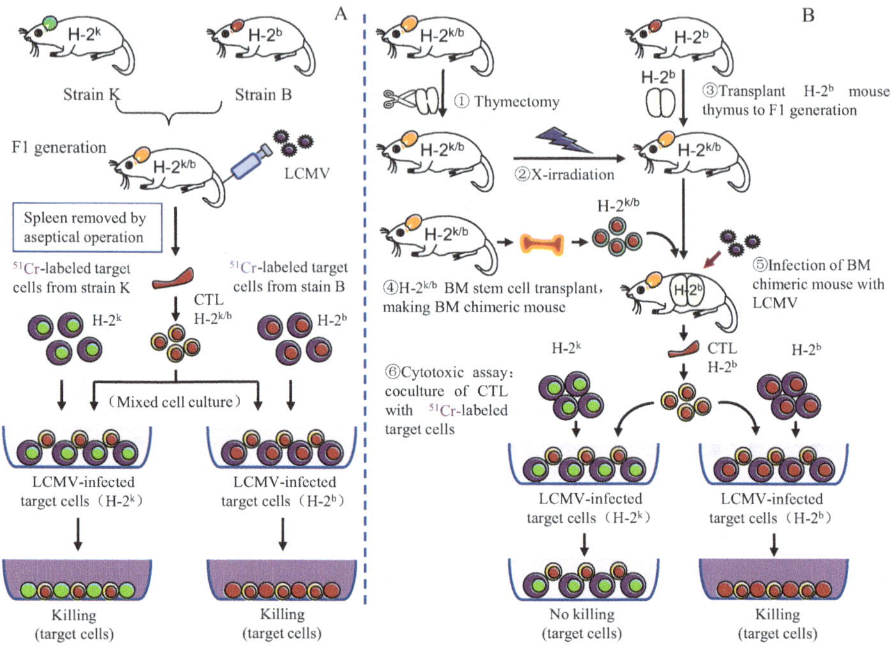

Fig. 4.5 Schematic diagram of the selective thymocyte maturation. (**a**) Inbreed strain of mouse H-2^k (or written as H-$2^{k/k}$, e.g., mouse CBA/H strain) were crossed with H-2^b (or written as H-$2^{b/b}$, e.g., mouse C57BL strain) to produce F1 generation (H-$2^{k/b}$). Infection of F1 generation mice with LCMV elicited immune responses, and spleen cells that were isolated from the spleen were used as CTL to co-culture with isotopically labeled target cells (peritoneal monocyte-macrophages infected with LCMV infected as target cells). If the target cells were from K strain (H-2^k) or B strain (H-2^b), the CTL of the F1 generation was able to recognize the K or B target cells and exerts a cytotoxic effect. (**b**) The thymus of the F1 generation (H-$2^{k/b}$) mice was surgically removed, and the thymectomized mice were irradiated with X-rays (killing all bone marrow stem cells and immune cells in the body) as thymus transplantation recipients. The thymus of H-2^b mice was transplanted into the recipient F1 generation mice; thus, this kind of F1-generation mice was characterized by the absence of bone marrow stem cells, the absence of immune cells, and the expression of only H-2^b by all tissue cells in the transplanted thymus. Then bone marrow stem cells from the F1 generation (H-$2^{k/b}$) mice were injected intravenously into the thymus-transplanted mice for bone marrow reconstruction to generate bone marrow chimeric mice. Bone marrow chimeric mice were infected with LCMV to elicit an immune response, and the spleen cells were taken for cytotoxicity assays. If the transplanted thymus in the chimeric mice was of strain B (H-2^b) background, the immune splenocytes (containing CTL) had a cytotoxic effect on the target cells from strain B but not from strain A. Conversely, if the transplanted thymus in the chimeric mice was of strain K (H-2^k), its immune splenocytes had a cytotoxic effect on the target cells from strain K but not from strain B

recognizes antigens presented by MHC, there would be T-cell clones in vivo that target various antigens, and that each T-cell clone should differ in receptor (TCR) structure, and that such differences would give different TCRs a unique antigenic signature (since the TCR itself is a protein complex that is antigenic to other species), whereby specific monoclonal antibodies could be used to differentiate the antigenic differences of TCR. Based on this speculation, in the early 1980s,

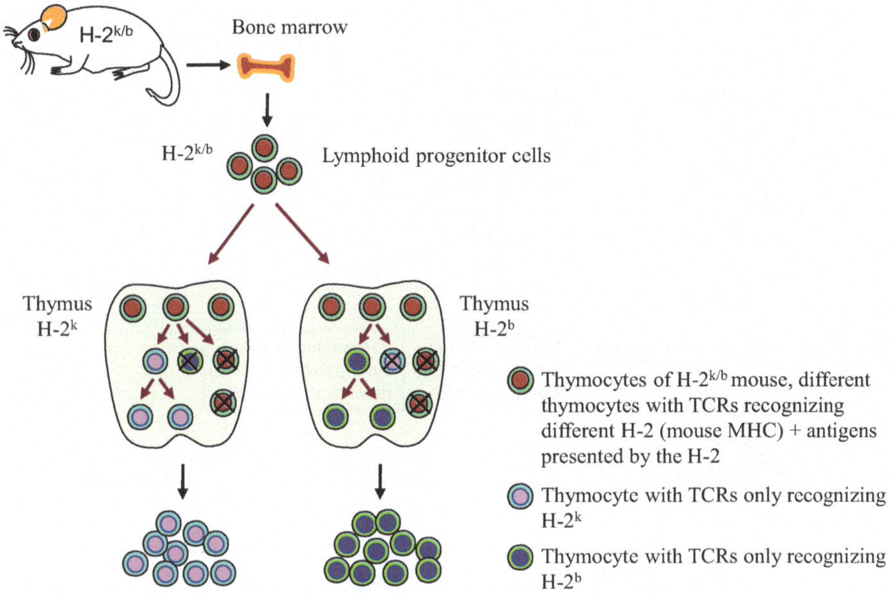

Fig. 4.6 Schematic diagram of the brief process for the development and maturation of selective thymocytes

immunologists produced monoclonal antibodies from different T-cell lines, searching for monoclonal antibodies that recognized and distinguished specific T-cell lines, i.e., clonotypic antibodies. It was found that some monoclonal antibodies recognized all TCRs, while others recognized only a specific TCR. It was assumed that TCRs, like antibodies, also had variable and constant regions. If the monoclonal antibody is directed against a constant region, the monoclonal antibody will recognize all T-cell lines. If the monoclonal antibody is directed against the variable region, the monoclonal antibody will only recognize specific T-cell lines. The isolation of TCRs using monoclonal antibody revealed that TCRs are heterodimers containing α and β chains, while a few T cells were found to express γδ heterodimeric TCRs.

In the early 1980s, Hedrick and Davis used DNA differential subtraction to determine different T-cell clones carrying different specific TCRs. Their approach was based on the following assumptions about the nature of TCR genes, briefly (1) that they should be expressed in T cells but not in B cells; (2) that the mRNAs for TCR proteins should be found on membrane-bound polysomes; (3) that like immunoglobulin genes, those that encode TCR proteins should be rearranged in T cells as a mechanism of generating diversity and consequently increasing the antigen-recognition repertoire; (4) that like immunoglobulin genes, they should have constant regions and variable regions, would confer the antigen-binding specificity. As expected, it was demonstrated that different T-cell clones had different specific TCRs (Fig. 4.7).

Both hepatocytes and B cells contain full-length TCR sequences that have not undergone random gene combinations, rearrangements, or mutations, so that

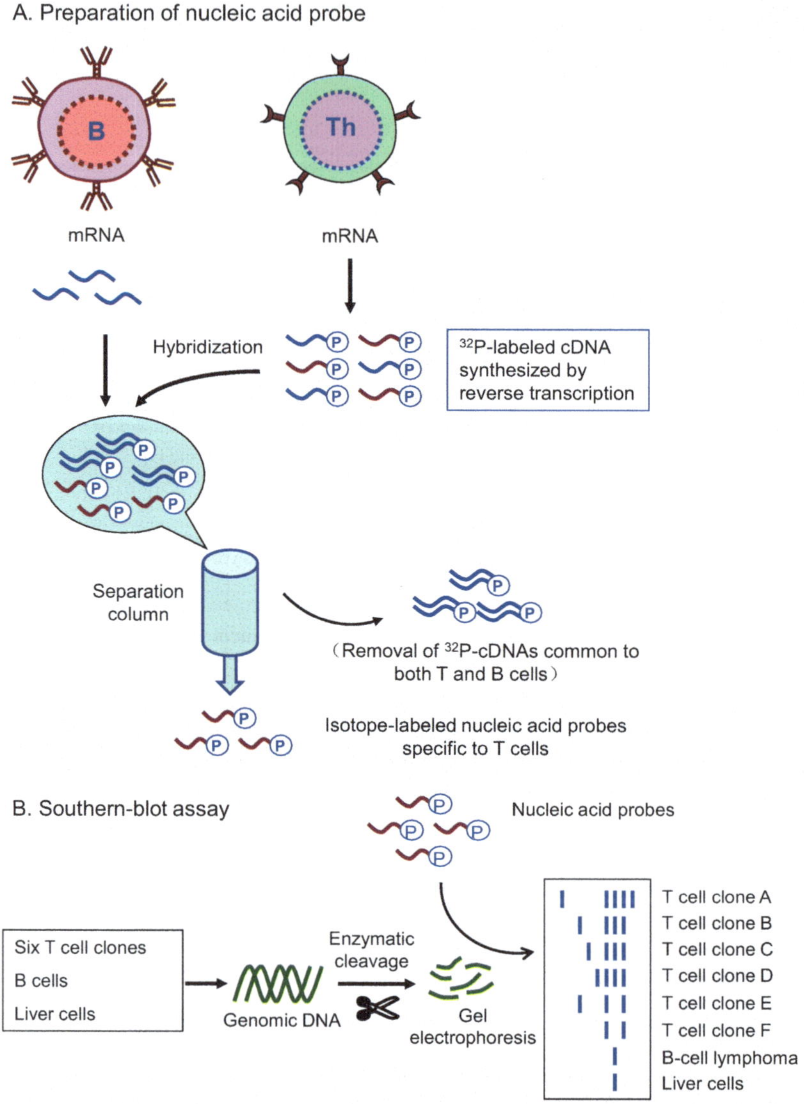

Fig. 4.7 Schematic diagram of TCR rearrangement discovered by DNA differential subtractive hybridization. (**a**) Total mRNA was extracted from helper T cells (Th) and reverse transcribed to synthesize isotope ^{32}P-labeled cDNA. This cDNA was hybridized to total B-cell mRNA, and the common component with hybridization was removed by separation columns to recover T-cell-specific ^{32}P-labeled cDNA as a probe. (**b**) Genomic DNA was extracted from B cells, hepatocytes, and different T-cell lines, enzymatically cleaved, gel electrophoresed, transferred, and hybridized with nucleic acid probes labeled with ^{32}P, exposed on photographic film, and analyzed. (Basis for hybridization test: Early studies by Davis showed that 98% of the genes expressed by lymphocytes are common to both B and T cells. TCR gene takes at least 2% of total genes. These genes can be used as probes to search for genes that rearrange in mature T cells. It was hypothesized that since the αβTCR has a constant and variable region, a process like B-cell immunoglobulin gene rearrangement would occur during T-cell development and maturation. As a result, six mature T-cell lines do show DNA rearrangements)

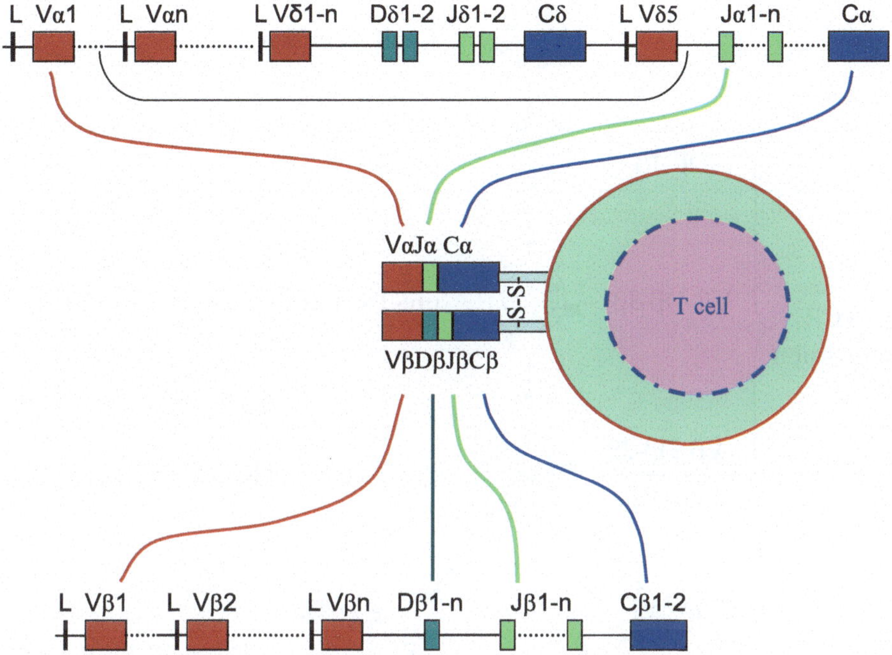

Fig. 4.8 Schematic diagram of T-cell receptor (TCR) gene formation

hepatocytes and B cells, like other somatic cells, contain the original TCR genomic DNA sequence with the same DNA fragments after enzymatic cleavage. However, when lymphoid progenitor cells migrate through the bloodstream to the thymus, the lymphoid progenitors proliferate and differentiate in the thymus, and during this process different fragments of the TCR gene (V-D-J) undergo random combination, rearrangement, and even mutation, forming DNA fragments of different sizes, generating TCRs with different specificities or different affinities for the same specificity (Fig. 4.8).

For a mature T-cell clone, the TCR gene sequence is identical. Taking the mouse αβTCR as an example, there are about 100 Vα genes ($n = 100$), 50 Jαgenes ($n = 50$), 25 Vβgenes ($n = 25$), 2 Dβ genes ($n = 2$), and 12 Jβ genes (n = 12). According to random combination, the number of variable regions in αβTCRs is approximately $V\alpha \times J\alpha = 100 \times 50 = 5000$, while the number of variable regions in β chains is approximately $V\beta \times D\beta \times J\beta = 25 \times 2 \times 12 = 600$. Random combination of α and β chains can produce $5000 \times 600 = 3 \times 10^6$ αβTCRs. In addition to random combination of gene fragments, there are gene rearrangements and even mutations (including point mutations and somatic mutations). Considering various factors, the number of TCRs is approximately 10^{13}.

After migration to the thymus, lymphoid progenitors proliferate and differentiate into mature T cells that recognize various antigens. Thymocytes go through multiple stages during maturation, and at each stage the cells display characteristic gene expression and specific cell surface markers, eventually maturing into CD4 or CD8 single positive T lymphocytes (Fig. 4.9). So how do thymocytes mature? What are

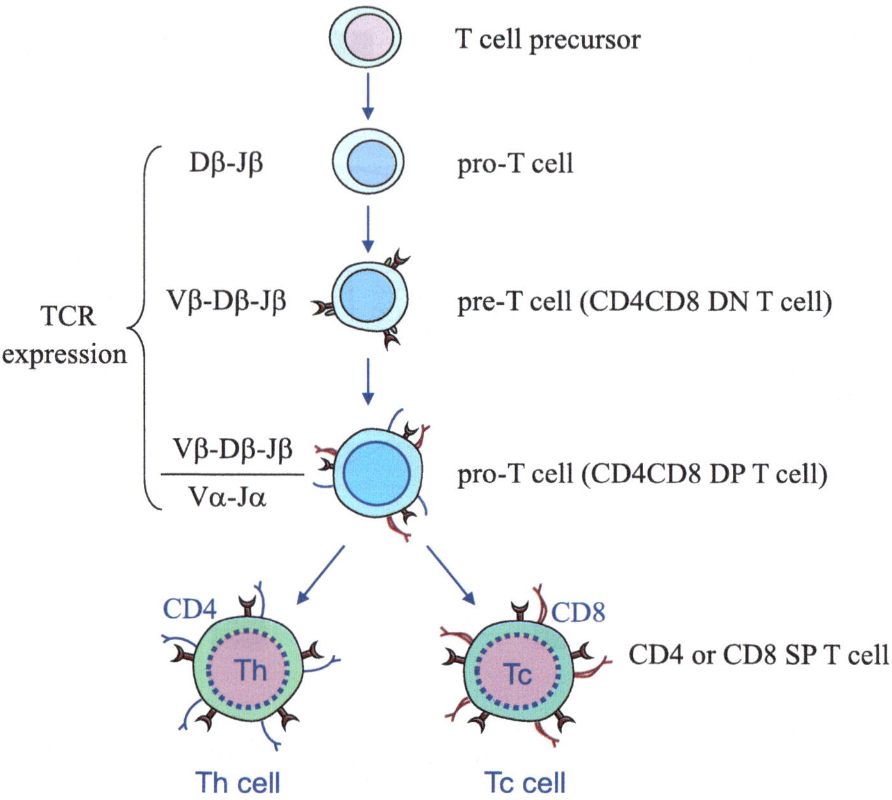

Fig. 4.9 Schematic diagram of the thymocyte maturation process

the exact molecular mechanisms? These questions have always been a hot topic in immunological research as well as a pyramid in life science research.

After migration to the thymus, lymphoid progenitors proliferate and differentiate, undergoing multiple stages such as T-cell precursor → pro-T cell(double negative, DN)) → pre-T cell → pro-T cell(double positive, DP). During the development of thymocytes, different fragments of TCR genes (V-D-J) undergo random combinations, rearrangements or even mutations to generate TCRs with different specificities or different affinities for the same specificity, and eventually thymocytes develop into mature CD4 or CD8 single positive T lymphocytes.

4.3.2 Structure and Function of TCR Complex

As mentioned earlier, TCRs are heterodimers and are divided into αβTCRs and γδTCRs. T cells as commonly described refer to αβT cells, whereas γδT cells are the main immune cells early in embryonic development, but their number decreases

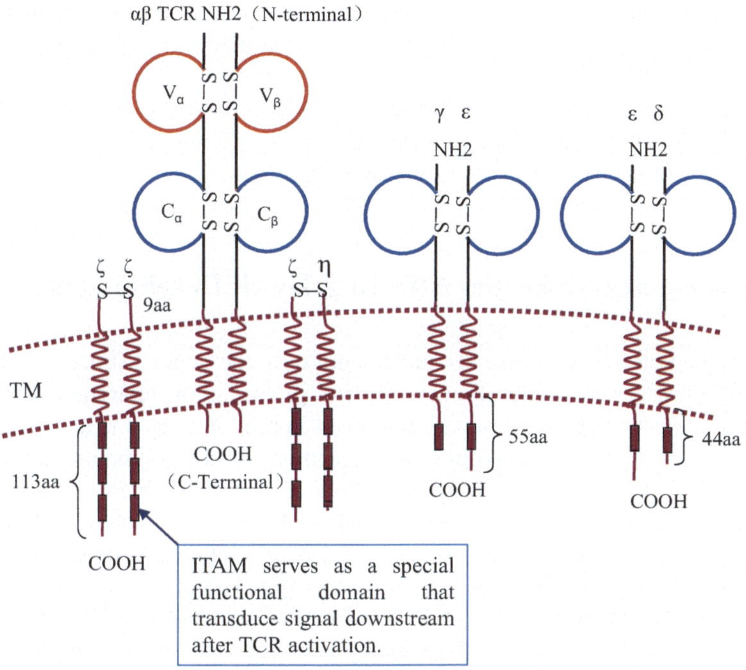

Fig. 4.10 Schematic diagram of the structure of TCR–CD3 complex. *ITAM* immunoreceptor tyrosine-based activation motif, *TM* transmembrane

significantly as the animals grow in age, and are mainly distributed in the liver and subcutis. γδT cells mainly recognize non-protein lipid antigens and are often considered to be innate immune cells that form the first line of defense against infection. The T cells mentioned here also refer to αβ T cells. A functional TCR actually consists of two parts: recognition of antigens and signal transduction. The α/β chain, which is in the extracellular region, is mainly responsible for recognizing MHC-Ag. As the intracellular region of the α/β chain is very short and cannot transmit signals downward by itself, there must be another helper molecule to transmit the signals generated by the engagement of α/β chain of TCR with Ag + MHC. CD3 is such a helper molecule for transduction of TCR-mediated signaling. Thus, TCR is actually the TCR-CD3 complex (Fig. 4.10).

CD3 is a complex of five different polypeptide chains that form four pairs of dimers, the γε heterodimer, the δε heterodimer, the ζζ homodimer, and the ζη heterodimer. η and ζ are encoded by the same gene but differ at the C-terminus during mRNA modification and splicing, so that η and ζ are actually homologous molecules. ζζ homodimer is the major component of the CD3 molecule, accounting for approximately 90% of the CD3 complex, followed by the ζη heterodimer. The intracellular regions of the γ, δ, and ε chains contain 40–50 amino acids, while the intracellular region of ζ contains 113 amino acids, forming the important intracellular immunoreceptor tyrosine based activation motif (ITAM). ITAM is present in the intracellular region of several immunoglobulin superfamily molecules, such as the Ig-α/β of BCR, the intracellular region of Fc receptor for IgE and IgG, etc. ITAM activation recruits and activates adaptor protein signaling downward to elicit an immune response in cells. Direct stimulation of T cells with anti-CD3 antibodies mimics the downward signaling process of antigen recognition by the TCR, which is widely used to activate T cells for research purposes.

4.4 Important Molecules Affecting T-Cell Development

As mentioned earlier, we currently understand only the general process of thymocyte development and maturation, while the precise molecular mechanisms of which have always been very esoteric questions in immunology and are not yet well understood. One of the most important questions is how immature thymocytes (CD4CD8DP) develop into CD4sp or CD8sp T cells and which original factors regulate the course of thymocyte development? These questions have not been satisfactorily answered up to date. As shown in Fig. 4.2, although the thymic proteasome of cTEC contains the thymus-specific catalytic subunit Psmb11 (β5t), which catalyzes to generate a unique peptide structure that selects CD4CD8DPT cells for maturation toward CD8T cell development via antigen presentation by MHC class I molecule, and the lysosome of cTEC contains the thymus-specific protease (cathepsin L Prss16) that degrades self-antigens to be presented via MHC-class II molecules, driving CD4CD8DPT cells toward CD4T cells, which suggest that certain molecules play a critical role in CD4 or CD8T cell development.

4.4.1 The Role of Major Histocompatibility Complexes and Antigens in the Development and Maturation of Thymocytes

Experimental evidence indicates that CD8T cells specifically recognize antigens presented by MHC class I (MHC-I) molecules, while CD4T cells specifically recognize antigens presented by MHC class II(MHC-II) molecules. However, if the MHC-I gene is knocked out, the animal does not produce CD8T cells. Similarly, if the MHC class-II gene is knocked out, the animal does not produce CD4T cells. Thus, MHC is required during CD4/CD8T cell development and maturation. Meanwhile, it also shows that during thymocyte development, only those T cells that recognize self MHC are able to develop and mature, further confirming the necessity of MHC during positive selection.

In H-Y transgenic mice, H-Y transgenic male mice lack mature CD8T cells because the TCR of male mouse recognizes Y protein (this protein is only expressed in males), whereas the female mice have normal CD8 T-cell development because they do not express Y protein. This phenomenon suggests that during thymocyte development, those T cells that recognize their own antigens are removed by a process of negative selection.

Since thymocytes undergo negative and positive selections during development, it seems that no thymocytes are able to develop to maturity, but in fact a small proportion (1%–3%) of thymocytes eventually develop into mature CD4 or CD8 T cells that leave the thymus for migration to peripheral tissues. If so, how to understand those T cells that survive and develop into mature T cells in the thymus? To explain this phenomenon, the affinity theory has been proposed, which states that the survival of those thymocytes that bind to their own MHC + Ag depends on the affinity strength, and that those thymocytes with a strong binding affinity of TCR-MHC or TCR-MHC + Ag will be induced to apoptosis, and only those T cells with low or moderate binding affinity will develop to maturity. The direct evidence in support of this theory is an affinity assay done with TAP (Transporter of antigenic peptide) knockout mice. TAP−/− mice were crossed with specific TCR-transgenic mice to obtain TCR-transgenic TAP−/− mice that were unable to make endogenous antigenic presentation due to the absence of TAP and whose CD8T cells expressed TCR that specifically recognized the same antigenic peptide due to the TCR transgene. Therefore, the affinity of TCR-MHC+Ag binding can be controlled by adding different concentrations of antigenic peptide to the system of in vitro mouse fetal organ (thymus) culture. This research system yielded excellent experimental results. The results showed that a small number of thymocytes developed into mature CD8T cells if high concentrations of short antigenic peptide were added to the cultures (Fig. 4.11a), likewise a small number of thymocytes developed if no antigenic short peptide was added (Fig. 4.11c), and only with the addition of appropriate concentrations of antigenic peptide could a normal number of mature CD8T cells be obtained (Fig. 4.11b).

4.4.2 *The Role of Transcriptional Regulators in Thymocyte Development and Maturation*

Transcriptional regulatory factors play an important role in gene expression and influence T-cell development and maturation. However, T-cell development and maturation is a complex process and many fine regulatory mechanisms remain to be investigated in depth.

4.4.2.1 Important Transcriptional Regulators Affecting Positive Selection

During thymocyte development, CD4CD8DPT cells in the cortical region undergo positive selection toward CD4SP or CD8SP T-cell subpopulations. The molecular mechanisms underlying the expression and regulation of CD4 or CD8 molecules are not well understood, but it is clear that CD4 and CD8 molecules are not expressed simultaneously in mature T cells. In CD8 T cells, the CD4 silencer is activated,

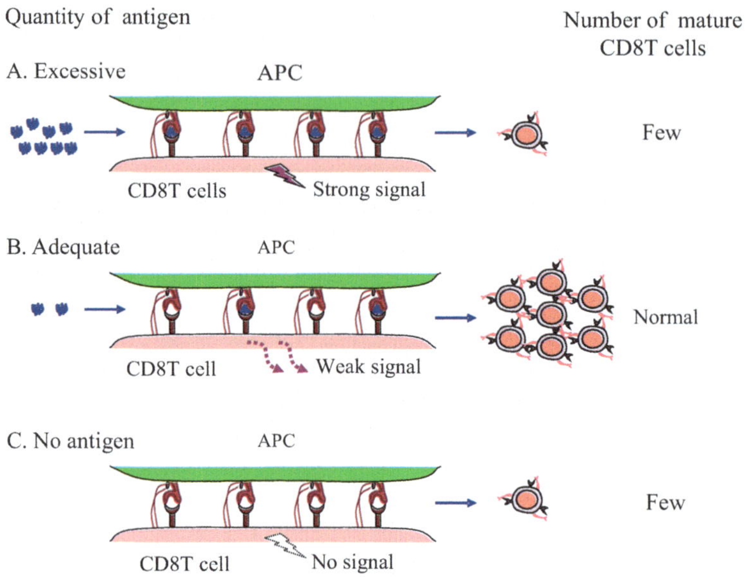

Fig. 4.11 Schematic diagram of the role of antigenic peptides in thymocyte development

suppressing CD4 expression. Two important transcriptional regulators have been identified that regulate CD4CD8DP thymocyte differentiation toward CD4SP or CD8SP T-cell subpopulations. One regulator is called ThPOK, and if this gene is mutated, animals lack CD4T cells, even thymocytes selected by MHC-II molecules also differentiate toward CD8T cells. If the gene is overexpressed, thymocytes selected by MHC-II molecules differentiate toward CD4T cells. Expression of ThPOK is regulated by the upstream gene GATA3 transcriptional regulator. The members of the Runx family of transcriptional regulators have a strong influence on thymocyte development, and among these members, Runx3 has a role in activating CD4 silencer, regulating differentiation of CD4CD8DP thymocytes to CD8SPT cells (see Fig. 4.12).

In thymic cortex, ThPOK determines the differentiation of CD4CD8DP thymocytes to CD4SP T-cell subpopulations, and Runx3 inhibition of ThPOK expression promotes the differentiation of CD4CD8DP thymocytes to CD8SP T-cell subpopulations, and ThPOK expression is regulated by the upstream gene GATA3 transcriptional regulator.

4.4.2.2 Important Transcriptional Regulators Affecting Negative Selection

In the thymic medulla, CD4 or CD8 T cells undergo a negative selection process, and the cells involved in performing the negative selection are mainly myeloid epithelial cells (mTEC) and dendritic cells (mDC), which, in addition to high expression of

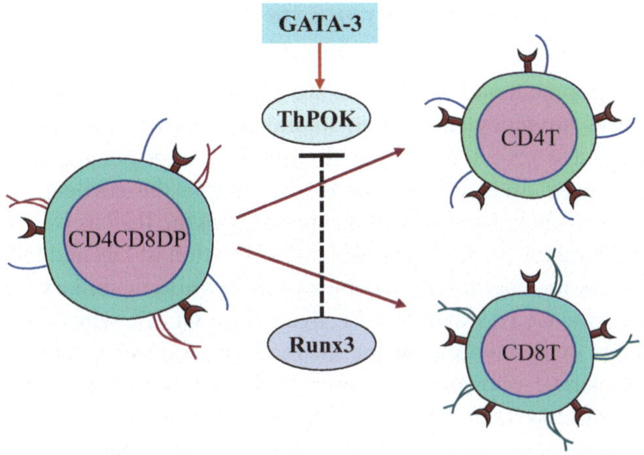

Fig. 4.12 Roles of transcriptional regulators ThPOK and Runx3 in T-cell development

CCL19 as well as CCL21 (ligand of CCR7) to promote CD4 and CD8 T-cell migration in the medulla, express a large number of self-tissue-restricted antigens (TRAs) that are presented by MHC-I and MHC-II molecules, and those SP T cells that bind strongly to self MHC or MHC+Ag are induced to apoptosis. AIRE (Autoimmune regulator) and Fezf2 (Fez family zinc-finger 2, also known as Zfp312 or Fezl) are important transcriptional regulators that regulate TRAs expression. AIRE affects the expression of TRAs in mTEC/mDC, regulates the expression of multiple self-antigens in myeloid APC, and directly affects the process of negative selection. AIRE is expressed exclusively in myeloid APC, and Aire$^{-/-}$ mice develop multiple autoimmune diseases. Fezf2 is another important transcriptional regulator expressed in thymic medullary APC that regulates the expression of multiple self-antigens and affects autoimmunity. Although transcriptional regulators have been found to play an important role in thymocyte development, the initial signals that regulate thymocyte differentiation and the underlying exact molecular mechanisms remain elusive.

Since thymocyte maturation undergoes both positive and negative selection processes, and the selection process is also the process by which a large proportion (97–99%) of thymocytes die by apoptosis, what is the original signal that triggers apoptosis? This is always an open question. Experimental evidence showed that certain molecules involved in apoptosis, such as tumor necrosis factor (TNF) and FASL, are not associated with apoptosis in thymocytes, while TRAIL is involved in the thymic apoptotic process only to a certain extent. While intracellular molecules such as BIM and AIRE play an important role in thymocyte apoptosis, it is always a mystery what their upstream molecules, i.e., the initial triggering signals, are. Elucidation of the molecular mechanism of the thymocyte maturation process is a prerequisite and basis for understanding immunity and immune tolerance.

4.4.3 Factors Regulating Thymocyte Migration

Thymocyte development and maturation are accomplished during migration in the thymus, and how thymocytes migrate in the thymus has always been a focus. Current studies have revealed that the thymus has a special microenvironmental structure that can be divided into two distinct parts: the cortical and medullary regions. Cortical epithelial cells (cTEC) express cytokine IL-7 as well as chemokine CXCL12, regulating the proliferation and differentiation of DN and DP thymocytes. SP T cells express the chemokine receptor CCR7, while medullary epithelial cells (mTEC) express the CCL19 and CCL21, ligands for CCR7, which drive the migration of SP T cells to the medulla during the maturation of SP T cells. Expression of S1PR1 (sphingosine-1-phosphate receptor 1), Foxo1 (Forkhead box protein O1), and Klf2 (Krüpple-like factor 2), which can be used as markers of maturation in studies, are molecules associated with migration of mature SP T cells.

4.5 Molecular Mechanisms of T-Cell Activation

Both humoral and cell-mediated immunity are associated with the activation of Th cells, so the studies on Th cell activation and the underlying molecular mechanism are one of the most important parts of immunology. Take Th cell activation as an example, the original trigger signal (first signal) for activation of naive T cells is TCR recognition of antigen presented by MHC-II class molecules ((TCR: MHC-II+Ag), and at the same time costimulatory signal (second signal) is required to achieve complete activation, followed by clonal proliferation, differentiation, and exertion of effects. The second signal for Th cell activation is mainly generated by the binding of CD28 on T-cell membrane to B7 of APC (CD28:B7 interaction), which induces T cells to overexpress IL-2 that provides growth signals to T cells via autocrine and paracrine secretion patterns and avoid the activation-induced cell death (AICD). If there is no costimulatory signal for T-cell activation, the cell will undergo apoptosis or anergy (a non-responsive state). Engagement of CD28 with B7 provides an important second signal for T-cell activation, and as T cell is activated, proliferate and differentiate, a variety of regulatory molecules (positive and negative regulators) are involved in this process to ensure that T-cell immune response proceeds within the physiological range. The activation process of CD8T cells is similar to that of CD4T cells and similarly requires the synergistic action of two signals, namely TCR binding to antigen presented by MHC-I class molecules and costimulatory signals.

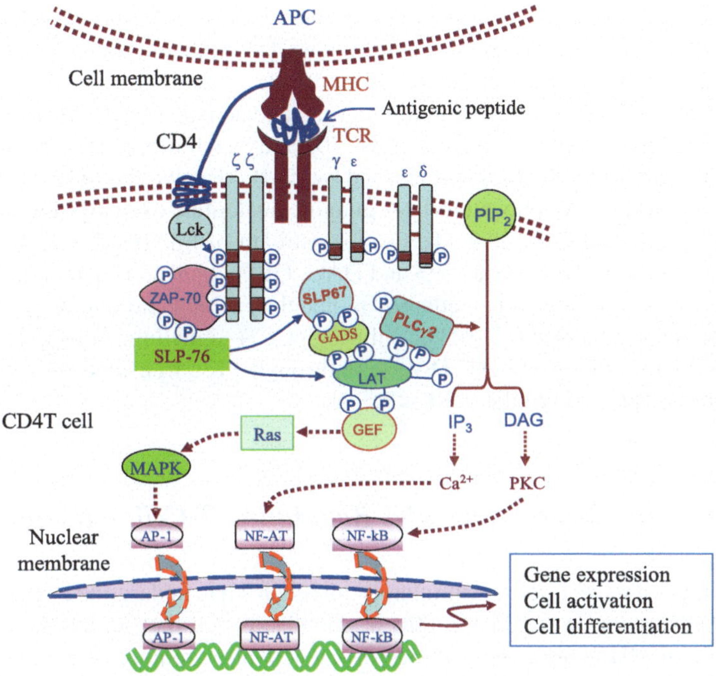

Fig. 4.13 Schematic diagram of the TCR-mediated signaling pathway. *Lck* tyrosine kinase, *ITAM* immunoreceptor tyrosine-based activation motif, *ZAP-70* zata associated protein of 70 kD, a protein kinase that binds to CD3-binding protein kinase, *SLP-76* SH2 containing leukocyte specific protein of 76 kD, a leukocyte-specific protein containing the SH2 structural domain with a molecular mass of 76 kD, *LAT* linker of activated T cells, a junction molecule for T-cell activation, *PLCγ* phospholipase Cγ, *GEF* guanine nucleotide exchange factor, *GADS* Grb2 like adaptor downstream of Shc, *PIP2* phosphoinositol biophosphate, *IP3* inositol 1,4,5triphasphate, *DAG* diacylglycerol, *PKC* protein kinase C, *AP-1* activator protein-1, *NF-AT* Nuclear factor of activated T cells, *NF-κB* nuclear factor kappa enhancer binding protein

4.5.1 Mechanism of Antigen Recognition and Signal Transduction by TCR–CD3 Complex Molecules

As previously described, the TCR only recognizes antigenic peptides presented by MHC, unlike the BCR that directly binds to free antigen, which indicates that T cells cannot directly neutralize antigens in vivo and only recognize specific antigenic peptides presented by MHC molecules on the cell surface (see the section on major histocompatibility complexes). The engagement of TCR with antigenic peptides presented by MHC generates activation signals to the cell, triggering an immune response through multiple intracellular signaling pathways (Fig. 4.13). In the case of CD4T cells, for example, their TCR recognizes antigenic peptides presented by antigen-presenting cell (APC) MHC-class II molecules, and after the CD4 helper molecule binds to the β2 domain of the MHC-class II molecule (whereas the CD8 T-cell helper molecule CD8 binds to the α3 domain of the APC MHC-class

I molecule), Lck (a protein tyrosine kinase) binds tightly to the intracellular portion of CD4 molecule, receiving the signal from the CD4 molecule, causing phosphorylation of the ITAM on the ζ chain of CD3 molecule.

Phosphorylation of the ITAM region recruits and activates ZAP-70, another protein tyrosine kinase. After activation, ZAP-70 catalyzes multiple membrane-bound proteins. For example, SLP-76 was cleaved by ZAP-70 activation to be SLP-67 and adaptor LAT, while LAT phosphorylates guanine nucleotide exchange factor (GEF), phospholipase C and Grb2-like adaptor downstream of Shc (GAD). PLCγ activation hydrolyzes PIP2 to generate IP3 and DAG, GEF signals downstream, and activates MAPK kinase and other adaptor molecules, which ultimately act through transcriptional regulator AP-1. IP3 acts through Ca2+ signaling pathway to activate transcriptional regulator NF-AT. DAG acts through PKC transmission pathway to activate transcriptional regulator NF-κB.

4.5.2 Costimulatory Molecules Regulating T-Cell Activation

The understanding of T-cell costimulatory molecules has undergone a research history of many years. In the late 1980s, it was discovered that, in addition to TCR recognition of MHC-antigenic peptides, T-cell activation requires the second signal generated by CD28 co stimulatory molecules for complete cell activation. Thus, T-cell activation not only requires recognition of MHC-presented antigenic peptides by the TCR, but also is regulated by costimulatory molecule signaling to keep the immune response in the proper range. Costimulatory molecules are members of the immunoglobulin superfamily (IgSF), the tumor necrosis factor receptor family (TNFR), or the integrin family (integrin F). For example, IgSF is also known as the B7 family, and costimulatory molecules of the B7 family include B7-1 (CD80), B7-2 (CD86), B7-H1 (PD-L1), B7-DC (PD-L2), B7-RP-1, and B7-H3, while OX40 is a member of the TNFR family and its ligand OX40L is a member of the TNF family. There are two major classes of costimulatory molecules that regulate T-cell activation, one acting as a positive regulator and the other as a negative regulator.

- Costimulatory molecules for positive regulation: this class of molecules binds to the corresponding ligands to generate costimulatory signals that have a role in promoting T-cell activation, e.g., CD28, ICOS, 4-1BB and OX40 bind, respectively, to membrane molecules B7-1 (CD80)/B7-2 (CD86), B7H2, 4-1BBL and OX40L of APC to generate positive regulatory signals to promote T-cell activation (see Fig. 4.14). Although there are multiple costimulatory molecules that promote T-cell activation, they may play different roles during T-cell activation. For example, CD28 is expressed in the initial T cells and generates a costimulatory signal (second signal) upon binding to B7, whereas the expression of 4-1BB, a costimulatory molecule, is induced after the initiation of T-cell activation, i.e., it is only expressed in activated T cells and NK cells, and the binding of 4-1BB to 4-1BBL of APC serves to enhance T-cell activation. In addition, 4-1BB binds to 4-1BBL to promote B7-1/B7-2 expression in APC and the production of pro-inflammatory cytokines such as IL-6, IL-12, and TNF-α. In addition, LIGHT and LTβ are also positive regulatory costimulatory mol-

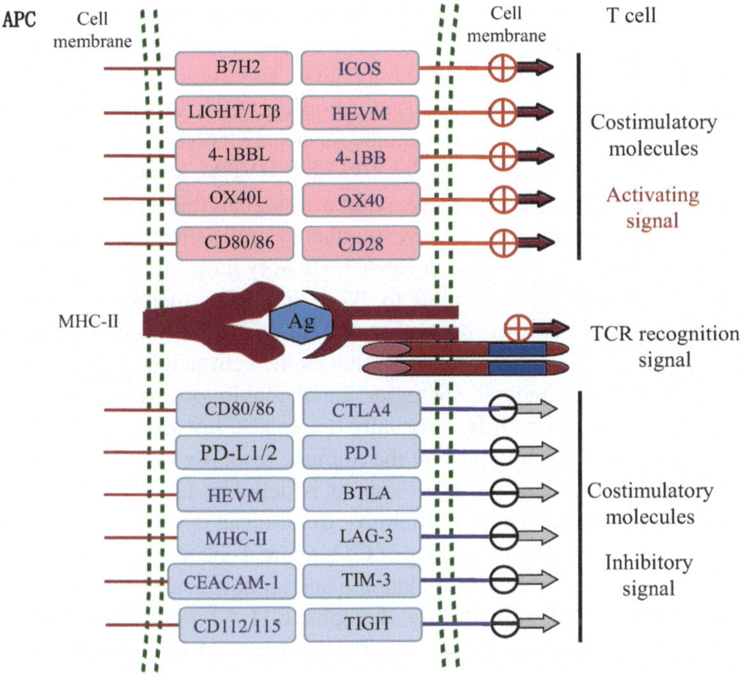

Fig. 4.14 Schematic diagram of the interactions of costimulatory molecules involved in positive or negative regulation of T-cell activation. Costimulatory molecules for positive regulation promote T-cell activation, mainly including CD28: B7-1/B7-2, ICOS: B7-H2, 4-1BB: 4-1BBL, OX40: OX40L, HVEM: LIGHT/LTβ, etc. Negative regulatory costimulatory molecules generate inhibitory signals, mainly including: CTLA4: B7-1/B7-2, PD1: PD-L1/PD-L2, HVEM: BTLA, LAG-3: MHC-II, TIM-3: CEACAM-1, TIGIT: CD112/CD115, etc. Abbreviated names and properties, *BTLA* B and T lymphocyte attenuator, also known as CD272, is expressed primarily in T cells within the tumor environment and is a receptor for HVEM, *CTLA4* cytotoxic T lymphocyte antigen-4, also known as CD152, is a receptor molecule whose expression is induced upon T-cell activation, and compete with CD28 to bind to ligand B7-1/B7-2; ICOS: inducible costimulator, *CEACAM-1* a T-cell-expressed costimulatory molecule that acts as a ligand to bind to self-membrane receptor TIM-3 or TIM-3 of other cells, *4-1BB* also known as CD137, is a costimulatory molecule that is mainly expressed upon in T and B lymphocytes and monocytes. *OX40* also known as CD134, is a costimulatory molecule only expressed in activated T cells, involved in regulation of cytokine expression, antigen presentation and signal transduction, *PD-1* programmed death-1, also known as CD279, is a receptor whose expression is induced upon T-cell activation, *TIM-3* T-cell immunoglobulin domain and mucin domain, is expressed in a variety of innate immune cells, also expressed during the T-cell activation and differentiation phases, mainly binds to T cell its self CEACAM-1 molecule or CEACAM-1 molecules of other cells, *LAG-3* lymphocyte-activation gene-3, a type I transmembrane protein, highly homologous to CD4, an inhibitory receptor for CD4T cells, mainly inhibits IL-2 production by CD4T cell, *LIGHT* lymphotoxin like, exhibits inducible expression and completes with HSV glycoprotein D for HVEM, a receptor expressed by T lymphocytes, also known as TNFSF14 or CD258, *LTβ* lymphotoxin-β, *PD-L1* program death ligand 1, also known as B7H1 or CD274, a ligand for PD-1, expressed in a variety of tissues and tumor cells, *PD-L2* program death ligand 2, also known as B7-DC or CD273, expressed primarily in dendritic cells and macrophages, *HVEM* herpes virus entry mediator, also known as CD270, a costimulatory molecule expressed by a variety of immune cells and a member of type I transmembrane protein TNF receptor (TNFR) superfamily, HVEM binds to BTLA to generate immunosuppressive signals, and if it binds to the ligands LIGHT or LTβ, it will promote immune response, and so it has a dual roles, *TIGIT* T-cell immune receptor with immunoglobulin and ITIM domain, is mainly expressed in T cell and NK cell, has a similar function to CTLA4, and its ligands are CD112/CD115

ecules that promote inflammatory responses upon binding to the corresponding receptors (e.g., HVEM).

- Costimulatory molecules for negative regulation: This class of molecules bind to the corresponding ligands to inhibit T-cell activation, e.g., CTLA4, PD-1, BTLA, LAG-3, TIM-3 and other costimulatory molecules for negative regulation, and they bind to membrane molecules B7-1/B7-2, PD-L1/PD-L2, HVEM, MHC-II, CEACAM-1 of APC, respectively, generating negative regulatory signals that inhibit T-cell activation (see Fig. 4.15). Among the negative regulatory molecules for T-cell activation, CTLA4 is the most representative negative regulatory molecule with the same ligand for CD28, and both of them bind to B7-1 and B7-2 membrane molecules of APC. Experimental evidence show that the naive T cells express a high level of CD28, while CTLA4 expression is little, but as the activated T cells proliferate, the expression of CTLA4 gradually increases, and the inhibitory signal generated by the engagement of CTLA4 with B7 gradually dominates, just like making the rushing train gradually slow down to prevent the immune response from being overreactive and causing immune damage. If an animal is deficient in CTLA4 (e.g., CTLA4 knockout mice), a lethal T-cell proliferative disease can develop. Clinically, tumors can be treated by blocking the activity of CTLA4, and encouraging results have been achieved for treatment of human melanoma and kidney cancer with CTLA4 monoclonal antibodies. In addition, the use of soluble CTLA4 molecules to block B7 can inhibit T-cell activation, and this approach is being used to explore the treatment of autoimmune diseases such as psoriasis and rheumatoid arthritis.

In recent years, PD-1, BTLA, TIM-3, and LAG-3 have been found to negatively regulate T-cell activation. The ligands of PD-1 are PD-L1/PD-L2. PD-L1 is expressed in a variety of tissues and tumor cells, whereas PD-L2 is expressed only in macrophages and dendritic cells. The ligand of BTLA is HVEM, which is expressed by a variety of immune cells and belongs to the member of the type I transmembrane protein TNF receptor (TNFR) superfamily, and BTLA-HVEM interactions produce immunosuppressive signals that may be associated with T-cell immune tolerance. However, if HVEM binds to the ligand LIGHT or LTβ, it will promote an immune response associated with autoimmune disease, thus HVEM has dual roles (Fig. 4.14). In addition, new immune checkpoint molecules have been discovered in recent years, such as VISTA (V-domain Ig suppressor of T-cell activation, also known as B7-H5, PD1-H or Gi24) and TIGHT (T-cell immunoreceptor with Ig and ITIM domains). If VISTA is knocked out in mice, the mice have an enhanced immune response and develop autoimmune disease, and blocking VISTA by administering ipilimumab (VISTA monoclonal antibody) to tumor patients in clinical practice improves antitumor immunity and has yielded positive therapeutic results, and TIGHT, similar to CTLA4, competitively binds to CD226 that can bind to the activating receptor, acting as an inhibitor.

It is worth recalling that the discovery of CTLA4 has a hard time history. In 1988, Dariavach et al. found that activated CD8 T cells express a molecule with high homology to CD28, termed CTLA4.In 1991, Ledbetter and Allison demonstrated that CTLA4 is a second receptor for B7, that its binding to B7 has an affinity more than 20 times that of CD28-B7, and that CTLA4 binding to B7 produced a stimulatory signal that strongly inhibited T-cell activation (Fig. 4.15). In the 1990s, James P. Allison and Tasuku Honjo discovered that blocking immune checkpoints

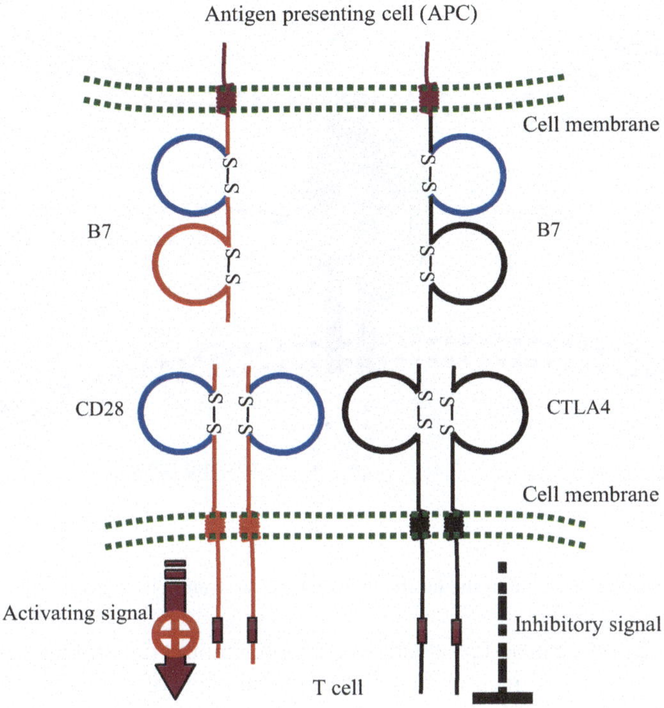

Fig. 4.15 Schematic diagram of the inhibitory effect of CTLA4

(inhibitory costimulatory molecules (CTLA4/PD-1) of T cells) by antibodies could restore the cytotoxic activity of effector T cells, and achieve a therapeutic effect on the tumor. As the clinical treatment of tumors by blocking immune checkpoints with monoclonal antibodies has achieved exciting efficacy, it has kindled new hope for tumor treatment. For this, James P. Allison and Tasuku Honjo were jointly awarded the Nobel Prize in Physiology or Medicine 2018.

As described above, multiple costimulatory molecules are required to generate positive and negative regulatory signals during T-cell activation, proliferation and effector phases to play coordinated roles in keeping the immune response within the appropriate physiological range. If the regulatory signals generated by costimulatory molecules go wrong, it can lead to apoptosis, anergy or autoimmune disease, and the quantitative investigation will definitely help to elucidate the exact regulatory mechanisms.

4.5.3 Activation of T Cells by Superantigens

Superantigens are toxic substances produced by insect, bacteria or viruses. These substances have high binding affinity for the alpha chain of MHC-II molecules and the Vβ chain of CD4T TCR, and therefore can bind them tightly together through the outer side of MHC-II and TCR (Fig. 4.16), which cause simultaneous activation

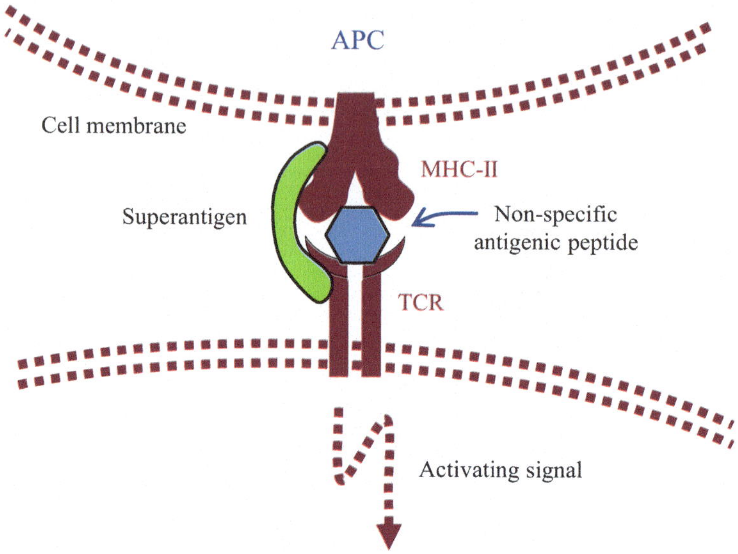

Fig. 4.16 Schematic diagram of the mechanism of T-cell activation by superantigens

of 5%-20% CD4T cells and production of a large amount of cytokines, resulting in immune damage, consequently displaying clinical signs of systemic toxicity, such as toxin-induced septic shock caused by staphylococci enterotoxins (SEA, SEB, SEC, SED, and SEE) contaminating food or by insect bites.

The superantigen binds directly to the outside of TCR Vβ chain and the outside of MHC-II α chain without restriction of specificity, which enables the simultaneous activation of a large number of T cells.

4.5.4 Regulation of T-Cell Activation and Tumor Therapy

Tumor immunotherapy has always been a concern, and currently there are two main directions of therapeutic approaches, one is to block immune checkpoints with monoclonal antibodies to restore the cytotoxic activity of effector T cells for tumor therapy, and the second is to use chimeric antigen receptor T-cell therapy (CAR-T), in which the tumor antigen-recognizing portion of an antibody is fused with the TCR-CD3 intracellular region to form a chimeric antigen receptor expressed in the membranes of CD8T cells, allowing T cells to target and kill tumor cells by recognizing tumor antigens.

4.5.4.1 Blocking Immune Checkpoints for Tumor Immunotherapy

Blocking the immune checkpoint CTLA4 and/or PD-1 with monoclonal antibodies to restore the cytotoxic activity of effector T-cell was the first immunotherapeutic technique to be used in the clinic, and in 2011 the FDA-approved anti-CTLA4 monoclonal antibody (trade name ipilimumab) for the clinical treatment of

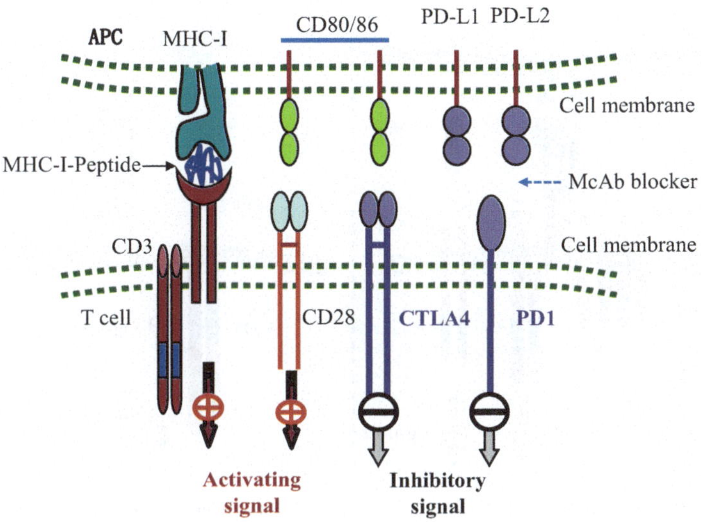

Fig. 4.17 Schematic diagram of blocking immune checkpoint CTLA4/PD-1 for tumor treatment

melanoma, followed by anti-PD-1 monoclonal antibody (trade name nivolumab) and anti-PD-L1 monoclonal antibody (trade name devalumab) blocking PD-1/PD-Ls) were also approved for tumor treatment (see Fig. 4.17). Typically, clinical tumor immunotherapy is used in combination with chemotherapy. In addition, the immunosuppressive costimulatory molecules LAG-3 and TIM-3 expressed by T cells and NK cells are also targets for tumor immunotherapy.

Tumor cells highly express ligands (CD80/86 as well as PD-L1/PD-L2) for T-cell immune checkpoints (CTLA4 and PD-1), inhibit T-cell activation, and put them in an anergic state. Blocking tumor cells from binding to immune checkpoint receptors with a monoclonal antibody restores T-cell cytotoxic activity.

4.5.4.2 CAR-T for Tumor Immunotherapy

CAR-T tumor immunotherapy technology embodies the most ingenious combination of techniques of molecular biology, cell biology, and immunology. This technology constructs a fusion expression vector by genetic recombination technology, combining the gene of antigen-recognizing portion (scFV, i.e., single-chain variable region), which specifically recognizes tumor antigen by antibody, with the gene sequence of CD3 activation region and intracellular region of the costimulatory molecule (CD28 or 4-1BB) of CD8T cells, packaging lentivirus, and expressing fusion proteins in CD8T cells through lentiviral expression system. This recombination receptor includes an extracellular region (scFV) + intracellular region (CD3 activation region + signal transduction region of costimulatory molecules), which constitutes the so-called chimeric antigen receptor (CAR) of T cells (Fig. 4.18), through which CD8T cells are activated upon recognition of tumor-specific antigens and subsequently release perforin and granzyme to target and kill tumor cells. In

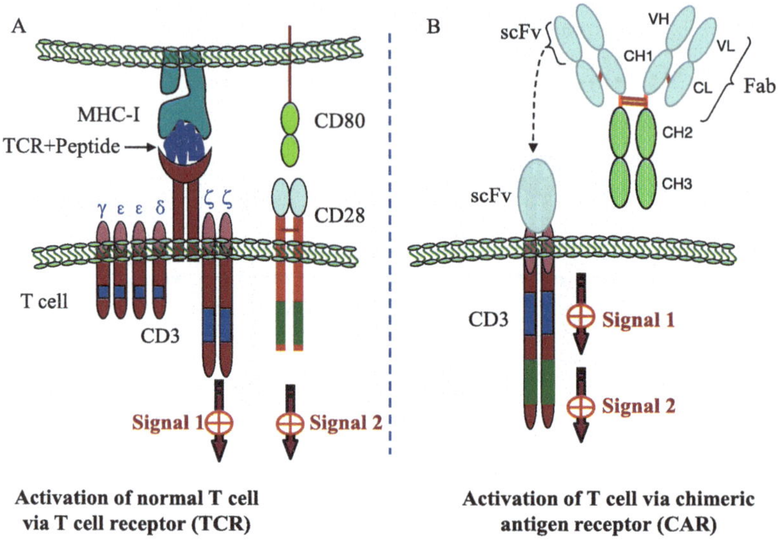

Fig. 4.18 Schematic diagram of the principle of CAR-T for tumor treatment. (**a**) Conventional T-cell activation process; (**b**) The chimeric receptor of T cells is a fusion protein consisting of an extracellular region (scFV) + an intracellular region (CD3 activation region + signal transduction region for costimulatory molecules). The extracellular antibody recognition region specifically binds to tumor antigens and generates signals transduced into intracellular portion of the receptor, which contains functional domains that generate the first and second signals required to cause T-cell activation. *scFV* single chain variable region, *CAR* chimeric antigen receptor

2012, a patient (a 7-year-old girl named Emily Whitehead) with acute lymphoblastic leukemia, was successfully cured with CAR-T, and up to now (April of 2024), it is close to 12 years, she is cancer free, which became a milestone in tumor immunotherapy and brought new hope to cancer patients. In August 2017 CAR-T technology has been approved by FDA for tumor treatment in the U.S. CAR-T is the most successful clinical tumor immunotherapy technology using specific tumor cell-related antigens as therapeutic targets, opening a new window for tumor treatment with broad application prospects. Similarly, CAR-NK and CAR-iPS technologies are also being evaluated in clinical trials for tumor immunotherapy, and different types of improved CAR-T technologies have achieved encouraging results. It is believed that the cure rate of tumor diseases will continually increase.

Study Questions

1. What is the process of positive and negative selection of thymocytes? What is the biological significance?
2. what do you think of the specificity and MHC restriction of CTL to kill target cells?
3. What is the mechanism of immune damage caused by superantigens?
4. What is the role of CD28 and CTLA4?
5. What is the immune checkpoint?
6. What is CAR-T?

Chapter 5
B-Cell Development, Maturation, and Activation

Overview:
B cells originate from hematopoietic stem cells in the bone marrow, which is the site of B cell development and maturation (bursa of Fabricius is the site of B cell development and maturation in avian species). In the bone marrow, large numbers of B cells are produced daily, but only a small number eventually develop to maturity. These mature B cells leave the bone marrow (or bursa of Fabricius in birds) and enter the circulation system, settling down mainly in the spleen and lymph nodes. During B-cell development, those B cells that bind to self antigens are induced to apoptosis, so that under normal conditions, mature B cells recognize foreign antigens and do not react with self antigens, i.e., self-tolerance. However, if self-reactive B cells are present in the body and are activated to produce autoantibodies, they would attack self tissues and cause autoimmune disease. This chapter focuses on the basics of B-cell development, maturation, and activation.

5.1 B-Cell Development and Maturation

Early in embryonic development, the yolk sac, liver, and bone marrow are important sites for the generation of blood cells, but as embryonic development progresses, the bone marrow gradually becomes the main central immune organ. Recent studies have revealed that B cells in mice are divided into two populations, B1 and B2 cells, and that those B cells differentiate, develop, and mature from hematopoietic stem cells of the liver early in embryonic development are called B1 cells. As the embryo develops, the bone marrow gradually replaces the liver in performing hematopoietic tasks, and those B cells differentiate, develop, and mature from bone marrow hematopoietic stem cells are called B2 cells. B1 cells do not need antigen stimulation to produce natural antibodies to protect the fetus, and the main types of natural antibodies produced are IgM and IgG3, so B1 cells also belong to innate like lymphocytes, which

S. J. Zheng, *Veterinary Molecular Immunology*,
https://doi.org/10.1007/978-981-99-8929-4_5

are further divided into two subpopulations, B1a (CD5+) and B1b (CD5-), based on whether B1 cells express CD5 or not. As the embryo matures and liver hematopoiesis is replaced by bone marrow, B1 cell production gradually ceases, but B1 cells continue to renew themselves in response to self-antigen stimulation and retain a certain number after birth and even throughout life. Most B1 cells reside in the abdominal cavity, thoracic cavity, and lamina propria of the intestinal mucosa, and mainly play the role of antigen presentation, but also have certain phagocytic functions. In terms of immune response, B1 cells mainly recognize thymus-independent antigens, do not generate immune memory, and secrete IL-10, whereas B2 cells are conventional adaptive immune response B cells that require continuous differentiation and generation from bone marrow hematopoietic stem cells, which are continuously transported to peripheral immune organs for residence and humoral immune response against specific antigens, usually requiring the involvement of CD4T cells to generate immune memory. The B cells described in this chapter refer to B2 cells.

After the birth of an animal, hematopoietic stem cells in the bone marrow are the progenitors of all blood cells. Stem cells have the ability of self-renewal and generate stem cells that differentiate into various types of cells. All lymphocytes are derived from lymphoid stem cells, and B cells need to go through different developmental stages in the bone marrow, such as pro-B cells, pre-B cells, immature B cells, and naive B cells. During development, pro-B cells need to come into contact with each other and stromal cells. Pro-B cells express VLA-4 (an adhesion molecule), which is able to bind to VCAM-1 expressed by stromal cells, and the interaction of VLA-4 with VCAM-1 promotes the interaction of receptor c-Kit on the surface of pro-B cells with the ligand SCF expressed by stromal cells. After SCF binds to c-Kit, c-Kit signals downward, prompting pro-B cells to express IL-7R. Stromal cells secrete IL-7, which, through IL-7 binding to IL-7R, drives pro-B cells toward pre-B cells, further developing into immature B cells and eventually into mature naive B cells (Fig. 5.1).

The stromal cells play a very important role in B-cell development. On the one hand, stromal cells are in direct contact with pro-B cells and pre-B cells, and on the other hand, stromal cells secrete a variety of cytokines (e.g., IL-7), which promote B-cell development and maturation.

5.2 B-Cell Clones and the Basic Structure of the B-Cell Receptor

An important function of the bone marrow is the selection of B-cell clones. Cell clone is an identical group of cells derived from the same cell. In the bone marrow, 10^7 cells are generated each day, but the vast majority of cells undergo apoptosis before they complete the developmental process, and only a small number (5% to 10%) of B cells eventually develop and become mature naive B cells that leave the bone marrow and migrate to local lymphoid tissue. There are approximately 1 billion B-cell clones in mammals, and different B-cell clones contain different membrane surface receptors (B-cell

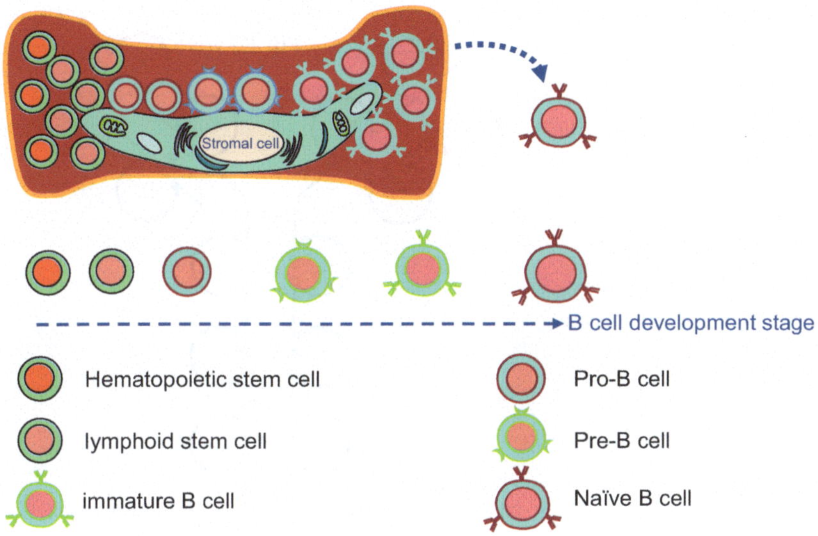

Fig. 5.1 Schematic diagram of the B-cell development process

receptor, BCR), and therefore recognize different antigens. However, for the same B-cell clone, the membrane surface antigen receptor BCR is identical, i.e., a B-cell clone can recognize one antigen only. Thus, 1 billion B-cell clones mean that there are sufficient numbers of BCRs that recognize 1 billion different antigens, so how does the diversity of BCRs develop during B-cell development? This is the fundamental question that needs to be answered in immunology. A complete BCR consists of two components: one component is the membrane immunoglobulin (mIg) molecule anchored in the B-cell membrane, consisting of two heavy chains (H chain) and two light chains (L chain), which are responsible for antigen recognition; the other one is the Ig-α/Ig-β heterodimer, which is responsible for downward signaling (Fig. 5.2).

The heavy and light chains of the BCR are divided into a variable region (V region) and a constant region (C region). The V regions of the heavy and light chains form an antigen-binding groove, and the amino acids in this region are complementary to the amino-acid sequence structure of the antigen, also known as the complementary-determining region (CDR). There are three CDRs in each V region, and the amino-acid sequences in CDRs are highly variable, hence named as high variability region (HVR). The structure of this region is related to the specificity and affinity of antibody-antigen binding. Similar to T-cell, B-cell development and maturation must undergo negative selection, but this process is accomplished in the bone marrow or bursa of Fabricius (BF, in birds). After negative selection, B cells mature, leave the bone marrow or BF (in birds), migrate to peripheral lymphoid tissue, and die after several weeks if no antigen is encountered. If encountered with antigen, B cells are activated with the help of Th cells and then proliferate and differentiate into plasma cells to produce antibodies and form memory B cells (the earliest memory B cells will be formed in about 2 weeks) (Fig. 5.3).

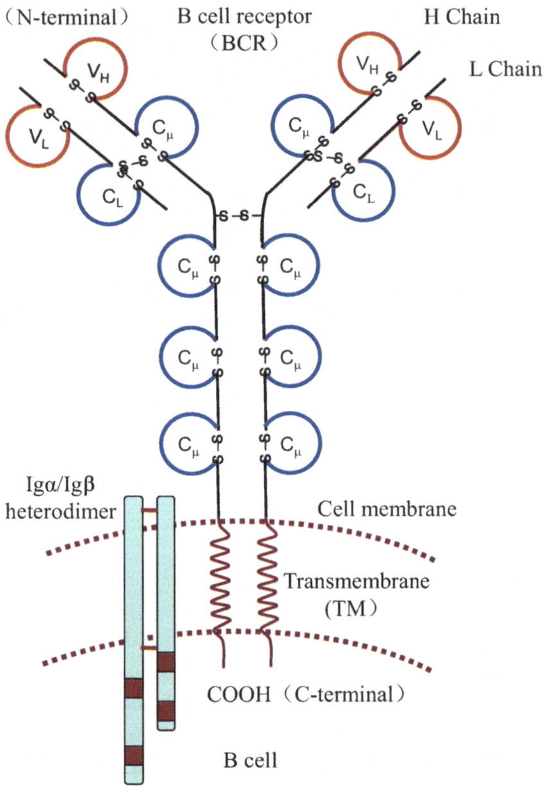

Fig. 5.2 Schematic diagram of the B-cell antigen receptor

The process of BCR formation is important for B-cell maturation because of the specificity and affinity of BCR binding to specific antigens. B-cell maturation depends on the rearrangement of immunoglobulin DNA during lymphoid stem cell differentiation. As shown in Fig. 5.2, each BCR molecule is composed of two identical light chains (L chain) and two identical heavy chains (H chain). There are two subclasses of κ and λ chains in the L chain, and either the κ or λ chain contains three gene fragments, V, J, and C. The VJ fragments combine to form the variable region of the light chain, and the C fragment encodes the constant region. The H chain contains V, D, J, and C gene fragments, with the VDJ rearrangement and combination to encode the variable region of the heavy chain and the C gene fragment encoding the constant region.

5.3 Molecular Mechanisms of B-Cell Receptor and Ig Formation

The B-cell receptor (BCR) is an immunoglobulin (mIg) anchored in the B-cell membrane, and immunoglobulin genes are assembled from multigene during the development and maturation of B cells. In the genomic DNA of hematopoietic stem

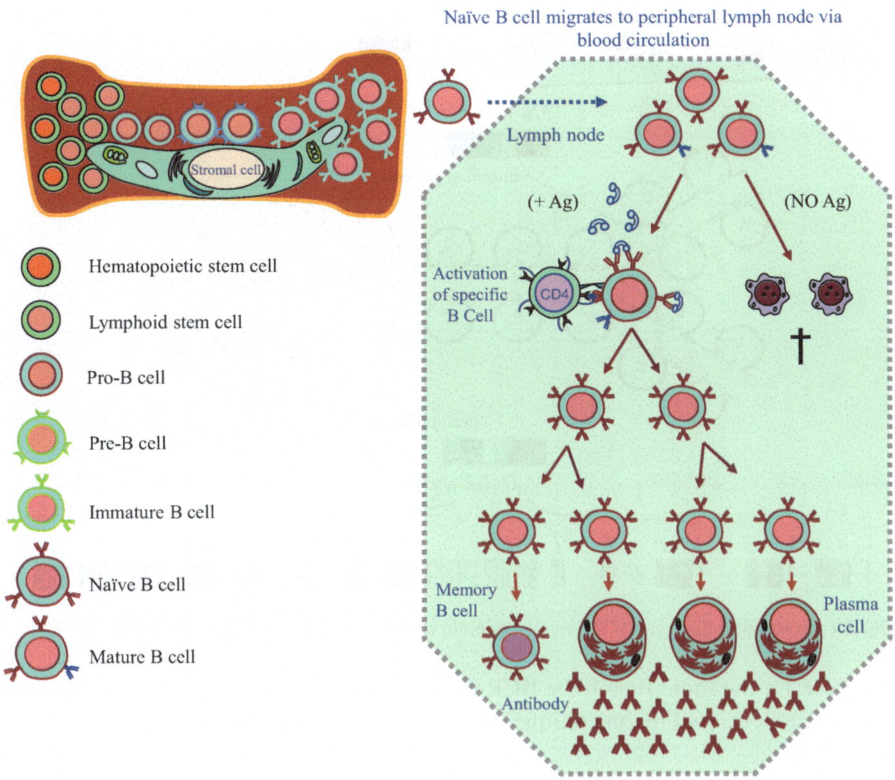

Fig. 5.3 Schematic diagram for B-cell development, maturation, proliferation, and differentiation

cell, these multigene segments are separated by genes of different sizes, each encoding a segment of protein, e.g., the mouse κ-L chain has 85 V genes, 5 J genes (1 gene is a pseudogene, which does not encode a protein), and 1 C gene (Fig. 5.4); the λ-L chain has 3 V genes, 4 J genes (1 gene is a pseudogene) and 4 C genes; the H chain has 51 V genes, 27 D genes, 6 J genes, and several different subclasses of C genes (Cμ, Cδ, Cγ, Cε, and Cα).

The discovery that the genes encoding the L and H chains were not only recombined from multiple gene fragments but also located on different chromosomes changed the once-held dogma that "one protein is encoded by one gene" and deepened the understanding of the life sciences, thus unveiling the mystery of antibody diversity and leading immunology into the era of molecular immunology. Susumu Tonegawa was awarded the Nobel Prize in Physiology or Medicine in 1987 for this outstanding work (see the Introduction section for more details). It should be emphasized that during B-cell development, gene fragments are randomly recombined, and each final mature B-cell clone contains only one defined L-VJC/H-VDJC combination and encodes only one BCR. Because each B-cell contains a unique L-VJC/H-VDJC combination, the BCR contained in one B-cell clone is specific.

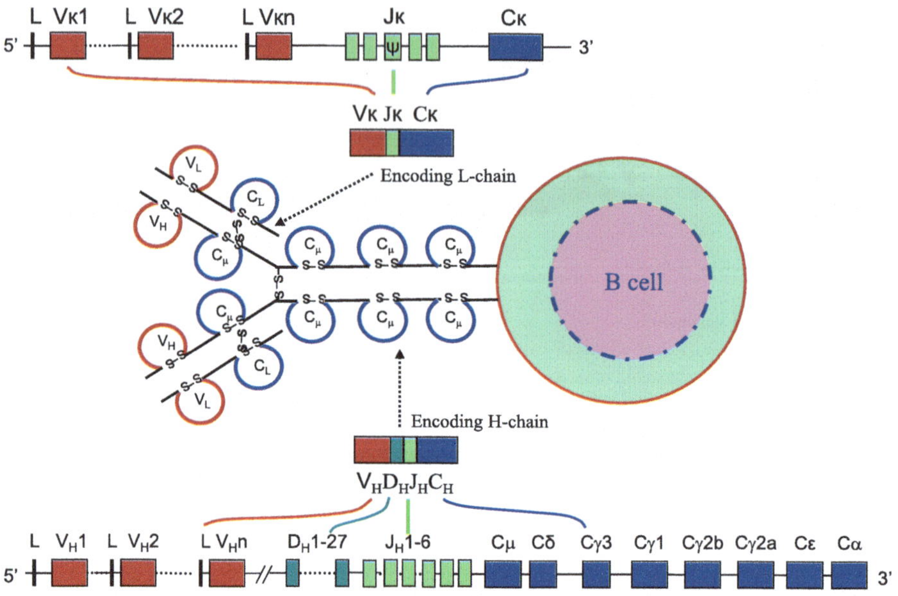

Fig. 5.4 Schematic diagram of the rearrangement of immunoglobulin L and H chains

However, for the same B cell, the BCR molecules on its cell membrane are all identical and are encoded by one unique combination of L-VJC/H-VDJC gene.

5.3.1 DNA Rearrangement of Ig Variable Region

The variable region of Ig is encoded by two gene fragments, one for the V gene in the L chain and the other in the H chain. Gene sequencing analysis revealed that in the germline, unique recombination signal sequences (RSS) exist on both sides of each V, J, and D gene fragment. Each RSS contains a palindrome structure consisting of 7 bases, an insertion fragment consisting of 12 or 23 bases, and a conserved AT-rich sequence. These RSSs contain recombinase cleavage sites that can be cleaved by recombinases, resulting in a V-J or V-D-J gene rearrangement. The enzymes that perform RSS sequence cleavage are collectively referred to as V(D)J recombinases. The first V(D)J recombinase genes identified were recombination-activating genes (RAG)-1 and RAG-2. RAG-1 and RAG-2 encode the RAG-1 and RAG-2 gene recombinant proteases as well as terminal deoxynucleotidyl transferase (TdT), which are enzymes specific to lymphocytes undergoing recombination of the V(D)J gene. RAG-1 and RAG-2 recognize RSS sequences for cleavage, cut the RSS sequence between V-J or between V-D-J, add nucleotides at the cleavage site, which is called P-addition (P-region nucleotide addition), and then the gene fragment encoding the protein is modified by a single-stranded endonuclease. During the H-chain gene recombination, in addition to the P-addition, 15 amino acids are added at the enzymatic cleavage end of the V, D, and J gene fragments by the TdT enzyme, which is a process called N-addition (N-region nucleotide addition). Finally, the gene fragment is repaired and ligated by the double-stranded break repair enzyme (DSBR) to form a complete variable region gene fragment encoding the L or H chain.

Although the chromosomes of animal somatic cells are diploid, each B cell expresses only one specific variable region of the BCR, with only one specificity, and does not express parental BCR molecules like the way antigen-presenting cells express MHC molecules of both parents (see Chap. 7), i.e., the H chain of Ig is derived from only one chromosome and the L chain is similarly derived from only one chromosome, a phenomenon called allelic exclusion. The molecular mechanism behind the phenomenon of allelic exclusion, which ensures that a mature B cell contains only one combination of V-J and V-D-J, i.e., expresses only one specific Ig and therefore recognizes only one antigenic epitope, is not yet clear. A theoretical model was proposed by G.D. Yancopoulos and F.W. Alt to explain allelic exclusion, by which they suggest that once a gene rearrangement is completed, the protein encoded by it acts as a negative regulatory signal to inhibit further rearrangements from similar genes. The model essentially explains the phenomenon that a mature B cell expresses only one specific Ig. The experimental evidence indicates that if transgenic mice highly expressed μ chain of one heavy chain or the κ chain of a light chain, then the corresponding endogenous heavy or light chains would not express, supporting the theory of allelic exclusion to some extent, but the exact molecular mechanism needs to be further investigated.

5.3.2 Combination of Light and Heavy Chains

Based on the rearrangement of H- and L-chain gene fragments, humans theoretically have 8262 H- and 320 L-chains, which, if randomly combined, would yield approximately 2,643,840 Ig. However, in practice, one H- and one L-chain are not all expressed with equal probability, and pairing with each other is not completely random. There are always dominantly expressed H- and L-chains, so the effect of L- and H-chain combinations on how much Ig diversity is affected is not well understood.

5.3.3 Somatic Hypermutation and Affinity Maturation

During B-cell maturation, V-J and V-D-J gene fragment combinations require P-additions and N-additions, which present the opportunity for random combinations of bases to form different gene fragments, and the addition of different gene fragments directly generates variable regions of L and H chains with different specificities. Although mature B cells already have a defined combination of V-J and

V-D-J gene fragments, upon encountering an antigen and being activated, they begin to divide and proliferate, and with each new generation, gene mutations occur at a rate of 1 in 1000, i.e., 1 base change per 1000 bases after one cell cycle. The phenomenon can be detected by analyzing mRNA change in the Ig variable region of different B-cell clones after the animals are immunized. The mutation rate of other somatic genes under normal conditions is 1 in 100 million, i.e., 1 replacement in 10^8 bases occurs for every generation of cells. In comparison, the rate of mutation of the Ig gene in B cells is 100,000 times higher than that of other normal somatic cells, and is thus called somatic hypermutation. Somatic hypermutation occurs during the division and proliferation of B cells after antigen recognition, and the mutations are concentrated in the complementarity-determining region (CDR) within the variable region of the H and L chains, which determines the affinity of the antibody for the antigen, so somatic hypermutation can produce antibodies with different affinities, and B cells with high-affinity antibodies (also BCR) will be preferentially activated and proliferate under the stimulation of low concentrations of antigen, gradually forming a dominant population of cells with high affinity, and the whole process of producing high-affinity antibodies through antigenic selection is called affinity maturation. Since the heavy and light chain variable region genes of Ig add up to about 600 bp (base pair), B cells will have at least one mutation per two generations of proliferation, which greatly increases the diversity of Ig.

5.3.4 Genetic Variation in Constant Region and Antibody Isotype Class Switching

Each B-cell H-chain has a defined combination of variable-region V-D-J gene fragments, which are formed and further combined with C (constant-region) fragment genes (mainly Cμ) to form a complete H-chain. However, when B cells are activated by recognition of antigens, their predefined V-D-J fragments may combine with different constant region genes to form antibodies of different isotypes, a phenomenon known as antibody class switching or isotype class switching. Although the constant region of the antibody is different after class switching, the variable region remains the same. The molecular mechanism that causes Ig isotype class switching is not well understood, but interleukin (IL)-4 can cause Ig isotype class switching under certain conditions, for example, IL-4 can induce isotype class switching of the B-cell Ig heavy chain constant region from μ to γ1 or ε chains, i.e., the Ig produced undergoes isotype class switching from IgM to IgG1 or IgE.

5.3.5 Synthesis, Assembly, and Secretion of Immunoglobulins

In the nucleus, the DNA fragments of the H and L chains of immunoglobulin are rearranged, combined, and transcribed into mRNA using the recombinant as templates. mRNA is translated into protein on the polyribosomes of the rough

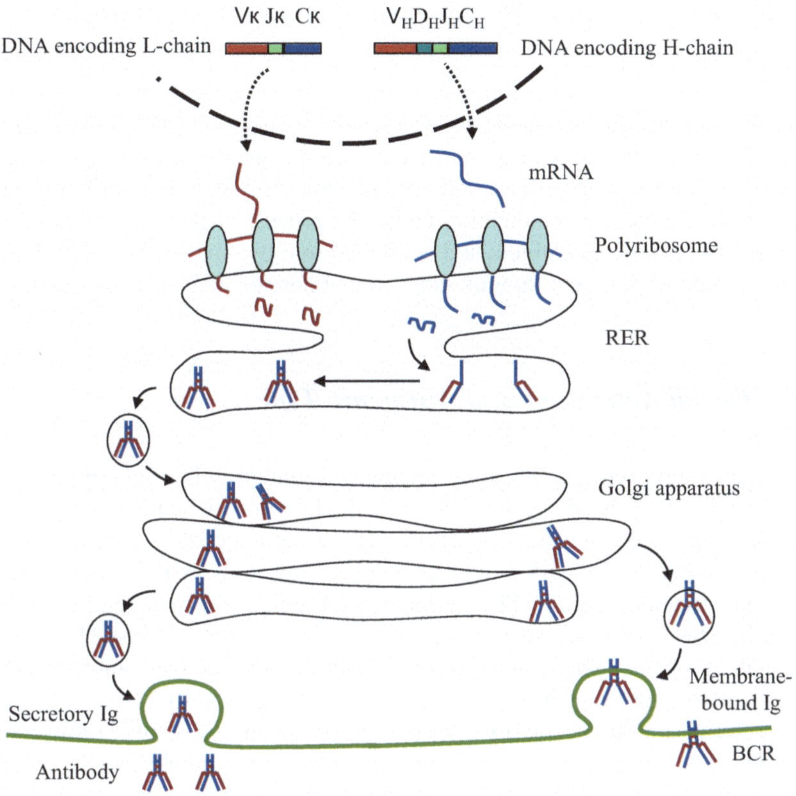

Fig. 5.5 Schematic diagram of the process of synthesis, assembly, and secretion of immunoglobulins

endoplasmic reticulum (RER), and the newly synthesized peptide chain contains the amino-acid terminal leader sequence that guides the newly synthesized peptide chain into the rough endoplasmic reticulum, in which subsequently the leader sequence is removed. The H and L chains are glycosylated within the RER and linked by disulfide bonds to form a complete Ig molecule, which is then transported to the Golgi apparatus and enters the secretory vesicle, which fuses with the cell membrane to release Ig. if the Ig molecule contains a transmembrane region, the Ig molecule is subsequently anchored onto the cell membrane when the secretory vesicle fuses with the cell membrane and becomes a membrane-bound immunoglobulin (mIg, BCR) (Fig. 5.5). Mature B cells express mIg, and upon antigen stimulation, proliferate and finally differentiate into plasma cells producing secretory Ig. The exact mechanism determining whether Ig is secretory or membrane-bound is not well understood, but it has been found that it may be related to the shearing and splicing during mRNA processing. If the transmembrane region of primary mRNA is retained during mRNA processing, mIg is produced, and vice versa, secretory Ig is produced. However, how the process of shearing off the transmembrane region of mRNA is regulated remains elusive.

5.4 B-Cell Activation, Proliferation, and Differentiation

After leaving the bone marrow, mature B cells circulate in body fluids into peripheral tissues and die after a few weeks if they do not encounter the appropriate antigen. If a B cell encounters and is stimulated with the appropriate antigen, it will be activated, proliferate, and differentiate into plasma cells secreting antibodies and a few of them differentiate into memory cells. The process of B-cell activation varies, depending on the type of antigen, and is divided into two main types of cell activation: thymus-dependent and thymus-independent antigen activation processes.

5.4.1 B-Cell Activation and Antigenic Type

If B-cell activation requires not only antigenic stimulation but also direct contact with helper T cells (Th), these antigens are called thymus-dependent antigen (TD). In comparison, if a B cell is directly activated by an antigen without Th-cell involvement, those antigens are called thymus-independent antigens (TI). There are two types of TI antigens, TI-1 and TI-2 antigens. TI-1 antigens is in general the components of bacterial cell wall, such as lipopolysaccharide (LPS). Most TI-1 antigens are B-cell mitogens, which at high concentrations can activate multiple B-cell clones simultaneously, with approximately 1/3 of the B cells being activated to secrete antibodies at the same time, without strict specificity, while at low concentrations can activate specific B cells to produce antibodies against specific antigenic epitopes. Immunization of nude mice without thymus with this type of antigen activates B cells to produce antibodies, and there is no significant effect on promotion of antibody production after transfusion with Th cells.

TI-2 antigen activates B cells by cross-linking multiple BCR molecules, and this type of antigen has three characteristics compared with TI-1 antigen: (a) it is not a B-cell mitogen and only activates specific B-cell clones; (b) it requires the involvement of cytokines produced by Th cells to fully activate B cells for production of antibodies and switch between different antibody isotypes under certain specific conditions; and (c) it only activates mature B cells and has an inhibitory effect on immature B cells. In contrast, TI-1 antigen has an activating effect on both mature and immature B cells. Immunization of TCR knockout mice with TI-2 antigen does not produce antibodies, and immunization of nude mice with the antigen stimulates antibody production by B cells because nude mice contain small amounts of T cells in their mucosal tissue.

The most striking features of the immune response induced by TI antigens are the lack of memory cells, the predominant type of antibody being IgM, and the weak immune response. These features suggest that Th cells play an important role in the formation of immune memory, affinity maturation, and antibody isotype class switching.

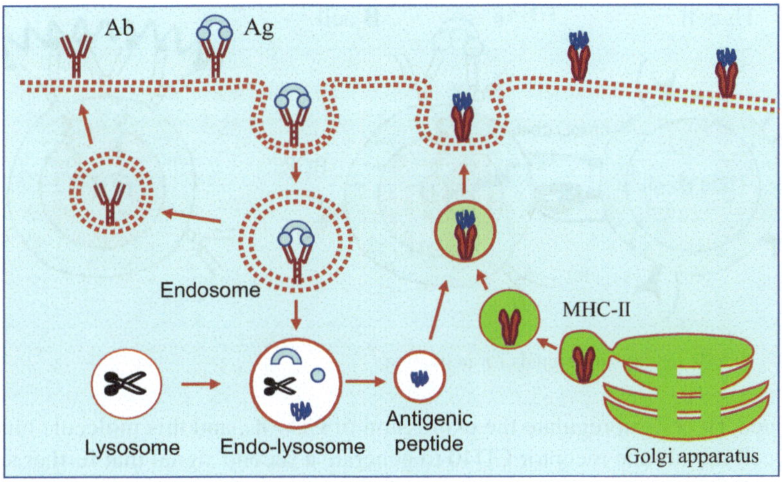

Fig. 5.6 B-cell recognition of antigen via the endosomal pathway presenting antigenic peptides

5.4.2 Molecular Mechanisms of B-Cell Activation

B cells recognize antigenic epitopes through B-cell receptor (BCR) on the cell membrane, which is chemically composed of membrane immunoglobulin (mIg) with a specific antigen recognition region. Because the BCR molecules are identical on the surface of each B cell, the BCR binds to antigen with strict specificity, so that a B-cell clone recognizes only one antigenic epitope. Exogenous TD antigen bound to the BCR is ingested into the cell to form the endosome, and under the low pH conditions of the endosome (pH 6–6.5), the antigen dissociates from the BCR and the BCR returns to the cell membrane surface, while the antigen passes through the acidic environment of the endosome and endo-lysosome (pH 5–6) (the endosome fused with lysosome (pH 4.5–5) to become endo-lysosome), where it is digested by various enzymes (including hydrolases, proteases nucleases, esterases, phospholipases, phosphatases, and more than 40 other digestive enzymes), the antigen is degraded into 13–18 amino-acid residues, and the short peptide is presented on the cell surface via MHC-II molecules to specific Th cells (Fig. 5.6).

As mentioned earlier, the activation of B cells by TD antigen requires direct contact with Th cells. The process of B-cell activation requires two basic signal stimuli: the first stimulating signal generated by antigen binding to the BCR and the second one generated by the interaction of CD40L on the Th-cell membrane with CD40 on the B-cell membrane. When the first signal is generated, B cells up-regulate the expression of MHC-II molecules and B7 co-stimulatory molecules and become antigen-presenting cells (APC), which present antigens via MHC-II molecules (Fig. 5.6). Th cells are activated by the recognition of antigens presented by MHC-II molecules through the T-cell receptor (TCR) and the binding of the co-stimulatory molecule CD28 on their cell membrane to the B7 molecule of B cells.

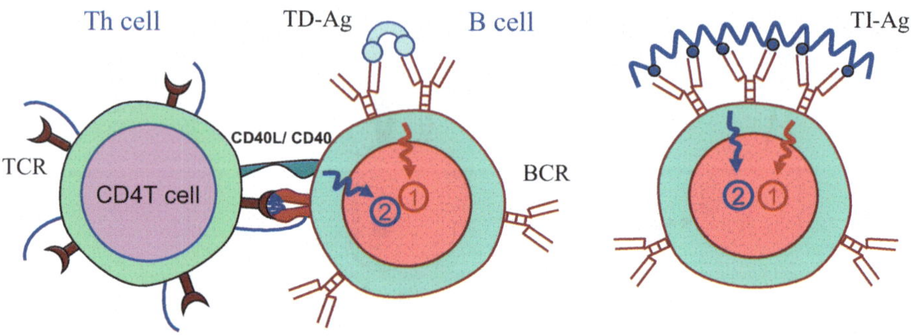

Fig. 5.7 B cells require two signals for activation

Activated Th cells upregulate the expression of CD40L, and this molecule binds to the B-cell membrane receptor CD40 to generate a second signal that further stimulates and activates B cells (Fig. 5.7), so that B cells activate Th cells first, and the activated Th cells in turn help B cells to be activated. For TI antigens, the presence of multiple repeats of antigenic epitope enables the simultaneous binding of multiple mIg molecules to form cross-linking that generate not only the first signal but also the second signal through the cross-linking.

Since the amino-acid sequence of the intracellular region of mIg is short, only a few to two dozen of amino acids, which is not sufficient to form a functional domain for signaling, how B cells transmit signals post-recognition of antigen was once one of the issues puzzling immunologists. mIg was later found to bind to the Ig-α/Ig-β heterodimer and transmit signals downward using the longer intracellular region contained in Ig-α/Ig-β. It is thus clear that the BCR actually consists of two parts: one part is mIg responsible for recognizing Ag, and the other is the Ig-α/Ig-β heterodimer that performs signal transduction (Fig. 5.2). The Ig-α/Ig-β heterodimer contains the immunoreceptor tyrosine-based activation motif (ITAM). Upon recognizing antigens by BCR, Src family tyrosine kinases (e.g., Fyn, Blk, Lyn, etc.) are activated, and the activated protein kinases phosphorylates tyrosine in the ITAM region, forming a docking site that recruits and activates the protein kinase Syk, which activates a series of kinases and adaptor proteins to form Signalosome, including Vav, Btk (Bruton's tyrosine kinase), PI3K (phosphoinositide 3-kinase), PLCγ2 (phospholipase C-γ2), BLNK (B-cell linker), etc. BLNK is an adaptor protein that, together with Btk, phosphorylates PLCγ2, and activated PLCγ2 hydrolyzes PIP2, a cell membrane phospholipid that, upon hydrolysis, produces DAG (diacylglycerol) and IP3 (inositol 1,4,5-triphosphate) as the second messengers that activate PKC (protein kinase C) and Ca2+-mediated signaling pathways (Fig. 5.8). The combined action of these molecules allows B cell into the initial stages of activation, including initiation and regulation of gene expression, cytoskeletal rearrangement movements, and endocytosis of BCR-bound antigens. Thus, the B-cell activation process is dependent on the coordinated action of multiple intracellular molecules.

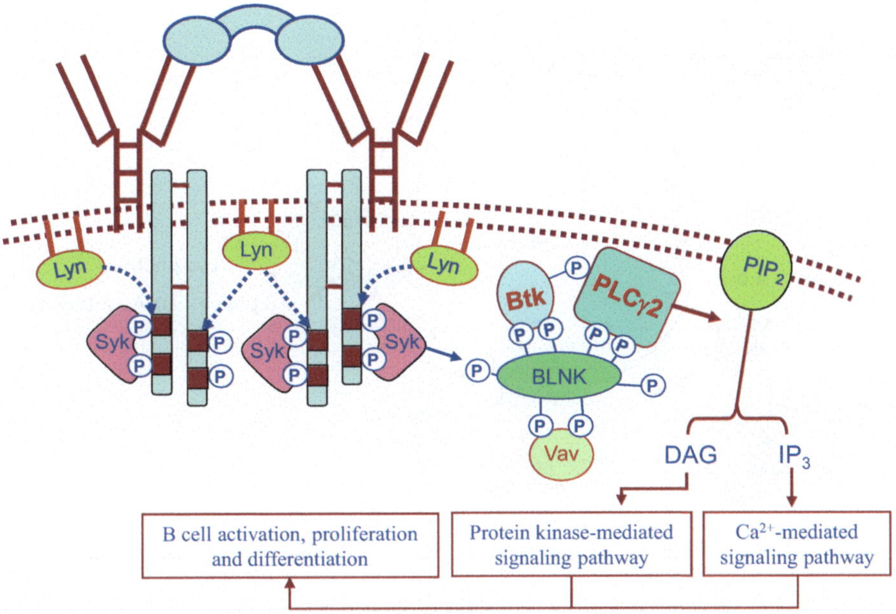

Fig. 5.8 Schematic illustration of BCR-mediated signaling. *Lyn* member of the Src family (tyrosine kinase family), which also includes Fyn, Blk, etc., *Syk* spleen tyrosine kinase, *Vav* a guanine nucleotide exchange factor, *Btk* Bruton's tyrosine kinase, *PI3K* phosphatidylinositol 3-kinase, *PLCγ2* phospholipase C-γ2, *Blnk (B-cell linker)* a B-cell ligand protein, *PIP2* phosphatidylinositol diphosphate, *DAG* diacylglycerol, *IP3* inositol 1,4,5-trisphosphate, *PKC* protein kinase C

5.4.3 Role of Co-Stimulatory Molecules in the Proliferation of B Cells

The BCR recognizes the antigen to generate the first stimulatory signal, and Th cells bind to CD40 of B cells via CD40L to form the second stimulatory signal. Both signals act together to activate B cells and bring them into a dividing and proliferating state (Fig. 5.9). The binding between CD40L-CD40 is critical to B-cell activation, and if the binding between CD40L-CD40 is blocked with the anti-CD40L antibody, the B cells cannot be activated. If B cells are stimulated with anti-CD40 antibody, the same effect of CD40L-CD40 interaction on B-cell activation could be achieved. In addition, the B-cell activating molecule BAFF (B-cell activating factor of the TNF Family) binds to BAFFR to signal downstream through the adaptor molecule TRAF (tumor necrosis factor receptor-associated factor), activating NF-κB signaling pathway to elicit immune response (Refer to Chap. 3 for the details of NF-κB signaling pathway). In addition to binding to BAFFR, BAFF also binds to BCMA (B-cell maturation antigen) and TACI (transmembrane activator and calcium-modulator and cyclophilin ligand). BCMA and TACI are both membrane molecules of B cells, and these molecules bind to APRIL (a proliferation-inducing

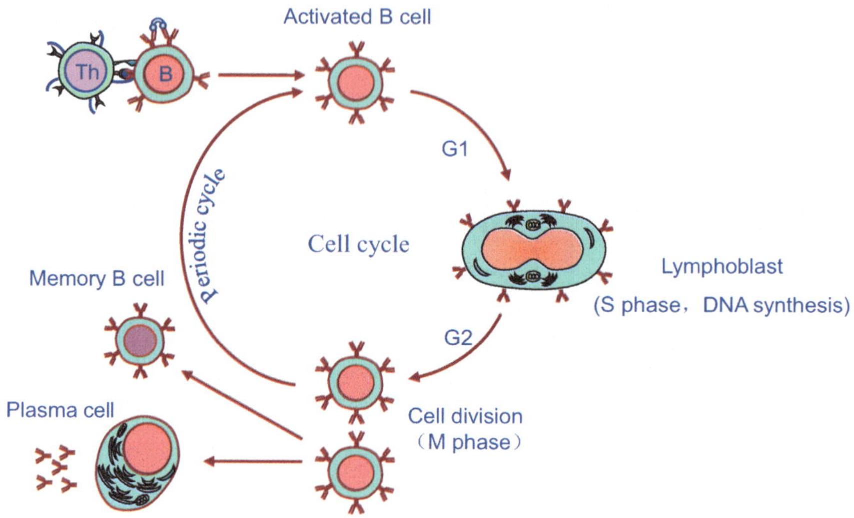

Fig. 5.9 Schematic illustration of B cells entering the cell dividing cycle after activation

ligand) besides BAFF. APRIL has a high homology with BAFF and is also a member of the TNF family. It has been demonstrated that transgenic mice overexpressing BAFF develop autoimmune diseases, while BAFFR$^{-/-}$ mice have a similar phenotype to BAFF$^{-/-}$ mice, both exhibiting a lack of mature B cells and lacking the ability to efficiently generate a humoral immune response.

BCMA, a membrane molecule of B cells, is associated with B-cell survival, and BCMA$^{-/-}$ mice exhibit reduced plasma cell survival. TACI is also a membrane molecule of B cells, and TACI$^{-/-}$ mice exhibit an increased number of peripheral lymphocytes, developing autoimmune diseases such as lupus erythematosus and lymphoma. Humans with mutations in the TACI gene exhibit varying degrees of immunodeficiency. Thus, TACI is hypothesized to have a negative regulatory effect on B-cell activation.

The B-cell activation process is subjected to inhibition by the negative regulatory molecule Act1 to ensure that the immune response is carried out within the appropriate range. If the negative regulatory mechanism is missing, the B-cell immune response will go out of control, leading to a series of immune damages caused by an over-reactive immune response. For example, Act1 knockout mice exhibit splenomegaly and lymph node enlargement, B-cell hyperplasia, immunoglobulinemia, autoantibody production, immune response hyperactivity, and multi-tissue inflammatory responses (Fig. 5.10). Although Act1 acts as a negative regulator within the cell, it is not well understood what the signal upstream is. The surface of the B-cell membrane contains a negative regulatory molecule CD22, which transmits an inhibitory activation signal downstream upon binding to CD22L (which is expressed by a variety of cells, such as dendritic cells and B cells), so it is currently thought that CD22 determines the threshold of B-cell activation. In any case, appropriate negative regulation of B-cell activation is important in maintaining the physiological homeostasis of the organism.

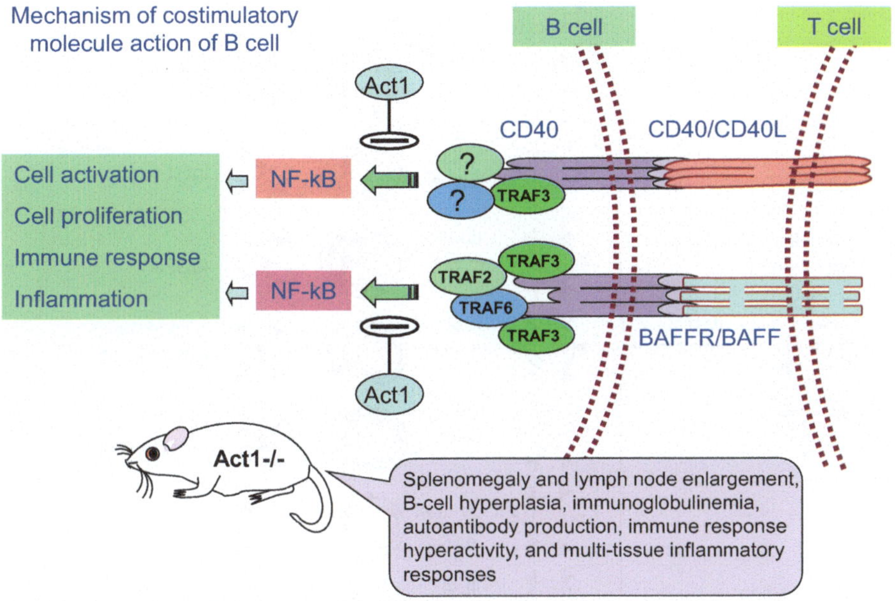

Fig. 5.10 Schematic diagram of the role of co-stimulatory molecules in B-cell activation. BAFF (B-cell activating factor of the TNF family): a TNF family B-cell activating factor; TRAF (tumor necrosis factor receptor-associated factor): a TNF receptor binding protein; BCMA (B-cell maturation antigen): a B-cell maturation antigen; BAFFR: a BAFF receptor; APRIL (a proliferation inducing ligand): a cell proliferation ligand

B-cell activation initiates the expression of multiple genes, including cytokine receptors on the surface of the cell membrane that recognize cytokines secreted by Th cells, such as IL-2, IL-4, and IL-5. These cytokines have a role in promoting the proliferation and differentiation of B cells into plasma and memory cells, and in addition, they also function in regulating antibody isotype class switching and affinity maturation.

5.4.4 Effect of Cytokines on B-Cell Immune Response

Mature B cells recognize specific TD antigens through BCR and are activated with the help of Th cells to differentiate into plasma cells for antibody production, a process that occurs mainly in the germinal centers of lymph nodes or the spleen. The cells, in addition to providing co-stimulatory signals through the engagement of CD40L with CD40, secrete appropriate cytokines to drive the immune response depending on the nature of the antigen.IL-4 and IL-12 are important cytokines that regulate the immune response. IL-4 promotes the Th2 immune response, driving the immune response towards humoral immunity. On the contrary, IL-12 drives the immune response toward cell-mediated immunity. For example, in the CD4T cell polarization assay, if IL-4 and anti-IL-12 antibodies are added to the naive CD4T

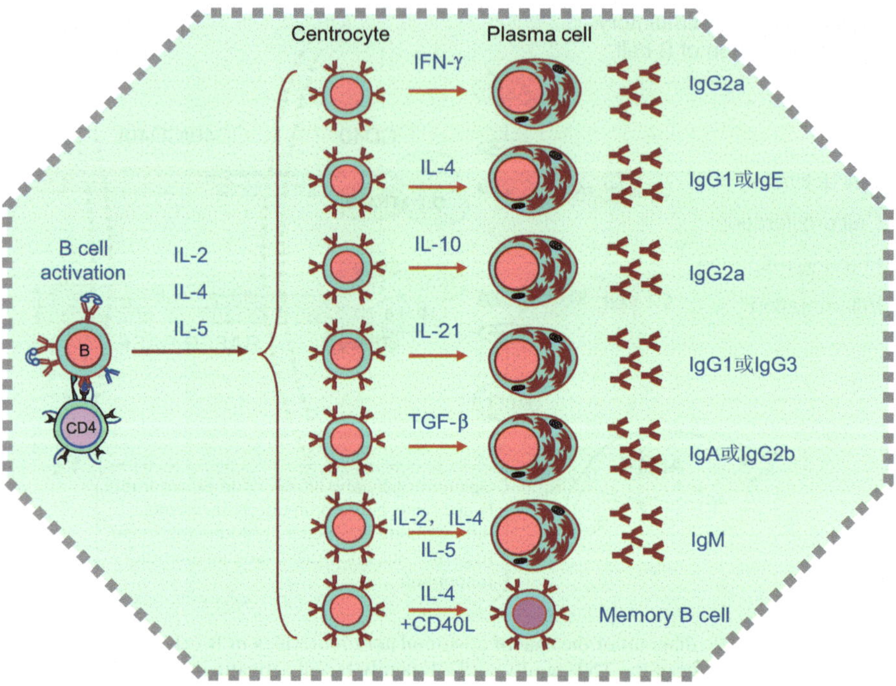

Fig. 5.11 Schematic illustration of the role of cytokines in antibody isotype class switching

cell culture medium, naïve CD4T cells will differentiate into Th2 cells, which mainly secrete IL-4 and IL-10. On the contrary, if IL-12 and anti-IL-4 antibodies are added to the naive CD4T cell culture medium, naïve CD4T cells will differentiate into Th1 cells, which mainly secrete IL-2 and IFN-γ. If there are more Th1 cytokines, the type of immune response is predominantly cell-mediated, while if there are more Th2 cytokines, humoral immunity becomes the dominant type of immune response. Cytokines regulate antibody isotype class switching (Fig. 5.11), and the exact regulatory mechanisms are not well understood.

5.4.5 B-Cell Differentiation and Antibody Affinity Maturation

In lymphoid tissue, antigen is captured by B cells or other antigen-presenting cells, and then presented by MHC-II molecules to activate Th cells. With the help of Th cells, B cells are activated, proliferate, and further differentiate into plasma cells to produce antibodies with a certain affinity (Ka). Antibodies produced early in the B-cell immune response generally have low affinity, but gradually with the affinity maturation, antibodies with high affinity would be produced. The affinity maturation of B-cell antibody results from point mutations in the genes encoding

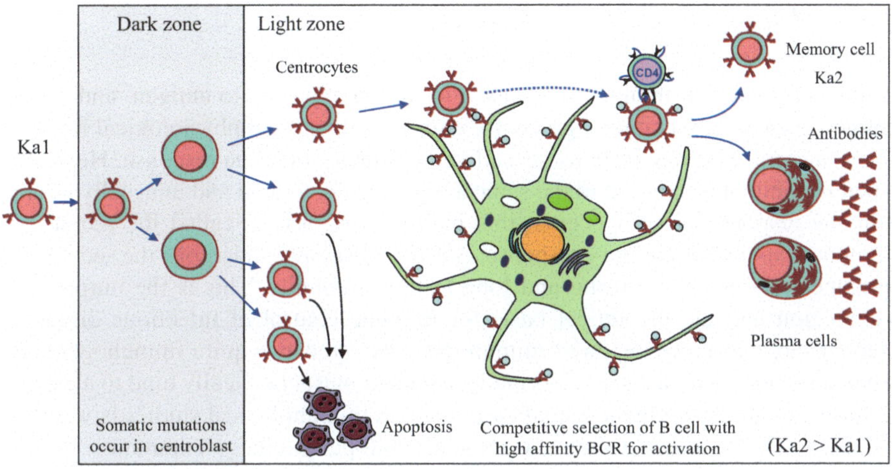

Fig. 5.12 Schematic diagram of the B-cell antibody affinity maturation process

the variable region of the antibody (see Sect. 5.3.3 of this chapter), with the involvement of dendritic cells. The antibody binds to the antigen to form an antigen-antibody complex, which binds to the surface of the dendritic cell membrane via the Fc fragment of the antibody (the dendritic cell membrane contains the Fc receptor), and the activated B cells divide and proliferate to form the germinal center. B cells in the proliferative phase are called centroblast and are characterized by increased cell size and cytoplasm, and lack of cell mIg. The area where these cells accumulate is called the dark zone of the germinal center. The centroblast divides and becomes centrocytes, which are small in size and contain mIg on the surface of the cell membrane. The centrocytes move from the dark zone to the light zone, where they come into contact with the Ag–Ab complex displayed on the surface of the dendritic cells and bind antigens competitively. The centrocytes with high-affinity mIg are able to bind antigens and are activated with the help of CD4T cells, and further differentiate into plasma cells producing antibodies and memory cells. Low-affinity cells are unable to compete for antigens, subsequently undergo apoptosis, and are ultimately cleared by macrophages (Fig. 5.12).

The affinity of B-cell antibodies increases progressively through the affinity maturation process, and the affinity is indicated as an affinity constant (Ka). In the germinal centers of the lymph nodes or spleen, B cells with low affinity (Ka1) give rise to central cells with different affinities during proliferation due to somatic mutations, and since antigen-antibody binding is a dynamic process of binding and dissociation (Ag + Ab $\rightleftharpoons$ Ag-Ab). B cells, whose BCRs with high affinity can competitively acquire Ag from the Ag-Ab complex of dendritic cells, are activated with the help of CD4T cells, and differentiate into plasma cells to produce high-affinity (Ka2) antibodies. Memory B cells and plasma cells have a higher affinity (Ka2 > Ka1) than the parental cells after the affinity maturation process in the germinal center.

5.4.6 *Hapten and Carrier Effects*

Animals produce an immune response after initial exposure to antigen, and as the antigen is cleared, the animal gradually returns to its normal physiological homeostasis, and this immune response is called the primary immune response. However, when the animal is exposed to the same antigen again, a rapid and unusually strong immune response process occurs; this immune response is called the secondary immune response or memory response. Animals that have undergone the secondary immune response have strong and long-lasting immunity. This is the purpose of vaccination and the theoretical basis for immune control of infectious diseases. Haptens are substances without immunogenicity but can acquire immunogenicity when combined with a carrier. Normally, a hapten must chemically bind to a carrier to form a complete antigen. A confusing issue in immunological studies is whether BCRs and TCRs recognize the same antigenic epitopes. Studies on immune response induced by immunization of animals with a hapten- carrier complex have unraveled this mystery. When animals were immunized with a hapten-carrier, the animals produced antibodies not only against the hapten but also against the carrier, but if the animals were immunized with the hapten and the carrier separately, the animals did not produce specific antibodies against the hapten. After the primary immune response is induced by immunizing the animal with the hapten-carrier, if the animal is immunized with another carrier conjugated with the antigen, the animal does not produce a secondary immune response against the hapten. Thus, a hapten must conjugate with the same carrier as the one that conjugates with the hapten to elicit the primary immune response can mount the secondary immune response, which is the phenomenon called carrier effect. Experimental evidence indicates that the carrier effect is associated with CD4T cells and that in a memory response, CD4T cells only recognize the carrier conjugated with the hapten at the time of the primary immunization so that they can help B cells in the memory immune response, the underlying exact molecular mechanism is still unclear.

5.5 Pioneering Experimental Findings in B-Cell Development and Maturation

B cells undergo negative selection during development and maturation, e.g., about 5×10^7 cells are produced daily in mouse bone marrow, but more than 90% are cleared via apoptosis, thereby less than 5×10^6 cells develop to maturity. B cells in the immature stage can become apoptotic if their BCR (mIgM) binds to an antigen or if their μ-chain (the constant region of the BCR H-chain) binds to an anti-μ-chain specific antibody. Nemazee and Burki (1989) prepared a transgenic mouse that encodes and produces an antibody (IgM) against its own MHC (H-2^k) molecule, K^k, and antibody H and L chains both recognizes K^k. Since all nucleated cells in mouse express the K^k molecule, B cell recognized and bound to the K^k molecule via its

BCR early during development, and these B cells are cleared in the bone marrow, there are no mature B cells in the peripheral lymphoid tissue of this mouse. This experiment confirms a negative selection process in the central immune organ. However, if the BCR of the transgenic mouse is for the K^b molecule and the type of MHC molecule expressed in the mouse is K^k, then the mouse has normal B-cell development and there are a large number of mature B cells in the peripheral lymphoid tissue. However, if this mouse is crossed with the liver cell-specific K^b transgenic mouse (which is genotyped as K^k, but has the liver-specific promoter linked to K^b, so that only the liver cell-expressed MHC is K^b), the double-transgenic mice had no mature B cells capable of binding K^b molecules in their lymph nodes. This assay suggests that B cells undergo the same negative selection process in peripheral tissues, which is distinct from T cells. Those B cells that are able to react with self-antigens are either cleared or in an unresponsive state (anergy) when they enter the peripheral circulation, but will be eventually cleared as time goes on.

Study Questions

1. How is the diversity of immunoglobulins generated?
2. What is the molecular mechanism of B-cell activation?
3. What is the effect of cytokines on the B-cell immune response?
4. What is affinity maturation and what is the process of B-cell-antibody affinity maturation?
5. What is the carrier effect?

Reference

Nemazee, D.A., and K. Burki. 1989. Clonal deletion of B lymphocytes in a transgenic mouse bearing anti-MHC class I antibody genes. Nature. 337:562–566.

Chapter 6
Antigens and Antibodies

Overview:
Substances that can be recognized by the B-cell immunoglobulin receptor (BCR) or by the T-cell receptor (TCR) after binding to the MHC are called antigens. Antibodies, the immunoglobulins that bind to antigens, are secreted by B cells post activation via recognition of antigen. B cells are activated by antigenic recognition through the BCR to produce antibodies that bind specifically to antigens for the purpose of neutralizing and clearing antigens.

6.1 Antigens

6.1.1 Basic Properties of Antigens

There are a variety of substances in nature, but not every substance is an antigen. Antigens have two basic properties: (1) immunogenicity, which means that the antigen stimulates an organism to produce an immune response (the animal produces antibodies or effector T lymphocytes) after immunization of the animal, and (2) antigenicity (also called reactivity), which means that the antigen can bind specifically to antibodies or immune effector T cells. Based on these two properties, antigens are classified as complete antigens, which are both immunogenic and antigenic, and haptens, which are only antigenic.

6.1.2 Antigenic Epitopes

An epitope, also known as a determinant, is a specific chemical structure present in an antigen that binds specifically to the TCR/BCR or the Fab portion of the antibody and is the basis required for eliciting immune responses.

S. J. Zheng, *Veterinary Molecular Immunology*,
https://doi.org/10.1007/978-981-99-8929-4_6

Due to the different structural complexity of antigen molecules, different types of antigenic epitopes are formed. As mentioned earlier, antigenic epitopes are also called antigenic determinants and are classified into sequential determinants and conformational determinants according to their structure. Sequenial determinant refers to a consecutive short peptide fragment of amino-acid sequences; conformational determinant is a structure formed by the spatial conformation of a molecule with a discontinuous sequence. Depending on their function, antigenic determinants are divided into hidden and functional determinants. Hidden determinants are those present inside the antigen, while functional determinants are those recognized by B cells or bound to antibodies. Hidden determinants can be exposed as functional determinants through processing (e.g., enzymatic digestion and degradation) (Fig. 6.1).

For a large molecule antigen, the immunogenicity of the antigenic epitopes on the surface of the molecule may vary in strength. Antigenic epitopes with more immunogenicity play an immunodominant role and induce the major part of the antibody production. In practice, it is often found that in serum antibodies against certain complex antigens, multiple antibodies exist in different concentrations and with different affinities.

The antigenic epitopes recognized by the TCR or BCR are not identical, and the differences between the two are listed in Table 6.1.

6.1.3 Haptens and Complete Antigens

Some small molecules, such as certain compounds, heavy metal ions, and hormones, are not immunogenic in themselves, but may become immunogenic if combined with a large molecule, which is called a carrier, thereby the small molecule is usually called a hapten. An hapten molecule is generally small, having only one antigenic epitope, but if combined with a large molecule, it can become a complete antigen, initiating immune response to produce antibodies, while the hapten is able to bind specifically to the antibody. Thus, a complete antigen can also be understood as a complex of a carrier with haptens. The number of functional antigenic epitopes that can bind to the

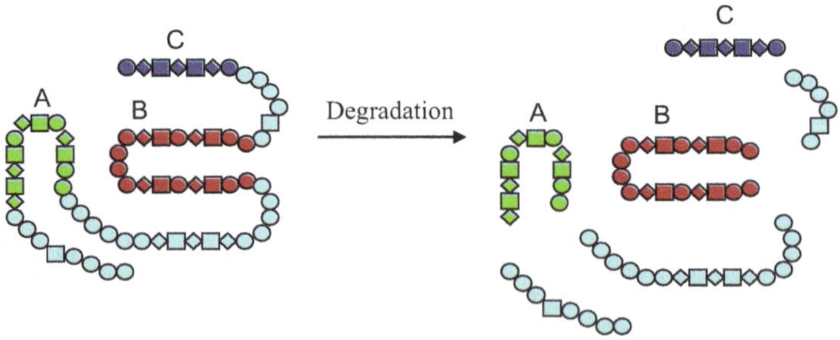

Fig. 6.1 Schematic diagram of antigenic epitopes. A, B, and C represent antigenic epitopes, where large molecular antigens are degraded to form multiple antigenic epitopes

Table 6.1 Comparison of epitopes recognized by T and B cells

Item	Epitopes recognized by T cells	Epitopes recognized by B cells
Epitope recognizing receptor	TCR	BCR
Involvement of MHC molecules	Required to present epitope	Not required
Chemical nature of epitopes	Mostly peptide, but some lipids, glycolipids, or microbial vitamin B metabolites presented on MHC-like molecules (CD1 or MR1)	Peptide, polysaccharides, lipopolysaccharides, lipid, or organic chemicals
Size of epitope	8~11 aa for $CD8^+$T cells 13~18 aa for $CD4^+$T cells	5~15 aa, or 5~7 sugars or nucleic acids for B cells
Epitope type	Linear peptides	Sequential or non-sequential (conformational) amino acids
Position of epitopes	Any location in antigen	On the surface of antigen (functional determinants)

antibodies is called the antigenic valence, and thus the hapten is a monovalent antigen, and monovalent antigens cannot directly activate B cells (Fig. 6.2). Some drugs and their metabolites, such as penicillin and certain sulfonamides, become complete antigens (allergens) post-binding to plasma proteins, gaining the ability to induce an immune response and therefore causing an allergic reaction.

Substances that are immunogenic are complete antigens, having both immunogenicity and reactivity. Factors affecting the immunogenicity of an antigen include the foreignness of the antigen, molecular size, chemical composition, and three-dimensional structure.

As mentioned above, the hapten needs to bind to the carrier to be immunogenic, and the immune system produces an immune response to the hapten as well as to the carrier. (For carrier effects, see Chap. 5)

6.1.4 Factors Affecting Immunogenicity

6.1.4.1 Foreignness

The immune system generates an immune response by recognizing foreign substances, and distinguishing self from non-self is a prerequisite for the immune system to function. Since B-cell clones capable of recognizing self-antigens are removed during development, mature B cells recognize only foreign antigens. In general, the greater the phylogenetic distance of the antigen is from the species, the better the immunogenicity. For example, immunization of cattle with bovine serum albumin (BSA) does not induce an immune response, i.e., BSA is not immunogenic to cattle, but it is somewhat immunogenic if sheep are immunized. However, BSA, if introduced into mice or birds, is more immunogenic because mice and birds are more distantly related to cattle.

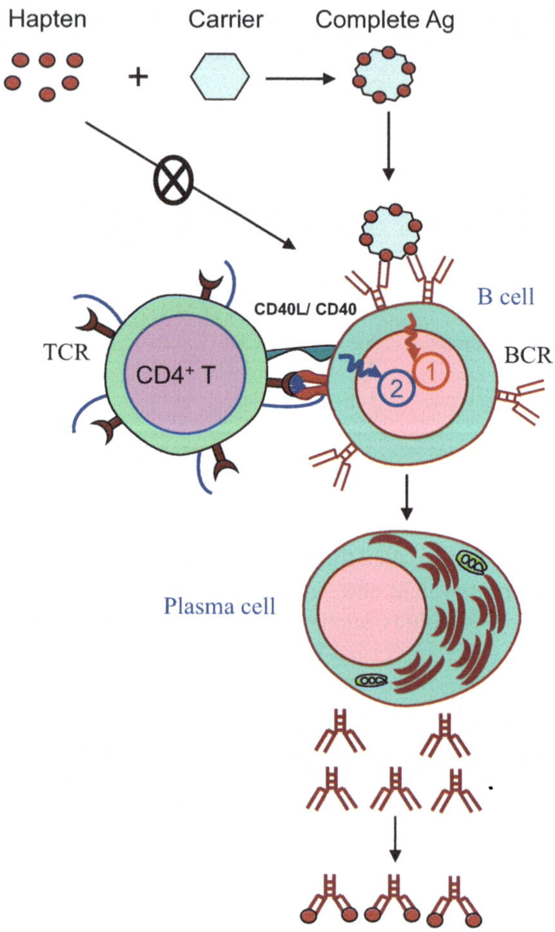

Fig. 6.2 Schematic diagram of antibody production by B cells post-activation with hapten–carrier complex. The hapten, when combined with a carrier, becomes a complete antigen. The hapten–carrier complex binds to B cells via cross-linking adjacent B-cell receptors with hapten (epitope) to generate the first activation signal, and with the involvement of helper T cells whose CD40L binds to CD40 on B cells to generate a second signal, which drives B cell completely activated, and the activated B cells subsequently proliferate and differentiate into plasma cells to produce antibodies that specifically bind to the hapten

6.1.4.2 Molecular Size

Immunogenicity correlates with molecular size. The molecules with a molecular mass larger than 100 kilo-daltons (kD) usually have good immunogenicity, and molecules between 5 and 10 kD have weaker immunogenicity, but for certain individual molecules, even smaller than 1 kD, they are still immunogenic.

6.1.4.3 Chemical Composition and Structure

Immunogenicity is not only related to foreignness and molecular size but also to the chemical composition and structure of the molecule. For example, simple repetitive amino-acid sequences and monosaccharides are less immunogenic, whereas complex protein molecules containing multiple amino acids are more immunogenic, and the primary, secondary, tertiary, and quaternary structures of protein molecules are all related to immunogenicity (Fig. 6.3).

A polypeptide with amino acids in the form of a single chain is a primary structure; if parts of the polypeptide undergo certain folding and helix, a secondary structure is formed; a tertiary structure refers to those that undergo further folding on the basis of the secondary structure to form certain functional regions; a quaternary structure refers to the combination of two or more polypeptide molecules to form a multimeric protein molecule.

As another example, certain compounds that have the same structural components but different molecular structures also develop different antigenicity, which is the property of antigen that was discovered during extensive studies of small-molecule antigenicity in the early twentieth century. Subtle structural differences in aminobenzene and aminobenzoic acid can lead to differences in antigenicity (Fig. 6.4).

There is no cross-reactivity of immune sera for these hapten molecules.

6.1.4.4 Non-protein Antigen

Certain complex carbohydrates, glycolipids, lipids, and other substances are immunogenic, but because the molecular structures of these non-protein antigens usually have repetitive sequences, they do not require the involvement of helper T cells to

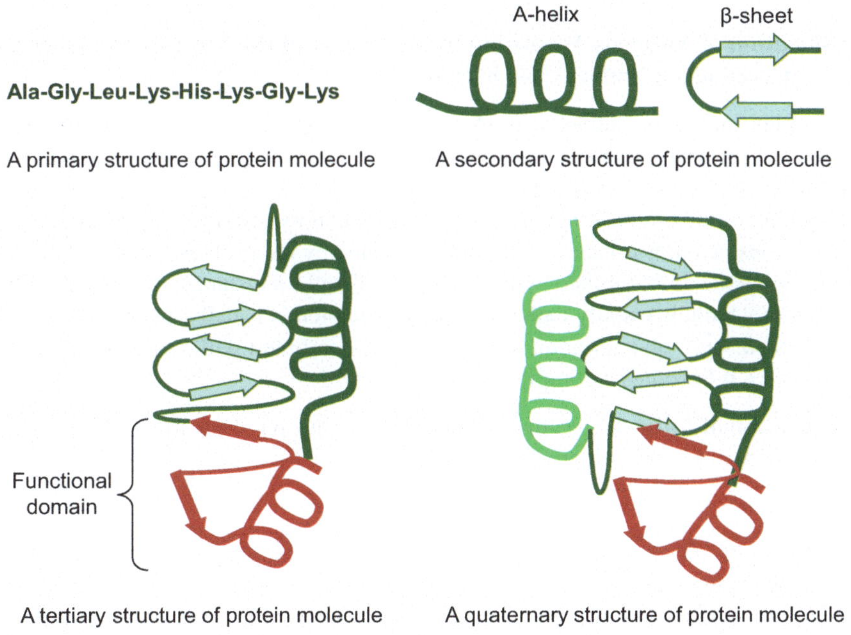

Fig. 6.3 Schematic diagram of the protein structures

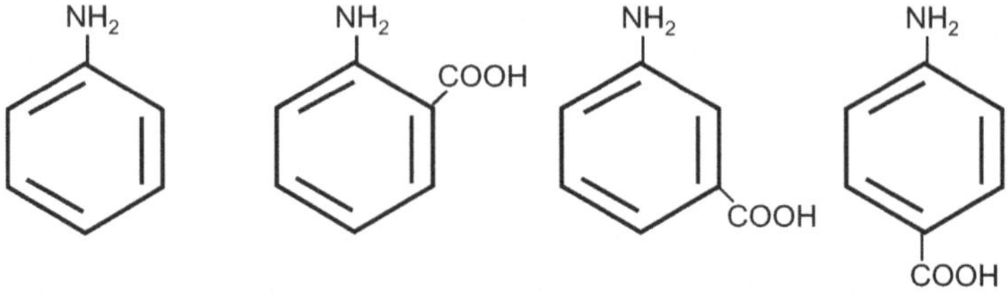

Fig. 6.4 Schematic diagram of the subtle differences in the structures of aminobenzene and aminobenzoic acid

activate B cells to produce IgM, such as the bacterial toxin lipopolysaccharide (LPS). In addition, glycans and glyco-containing molecules on cell membranes, as well as steroids, steroid hormones, and fatty acids, can become antigens under certain conditions. Immunoassays are often used to detect these substances in clinical practice, and blood group antigens on red blood cell membranes are molecules that contain glycans. Examination of these blood group antigens can provide guidance for clinical blood transfusions. Lipid antigens are usually presented by CD1 molecules rather than the conventional MHC, and their effector cells for recognizing the presented antigens are γδ T cells and NKT cells, both of which are innate immune cells.

6.1.4.5 Classification of Antigens

Classification of Antigens According to the Nature of the Antigen Antigens are classified as complete antigens and haptens.

1. Complete antigens: Substances that are immunogenic are complete antigens, such as proteins, glycoproteins, and certain lipids. Complete antigens have both immunogenicity and reactivity.
2. Haptens: Some small-molecule compounds, such as certain compounds, heavy metal ions, and hormones, are not immunogenic by themselves, but may become immunogenic if combined with a large molecule substance (a carrier). This small molecule is usually called a hapten. A hapten is usually a very small molecule with only one antigenic epitope, but if combined with a macromolecular substance it becomes a complete antigen.

Classification of Antigens According to the Way they Are Synthesized Antigens are classified as natural and artificial antigens.

1. Natural antigens include non-self and self-antigens.
   - Non-self-antigens: Also known as foreign antigens, antigens that are not their own components.
   - Self-antigens: Certain self-tissues do not come into contact with the immune system during embryonic development, and these substances are segregated

from the immune system by various barriers. If this segregation is disrupted, these tissues become antigens, causing an immune response. These substances are also called autoantigens, such as eye ball proteins and sperm proteins.

2. Artificial antigens include artificially combined antigens and synthetic antigens.
   - Artificially combined antigens: The antigens are chemically bound together to become complete antigens, e.g., certain small-molecule drugs are chemically bound to carriers to become complete antigens.
   - Synthetic antigens: A short amino-acid peptide that is synthesized by equipment (synthesizer) based on its amino-acid sequence can be used as a vaccine post-mixture with appropriate adjuvants to prevent and control diseases. In most cases, synthetic antigens are used to prepare antibodies.
   - Genetically engineered antigens: Genes encoding antigens are inserted into the expression vector by gene-engineering technique to express antigens by prokaryotic or eukaryotic expression systems, which is often used in molecular biology research.

Classification of Antigens According to Genetic Relationship to the Organism Antigens can be classified as xenogenic antigen, allogenic antigens, and autoantigens.

1. xenogenic antigens: Antigens from different species, for example, antibodies from poultry are antigens for rabbits, and antibodies (also called secondary antibodies) are usually obtained by immunizing one species with antibodies from another species in the experiment.
2. Allogenic antigens: Antigens that come from different individuals of the same species, for example, allogenic antigens are major antigens responsible for organ transplant rejection.
3. Autoantigens: Antigen from self-tissue is called autoantigen.

Classification of Antigens According to the Dependence of the Antigen-Induced Immune Response on T Cells

1. Thymus-dependent antigen (TD-Ag): This type of antigen is recognized by specific B cells and must require the involvement of CD4 T cells to completely activate B cells in immune response, this type of antigen is called thymus-dependent antigen, such as most protein antigens.
2. Thymus-independent antigen (TI-Ag): These antigens can directly activate B cells without the involvement of T cells. The common feature of these antigens is that a single molecule contains multiple antigenic epitope repeats, which can cross-link multiple BCRs to form a powerful stimulatory signal to directly activate B cells, such as bacterial lipopolysaccharide (LPS), polymeric flagellin, etc. (see Chap. 5).

Blood Group Antigens In 1901–1903, Kar Landsteiner discovered that blood transfusion between different peoples could cause jaundice, shock, or even death, and in 1909, classified human blood into four blood groups, A, B, AB, and O (Table 6.2), and clarified the principles of blood transfusion. Kar Landsteiner was also awarded the 1930 Nobel Prize in Medicine for this exceptional work.

6.2 Antibodies

Antibodies are immunoglobulins present on the surface of B cells or secreted by plasma cells that bind to antigens. When an antigen enters the body, it is recognized by the corresponding specific B cell, and this recognition is done through the B-cell receptor (BCR). After activation of the specific B cells that recognize the antigen, they proliferate and differentiate into plasma cells that secrete antibodies, which specifically bind to the antigen for neutralization and removal of the antigen. This is the basic process of antibody immune response, however, three issues should be clarified: (a) each B cell contains multiple BCRs on its surface, but for the same B cell, its BCRs are identical and recognize only one antigenic epitope; (b) BCRs recognize only antigenic epitopes, and since an antigen usually has multiple antigenic epitopes, such antigen is called compound antigen, thus one antigen often activates multiple B cell clones by varied epitopes simultaneously (a group of identical B cells originating from the same B cell is called a B cell clone), so the antibodies produced by multiple B cell clones are polyclonal and inhomogeneous. (c) plasma cells are short-lived, cannot live long, and so antibody levels will decrease gradually after removal of antigens, and specific antibodies will eventually vanish.

6.2.1 Basic Structure of Antibodies

Antibodies are present in body fluids (blood, lymphatic fluid, mucosal surface secretions, etc.) and reach all parts of the body through the blood circulation. The basic structure of mammalian antibodies consists of two heavy (H) chains and two light (L) chains, with peptide chains bound to each other by disulfide bonds. Two H chains and two L chains are combined to form the variable region (V-region) and the constant region (C region) (Fig. 6.5).

Table 6.2 Classification of the human ABO (H) blood groups

Blood type	Genotype	Antigen (RBC)	Antibody (serum)
A	AA or AO	A antigen	Anti-B antibody
B	BB or BO	B antigen	Anti-A antibody
AB	AB	A&B antigens	No anti-A or anti-B antibody
O	OO	No A or B antigen	Anti-A & anti-B antibodies

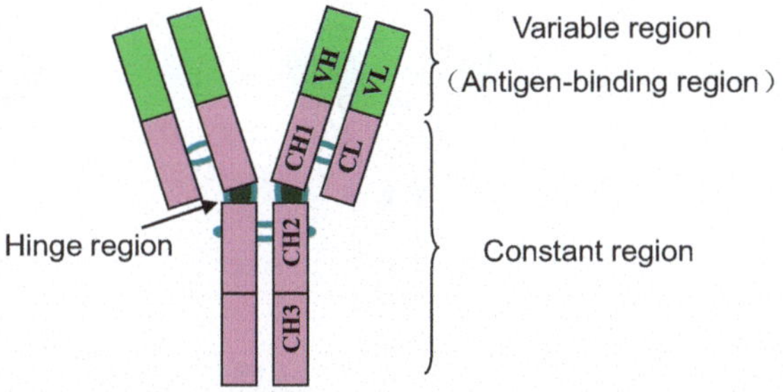

Fig. 6.5 Schematic diagram of the basic structure of immunoglobulins. Monomeric Ig consists of two H chains and two L chains, VH (1/4 N-terminal of the H chain) and VL (1/2 N-terminal of the L chain) form the variable region, which is the part of the antibody that recognizes and binds to the antigen. The rest of the H and L chains (CH1 to CH3, and CL) are the constant regions, and the part where CH1 and CH2 are connected is the hinge region

6.2.2 Types of Immunoglobulins

The immunoglobulins (Ig) of humans and mice are well understood and are broadly divided into five isotypes. The isotypes refer to the antigenic specificity of immunoglobulins that each individual of the same species has, with the antigenic determinants mainly present in the C region of Ig. Five commonly seen isotypes are IgG, IgD, IgE, IgA, and IgM (Fig. 6.6).

Human and mouse Ig were classified into classes and subclasses based on the differences in the antigenicity of the constant region of the H chain.

Ig Classes.

IgG - γ (gamma).
IgA - α (alpha).
IgM - μ (mu).
IgD - δ (delta).
IgE - ε (epsilon).

Subclasses.

IgG: IgG1, IgG2, IgG3, IgG4.
IgA: IgA1, IgA2.
IgM: IgM1, IgM2.

Immunoglobulins are present in both membrane (mIg) and secretory (sIg) forms. mIg is present on the surface of the B cell membrane and is a monomeric structure of IgM, the B cell receptor (BCR) for antigenic recognition. A small amount of IgD is also present in mature B cells. Secretory Ig are antibodies that are secreted by plasma cells into body fluids (blood, lymphatic fluid, and secretory fluid) to generate humoral immune responses, and these antibodies are mainly IgG, IgM (pentameric), IgA (monomers in circulating body fluids and dimers in secretory fluids), IgE, and IgD.

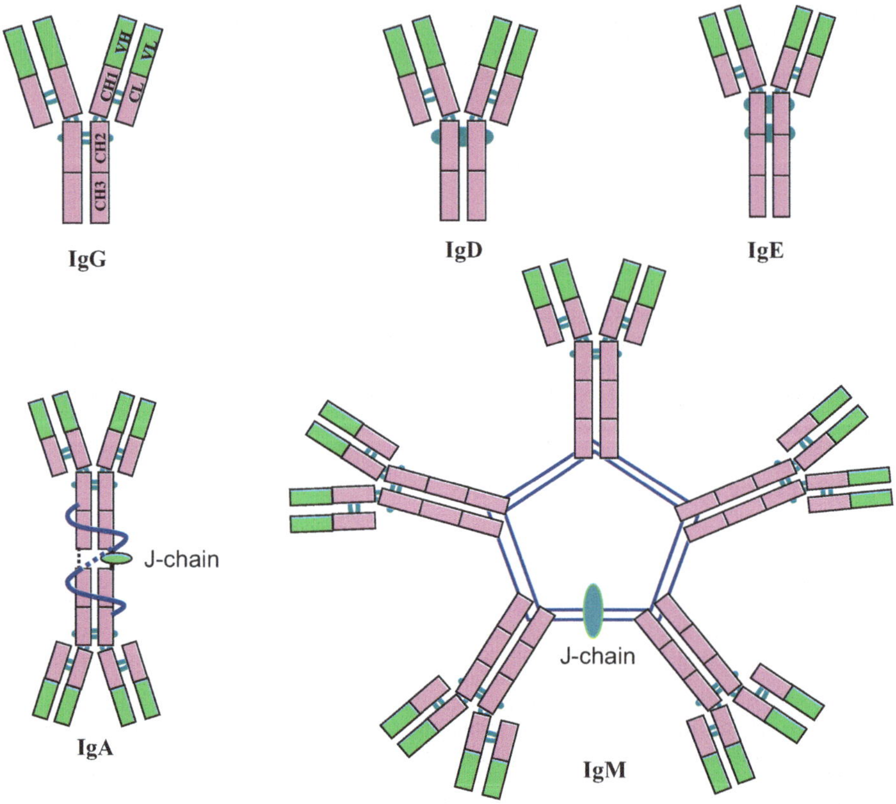

Fig. 6.6 Schematic diagram of the basic structures of five immunoglobulins

6.2.3 Variable Regions and Diversity of Antibodies

In the structure of Ig, the H chain is larger, with a molecular mass of about 50 kD, while the L chain is smaller, with a molecular mass of about 25 kD. One H chain and one L chain are bonded by disulfide bonds to form a heterodimer, and two identical heterodimers, 2X (H-L), are bonded together by disulfide bonds to form a single Ig molecule. Thus, Ig is also known as a dimer of dimmers. The variable region (VH-VL) determines the specificity of the antibody. This region is divided into two parts: the hyper-variable region (HVR) and the framework region (FR). The hyper-variable region, also known as the complementarity determining region (CDR), has three amino-acid sequence high-variable regions, called CDR1, CDR2, and CDR3, and changes in the amino-acid sequence and space of these parts determine the specificity and diversity of the antibody in recognition of antigen (Fig. 6.7).

As mentioned above, VH and VL together form an antigen-binding region, yet in some animals, such as camels and sharks, there are a large amount of H-chain-only antibodies in the body fluids (Fig. 6.8), which are equally effective in neutralizing antigen, and this feature of camel H-chain antibodies has been employed to produce specific antibodies. Importantly, genetic engineering technique has been employed to clone and express the VHH portion of camel antibody. The expressed VHH portion can directly bind to antigen, with highly stable structure, is the minimal unit for antigen-binding, and its MW is 15KD, within the range of nanometer, thus called nanobody(Nb). And so, what role does the L chain actually play? Is it necessary in antigen recognition? Why do most mammals and avian species have L chains? These questions cannot be satisfactorily addressed at present time.

Different species of animals may have varied classes and subclasses of immunoglobulins (Table 6.3), and some other animals have evolved to express some specific immunoglobulins (see the section on animal antibody evolution in this chapter).

6.2.4 The Role of Different Regions of the Antibody Molecule

The VH and VL of the antibody together form the antigen-binding site, while the main role of each domain in constant region is as follows.

1. CH1–CH3 and CL are associated with Ig genetic markers.
2. CH2(IgG) and CH3(IgM) are the binding sites for complement C1q.
3. CH2–CH3(IgG) have the ability to bind to and pass through the placenta.
4. CH3(IgG) is the binding site for FcγR on cells (e.g., phagocytes, NK cells).
5. CH4(IgE) is the binding site for FcεR on granulocytes (mast cells, eosinophils, basophils).

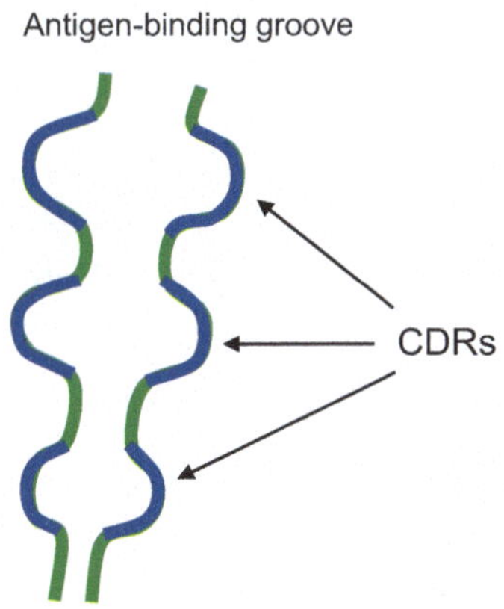

Fig. 6.7 Schematic diagram of antibody-antigen binding regions. The main regions of antibody recognizing and binding to antigen are the three complementarity determining regions of the variable region (CDR1-3), the amino acid sequence and three-dimensional structure of the CDR form complementary binding to the antigenic epitope, thus the CDR determines the specificity of antibody recognizing and binding to antigen. The other parts of variable region are framework regions to maintain the structure of the polypeptide molecule

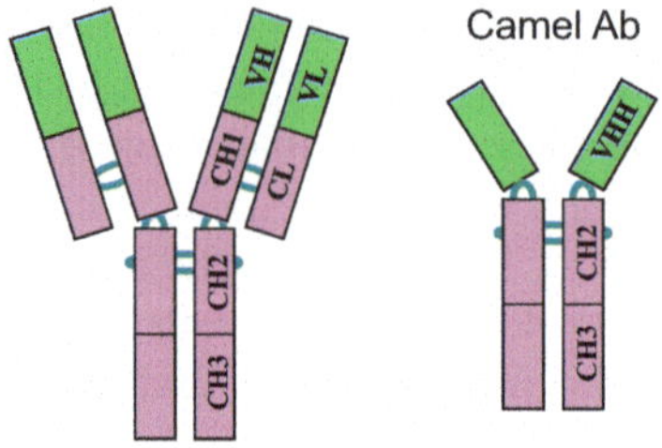

Fig. 6.8 Schematic diagram of camel antibody. (**a**) The general structure of monomeric antibody. (**b**) H-chain antibody, there are abundant H-chain antibody in camel body fluids, and the variable region of the camel H-chain antibody is called VHH (variable domain of heavy chain of heavy-chain antibody), which has no L-chain

Table 6.3 Commonly seen classes and subclasses of immunoglobulins in animals

Animal species	Immunoglobulin classes and subclasses				
	IgG or other Ig	IgA	IgM	IgE	IgD
Porcine	IgG1, IgG2a, IgG2b, IgG3, IgG4, IgG5, IgG6	IgA1, IgA2	IgM	IgE	IgD
Cattle	IgG1, IgG2a,IgG2b, IgG3	IgA	IgM	IgE	IgD
Sheep	IgG1, IgG2, IgG3	IgA1, IgA2	IgM	IgE	?
Horse	IgGa, IgGb, IgGc, IgG(B), IgG(T) (sometimes written as IgT)	IgA	IgM	IgE	?
Chicken	IgY	IgA	IgM	Not found	Not found
Duck	IgY. There is an IgG in the duck blood that lacks fuctional CH2 and CH3 domain, called IgY (Δ Fc)	IgA	IgM	Not found	Not found
Dog	IgG1, IgG2a, IgG2b, IgG2c	IgA	IgM	IgE	?
Cat	IgG1, IgG2	IgA	IgM	IgE	?
Mouse	IgG1, IgG2a, IgG2b, IgG3	IgA1, IgA2	IgM1, IgM2	IgE	IgD
Rabbit	IgG	IgA1-13 (13 subtypes)	IgM	IgE	Not found
Human	IgG1, IgG2, IgG3, IgG4	IgA1, IgA2	IgM1, IgM2	IgE	IgD
Gorilla	IgG1, IgG2, IgG3.	IgA	IgM	IgE	IgD
Giant panda	IgG1, IgG2	IgA	IgM	IgE	IgD

Note: ? indicates currently unknown

6.2.5 *Hydrolysis Fragments of Antibodies*

Immunoglobulins can be degraded by papain into three parts, two Fab fragments and one Fc fragment, while degraded by trypsin into two parts, F(ab)2 and Fc fragments. The antigen-binding site is located in the Fab fragment (Fig. 6.9).

6.2.6 Functions of Antibodies

The functions of the variable region of the antibody include (a) specific recognition and binding of antigen, the number of antigen-binding sites of the antibody is called the valence of the antibody, for example, monomeric (IgG, IgE) antibodies are 2-valent, secreted IgA is 4-valent, and pentameric IgM is 10-valent. (b) neutralization, neutralizing substances such as bacterial toxins, snake venom, and viruses. (c) promoting phagocytosis, the antibody binds to antigen to form Ab-Ag complex and then binds to Fc receptors of phagocytes through its Fc fragment, promoting phagocytosis of pathogens, which is the process known as antibody-mediated opsonophagocytosis (Fig. 6.10).

The functions of the constant region of antibody include (a) activation of the complement system, upon Ab (IgM, IgG) binding to antigen, the classical pathway of complement activation is triggered by activating C1q, while the aggregates of IgG4, IgA, and IgE activate the alternative pathway of complement activation. (b) Mediation of immune cell activity, such as promotion of phagocytosis (oposonization), antibody-dependent cell-mediated cytotoxicity (ADCC) (Fig. 6.11) (ADCC, referring to the release of substances such as perforin and granzyme to kill target cells by NK cells after binding to antibody via Fc receptors that binds to antigen on target cells), and mediating hypersensitive reactions (type I, II, and III hypersensitivity). (c) Crossing the placenta and mucosa.

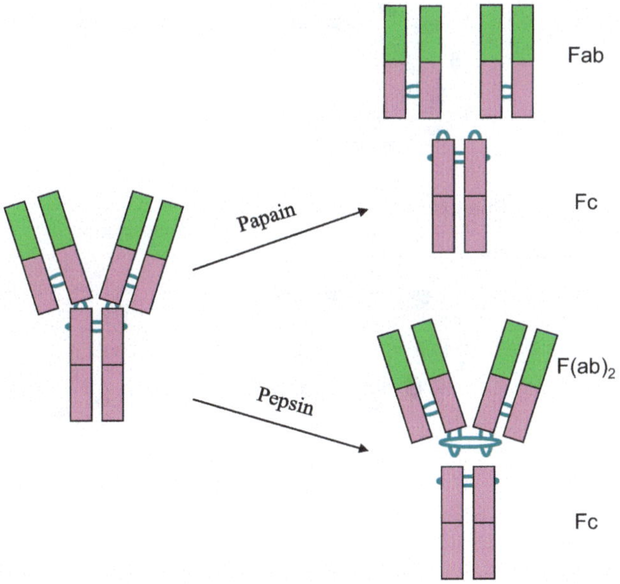

Fig. 6.9 Schematic diagram of different hydrolyzed fragments of immunoglobulins. *Fab* fragment, antigen-binding, *Fc* fragment, crystallizable

6.2.7 Evolution of Animal Antibodies

Currently, lines of experimental evidence indicate that Ig emerged in the process of biological evolution 5.5–4.6 million years ago (Fig. 6.12). During evolution, the immunoglobulin H chains of most vertebrates (mammals, birds, amphibians, some fishes) tended to be relatively conserved, but there are still some differences in the Ig of different organisms. For some organisms, the H-chain Fc fragments of Ig vary considerably, and the Ig with larger differences is indicated by different names accordingly, such as IgF (amphibians), IgT (horses, trout), IgX (amphibians), IgY (birds, amphibians, reptiles), IgZ (trout), etc.

6.2.8 Monoclonal Antibodies

Typically, a complete antigen often has several different antigenic epitopes, and each antigenic epitope is recognized and bound by a specific B-cell receptor (of note, although a B cell contains numerous identical BCRs, it recognizes only one particular antigenic epitope), so a single large antigen with multiple epitopes can be

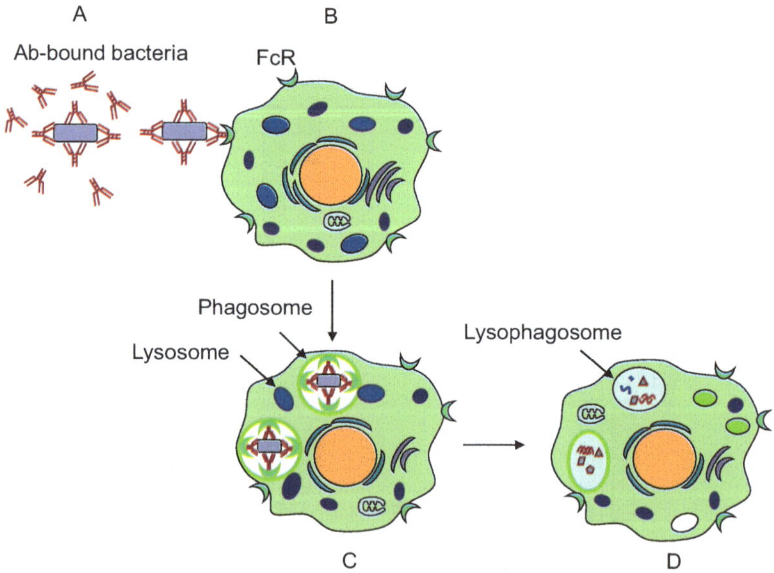

Fig. 6.10 Schematic diagram of antibody-mediated opsonophagocytosis of bacteria. (**a**) Antibody specifically binds to pathogenic bacterial antigens; (**b&c**) Macrophage Fc receptor (FcR) enhances phagocytosis of the opsonized- bacteria by binding to the Fc fragment of the antibody, Bacterial was phagocytosed, forming phagosome, and the phagosome subsequently is fused with lysosome to form lysophagosome. (**d**) phagocytosed bacteria are degraded and killed by various enzymes in the lysophagosome

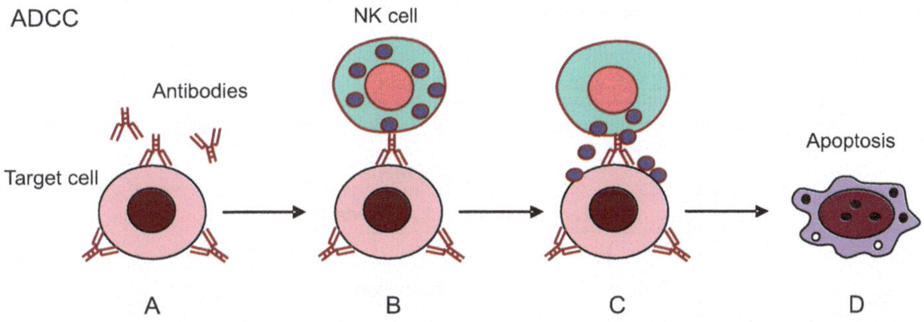

Fig. 6.11 Schematic illustration of the process of ADCC. (**a**) Antibody binds to antigen on cell membrane. (**b**) Fc receptor (CD16) on NK cell membrane binds to Fc of antibody. (**c**) Activation of NK cell releases perforin, granzyme, and other substances to kill target cells. (**d**) Apoptosis in target cell

recognized by several B cell clones, each derived from a B cell that recognize a particular epitope. With the help of Th cell, one B cell that recognize particular antigenic epitope is activated, proliferate and differentiate into plasma cells to produce antibodies, which are called monoclonal antibodies, indicating antibodies derived from a single B cell clone (a clone, a group of cells derived from a single progenitor) recognizing a particular epitope. In comparison, antibodies that are derived from multiple B-cell clones are called polyclonal antibodies, such as anti-serum.

Basic Principle This technique takes advantage of the fact that B cells produce antibodies but cannot survive long in vitro cell culture, while myeloma cells do not produce antibodies but can proliferate indefinitely, and the two cells can be fused to form hybridoma cells by fusion agent, which possesses the immortal growth properties and secretion of antibody. In general, the procedure is as follows: mice are immunized with antigen, and spleen cells are prepared and fused with myeloma cells (e.g., SP2/0) by fusion agent (e.g., PEG). As myeloma cells are hypoxanthine guanine phosphoribosyltransferase deficient (HGPRT−/−) and sensitive to aminopterin, cannot synthesize purines, and cannot survive in the selective HAT culture medium containing hypoxanthine (H), aminopterin (A) and thymine (T), they must have HGPRT in the HAT culture medium to survive, and HGPRT can be obtained by fusion with B cells. Thus, hybridomas can be screened by using the HAT selective culture medium. Individual cell-grown clones can be obtained by the limiting dilution, and the positive clones are screened based on detection of antibody in the culture medium. To ensure that the secreted antibodies are monoclonal, 3–4 rounds of cloning and screening are often performed.

A population of cells derived from a single B cell after activation and proliferation is a clone, and a population of cells formed by the activation and proliferation of multiple B cells is called polyclone. Monoclonal antibodies are antibodies secreted by one B-cell clone and are extremely uniform and homogeneous, while

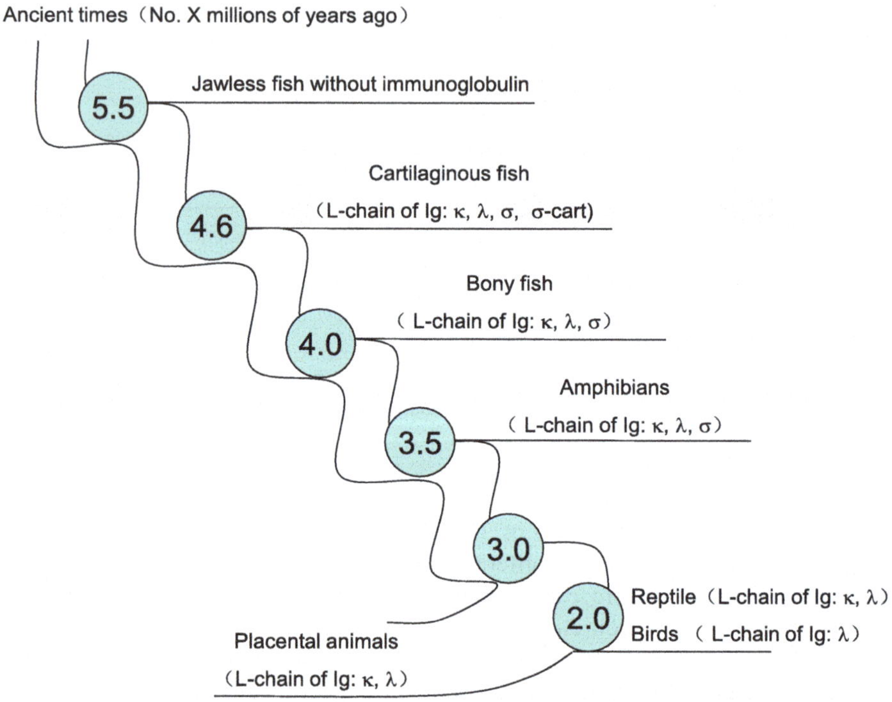

Fig. 6.12 Schematic diagram of the chronological history of immunoglobulin evolution. Ig emerged around 5.5 million years ago, and so far most placental animals have both κ and λ subclasses of L-chain, while birds have only λ L-chain

the antibodies secreted by multiple B cell clones are polyclonal antibodies, which are heterogeneous. Therefore, all antisera in animals are polyclonal antibodies, and these antibodies are identical or similar except for different antigen-binding sites. As it is impossible to directly divide polyclonal antibodies into monoclonal antibodies by biochemical methods, it becomes a dream to obtain monoclonal antibodies. In 1975, British scientists George Kohler and Cesar Milstein published a paper in Nature entitled "Continuous Cultures of Fused Cells Secreting Antibody of Predefined Specificity," which reported the establishment of a technique for the production of monoclonal antibodies by fusing B cells with myeloma cells to obtain hybridomas (Fig. 6.13). This technique allows the production of highly pure monoclonal antibodies. This technique became a milestone in the history of development of immunological techniques and a major impetus to life science research in general, for which Koler and Milstein were awarded the 1984 Nobel Prize in Medicine.

6.2.9 *Antigenicity of Immunoglobulins*

Immunoglobulins themselves are proteins or glycoproteins and are therefore antigenic to induce an antibody response. The antigenic determinants or epitopes on immunoglobulin molecules can be classified into three categories: isotypic, allotypic, and idiotypic determinants according to their location on the molecules.

1. **Isotype:** The isotype refers to the same immunoglobulin antigenic specificity that is shared by every individual of the same species, with the antigenic determinants found mainly in the C region of Ig, such as the difference between IgG and IgM (Fig. 6.14).
2. **Allotype:** Even in the same isotype antibody (e.g., IgG), the peptide chains encoding the amino acids may differ slightly, and the antigenic determinants formed by such subtle differences are called allotypes, e.g., the amino-acid sequences forming the constant region differ slightly between IgG1 from different strains of mice. The γ-chain (the constant region of IgG H chain) allotype is referred to as the genetic marker (Gm), and at least 25 Gm have been identified in human, usually named by the class and subclass followed by the allele number, such as G1m(1), G2m(12), G3m(7),G4m(11). Allotype is not only reflected in the difference of antigenic determinants of the same isotype antibody H chains

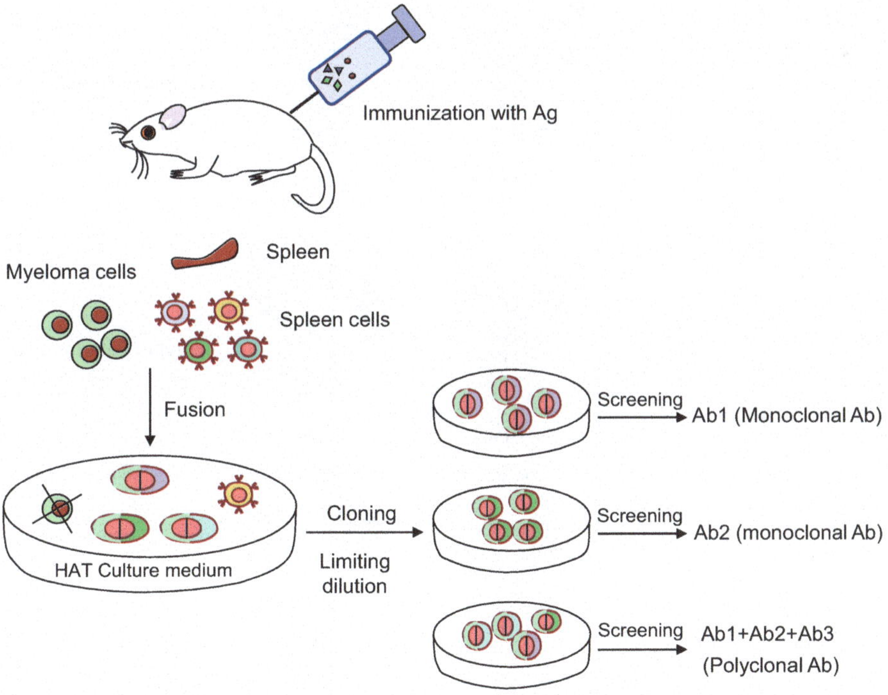

Fig. 6.13 Schematic diagram of the principle of monoclonal antibody preparation

but also in the antigenicity of L chains, for example, there are three allotypes of κ chains of L chains, named κm(1), κm(2), and κm(3), respectively.

3. **Idiotype: The specific antigen-binding site formed by the variable region of L chain (VL) and H chain (VL) of the antibody has** a unique amino-acid sequence and conformational structure that not only binds antigenic epitopes but also has antigenic properties. Each individual antigenic determinant of the variable region of the antibody is referred to as an idiotype. As an antibody has three hypervariable regions, there may exist multiple idiotypes in the variable region of antibody. The sum of the individual idiotypes is called the idiotype of the antibody. Because of the antigenicity of idiotype formed by the unique amino-acid sequence and conformational structure of the immunoglobulin hypervariable region, the immune system will also produce an immune response to the idiotype. The corresponding B cells will be activated to produce antibodies against the idiotype, which is called anti-idiotype antibody, and so on, forming an interesting network of regulation.

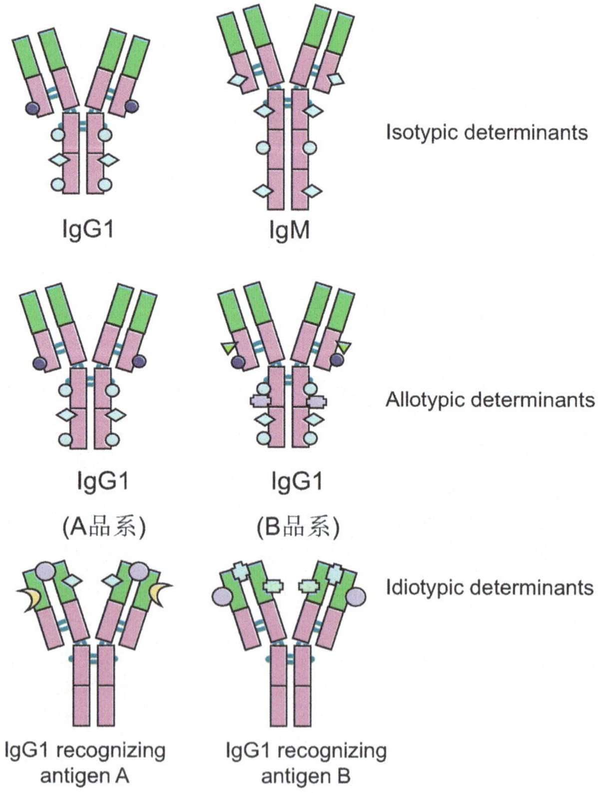

Fig. 6.14 Schematic diagram of isotype, allotype, and idiotype

6.2.10 Clinical Applications of Antibodies

Antibodies are used clinically for diagnostic and therapeutic purposes. As early as 1898, Behring and Wernicke discovered that immunization of animals with diphtheria toxin-antitoxin complexes could prevent and treat diphtheria infections, and antitoxin serum was later produced in large quantities for clinical use. Since then, various antisera have been widely used in clinical practice. Antibodies are generally used in following areas:

6.2.10.1 Detection of Antigens

Tumor diagnosis: Detection of tumor cell-associated antigens by monoclonal antibodies can make a specific tumor diagnosis.

Detection of trace small molecules: Monoclonal antibodies were used for monitoring certain drug residues, chemical contaminant, etc.

Surveillance for pathogens: Analyzing the feasibility of epidemics or diagnosing infectious diseases by monitoring pathogens.

6.2.10.2 Treatment of Diseases

Anti-snake venom antibodies: The most common potent treatment for venomous snake bites is the use of anti-snake venom antibodies.

Tumor-targeted therapy: Monoclonal antibody against tumor-associated antigen is conjugated with drugs targeting tumors for the treatment of tumor, and tumor immunotherapy with monoclonal antibodies blocking immune checkpoints (see Chap. 4) has been clinically used with encouraging outcome and shows good promise.

Prevention and control of infectious disease: Immune sera or yolk antibodies with high titers have been widely used in practice to prevent and control infectious diseases in animals, such as high-titer immune sera for the treatment of classical swine fever, yolk antibodies for the treatment of duckling viral hepatitis and infectious bursal disease, etc.

6.3 Antigen–Antibody Interactions

The antigen binds to the antibody through the antigenic epitope, and this binding is dynamic. The binding force between the antigen-binding site of the antibody and the antigenic epitope is composed of four forces of non-covalent bonding, which are hydrogen bonding, ionic bonding, hydrophobic bonding, and van der Waals force. The sum of the various forces during the binding between the antigen-binding site of the antibody and the antigenic epitope is called the affinity of the antibody for that antigen, which is usually expressed as association constant of affinity (K_a), but the strength of affinity is usually expressed as the dissociation constant of affinity (K_d). $Ag + Ab \underset{K_d}{\overset{?}{?}} Ag?Ab$

$$K_{d=} \frac{1}{K_a} = \frac{[Ag][Ab]}{[Ag?Ab]}$$

where [Ag], [Ab], and [Ag-Ab] refer to the concentration at equilibrium of the reaction.

Antigen–antibody binding is specific, and if the ratio of antigen and antibody is appropriate, an antigen–antibody complex is formed (Fig. 6.15), so people have used this basic principle to establish various antigen–antibody assays, which are widely used for disease diagnosis and epidemic surveillance.

6.3.1 Agar Gel Diffusion Test

Formulating agar to different concentrations yields support matrices of different softness and pore size, a feature that can be exploited to perform agar gel diffusion assays for the detection or analysis of antigens or antibodies (Figs. 6.16, 6.17 and 6.18).

6.3.1.1 Single-Phase Single Diffusion

The antibody is diluted to a certain concentration and pre-warmed at a water bath at 50–55 °C to mix well with dissolved agar (1.5% in saline) at 50–55 °C. The agar suspension containing the antibody is poured into the plate, cooled, and punched, and the antigen to be tested is added to the agar wells. The diameter of the ring of antigen–antibody precipitation line is proportional to the antigen concentration.

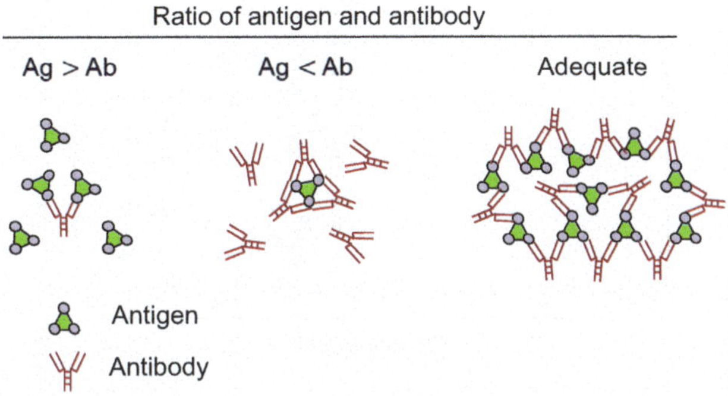

Fig. 6.15 Schematic diagram of antigen-antibody binding. When there are more antigens than antibodies or fewer antigens than antibodies, antigen–antibody binding does not form a visible complex. When the ratio of antigen–antibody is appropriate, antigen–antibody binding forms a complex visible to the naked eye

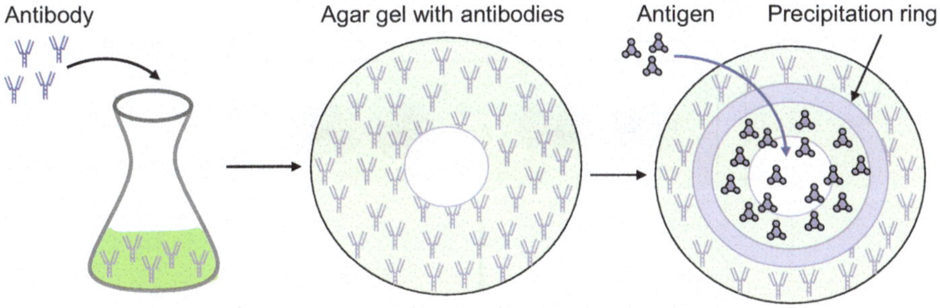

Fig. 6.16 Schematic diagram of the single-phase single diffusion principle

6.3.1.2 Double Diffusion

6.3.1.3 Immunoelectrophoresis

6.3.2 Hemagglutination Assay and Hemagglutination Inhibition Assay

Certain pathogens (e.g., Newcastle disease virus, avian influenza virus, and parvoviruses) have the property of agglutinating erythrocytes in the blood, and such pathogens contain hemagglutination antigens, and based on this characteristic, host immunity can be measured by monitoring antibodies against viral hemagglutination antigen. In the case of Newcastle disease virus (NDV), for example,

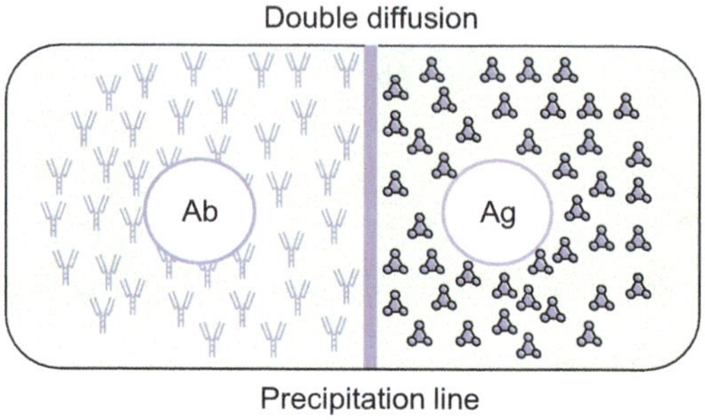

Fig. 6.17 Schematic diagram of the principle for double diffusion test. A prepared agar plate is punched, and either antigen or antibody is added to each of the two wells and placed in a wet box to diffuse, where the antigen and antibody encounter and interact, forming a line of precipitation visible to the naked eye at the optimum ratio of antigen and antibody. This method can be used for antigen or antibody detection

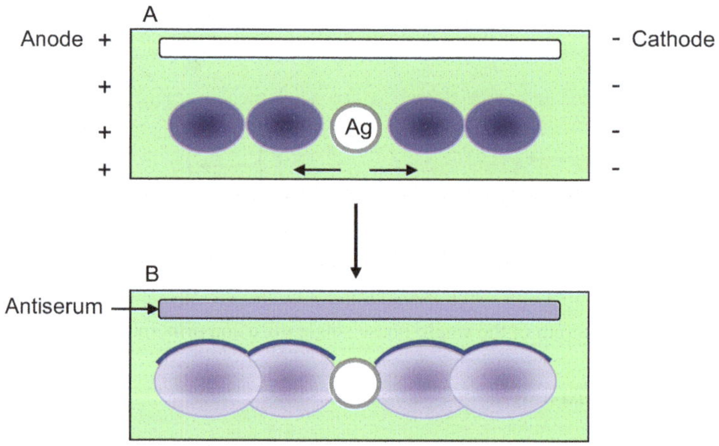

Fig. 6.18 Schematic diagram of the principle for immunoelectrophoresis. Immunoelectrophoresis is a two-step process in which antigen is first added to the agar gel wells and electrophoresis is performed by applying electricity (**a**), where the individual components are separated by the electric field due to the different sizes and charges of the antigen molecules, and then under removed electric field conditions, the serum is added to the gel rectangular well in parallel to the electric field for diffusion. Antigen encounters with antibody at diffusion to yield a visible precipitation line at the appropriate ratio (**b**). This method can be used to analyze the components of complex antigens

hemagglutination (HA) test is first performed to determine the hemagglutination titer of the viral fluid (Fig. 6.19). Hemagglutination inhibition (HI) test is then performed based on the hemagglutination titer parameters obtained from the HA test. The prepared antigenic solution is used to monitor the titer of the anti-hemagglutinin antibodies in chicken blood, obtaining the antibody titer (also called hemagglutination inhibition titer) (Fig. 6.20).

50 μL of diluent (sterilized saline or PBS) was added to the microplate wells, 50 μL of chicken serum was taken for a two-fold series of dilutions in wells 1–12, and 50 μL was discarded in the last well after dilution, and resulting serum dilutions in each well are: 1:2, 1:4, 1:8, 1:16, 1:32, 1:64, 1:128, 1:256, 1:512, 1:1024, 1:2048, and 1:4096, respectively, and 50 μL of diluted serum is in each well. 25 μL of NDV antigen with 4 hemagglutination units will be added to the diluted serum wells (if the antigen hemagglutination titer is 1:16, this means that the antigenic solution diluted 16 times can still agglutinate 0.85–1% of the equal volume chicken red blood cells, viral solution more than this dilution cannot form a complete agglutination, so according to the antigen agglutination titer, the antigen solution diluted at the antigen agglutination titer is therefore 1 agglutination unit. On this basis, if the antigen solution with the agglutination titer of 16 is diluted 1:4, the resulting antigen solution is 4 agglutination units). After mixing, the reaction wells are incubated for 10 min, followed by addition with 25 μL of 0.5% chicken fresh red blood cell (RBC) suspension to each well and mixing well. Half an hour after addition with chicken RBC, the results are determined by observing the inhibition of RBC agglutination. The total volume of the mixture in the reaction system at this point are 100 μL (the virus antigen was diluted to exactly 1 agglutination unit in the reaction). The highest dilution of the serum at which RBC agglutination is completely inhibited would be determined to be the hemagglutination inhibition titer, according to which the HI titer was 1:128 (well 7).

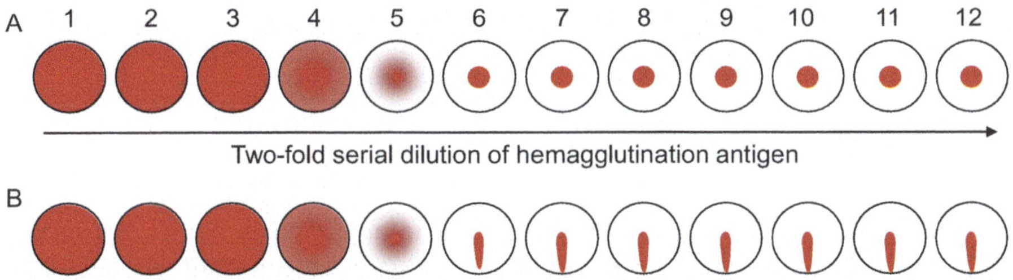

Fig. 6.19 Schematic diagram of hemagglutination assay for titration of Newcastle disease virus. (**a**) Vertical observation of aggregate of red blood cell (RBC) in the wells of microplate that is placed flat on the table. (**b**) Horizontal observation of aggregate of red blood cell (RBC) in the wells of microplate that is placed vertically on the table. The NDV virus antigen is diluted in a two-fold series: 50 μL of dilution solution (sterilized saline or PBS) is added to the microplate wells, followed by adding 50 μL of viral hemagglutination antigen to the first well and mixing well, and then 50 μL of the mixture is taken out and added to the second well, and so on, for a two-fold serial dilutions, and 50 μL is discarded in the last well after dilution. The viral hemagglutination antigen dilutions in each well are: 1:2, 1:4, 1:8, 1 :1:16, 1:32, 1:64, 1:128, 1:256, 1:512, 1:1024, 1:2048, 1:4096. An equal volume (50 μL) of 0.85–1% chicken fresh erythrocyte suspension is added to each well and mixed well, followed by half an hour incubation to observe the results. Criterion: The highest dilution of the viral antigen with complete erythrocyte agglutination was taken as the hemagglutination titer. The hemagglutination titer in this diagram is 1:16 (well 4). (**a**). The microplate was placed flat and the hemagglutination results were observed vertically from top to bottom. (**b**) The hemagglutination results were observed horizontally after the plate was placed vertically on a flat surface.

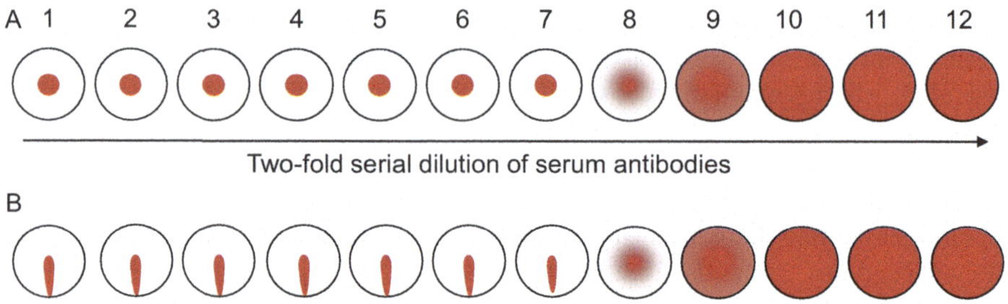

Fig. 6.20 Schematic diagram of monitoring hemagglutination inhibition titer of antibodies against Newcastle disease virus. (**a**) Vertical observation of aggregate of red blood cell (RBC) in the wells of microplate that is placed flat on the table. (**b**) Horizontal observation of inhibiting aggregate of red blood cell (RBC) in the wells of microplate that is placed vertically on the table

6.3.3 Indirect Hemagglutination Test

Soluble antigens bind to antibodies to form antigen–antibody(Ag–Ab) complexes, and because such molecular complexes are small, it is difficult to show a Ag–Ab reaction visible to the naked eye, especially when the amounts of antigen and antibody are small. If the antigen or antibody is bound to the surface of the red blood cell by a linker and then react with the corresponding antibody or antigen, red blood cell agglutination will be visible to the naked eye (Fig. 6.21), and this type of hemagglutination test using the red blood cell as the carrier is called an indirect hemagglutination test (IHA) or passive hemagglutination test.

6.3.4 Agglutination Inhibition Test

A small-molecule antigen is chemically linked to a particulate carrier and mixed with a specific antibody to form a naked-eye visible antigen–antibody complex, which is used as an indicator system. If the sample to be tested contains antigen, the antigen will bind to the antibody, and then the addition of the antigen-bound particulate carrier will not result in agglutination visible to the naked eye; this method is called the agglutination inhibition assay (Fig. 6.22).

6.3.5 Enzyme-Linked Immunosorbent Assay

The enzyme-linked immunosorbent assay (ELISA) is an assay in which a soluble antigen or antibody is adsorbed onto a solid-phase carrier and an enzyme-labeled immune complex is formed by an enzyme-labeled antibody or antigen to bind specifically to the antigen or antibody, adding the corresponding enzymatic substrate

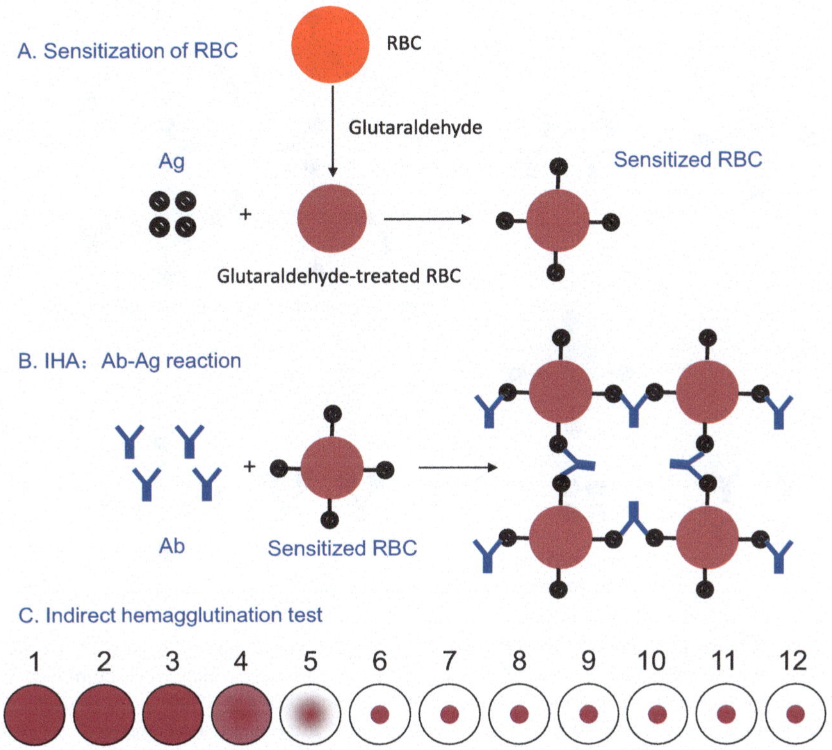

Fig. 6.21 Schematic diagram of the indirect hemagglutination test (IHA). (**a**) Fresh 10% red blood cells are slowly added to an equal volume of 1% glutaraldehyde (linker) and incubate at 4 °C for 30–60 min, during which mixture needs to be shaken continuously, and subsequently glutaraldehyde-fixed red blood cells were sensitized with antigen (the principle is: the aldehyde groups at both ends of glutaraldehyde bind to the amino group of red blood cell membrane protein and amino group of the antigen molecule, forming a strong Schiff base that connects the antigen with the red blood cell); (**b**) serial dilutions of the serum to be tested (1:10, 1:20, 1:40, 1:80, 1:160, 1:320 ......) are performed, and if the sample contains the antibodies to be tested, the antigen of the sensitized erythrocytes binds specifically to the antibodies to form a complex, and a visible hemagglutination occurs; (**c**) the highest dilution of the serum at which complete agglutination of the erythrocytes occurs is taken as the indirect hemagglutination titer of the sample to be tested, which in this figure is 1:80 (fourth well). This method can be used for diagnosis of infectious diseases and epidemiological surveillance

for color development, and the determination of the antigen–antibody reaction is based on the color of the reaction solution (Fig. 6.23). This method is a technique that combines the specificity of the antigen–antibody reaction with the efficiency of the enzymatic reaction, and is widely used in research and practices for the qualitative or quantitative detection of semi-antigens, antigens, and antibodies in body fluids, in addition to localization of antigen–antibody at the cellular and subcellular levels.

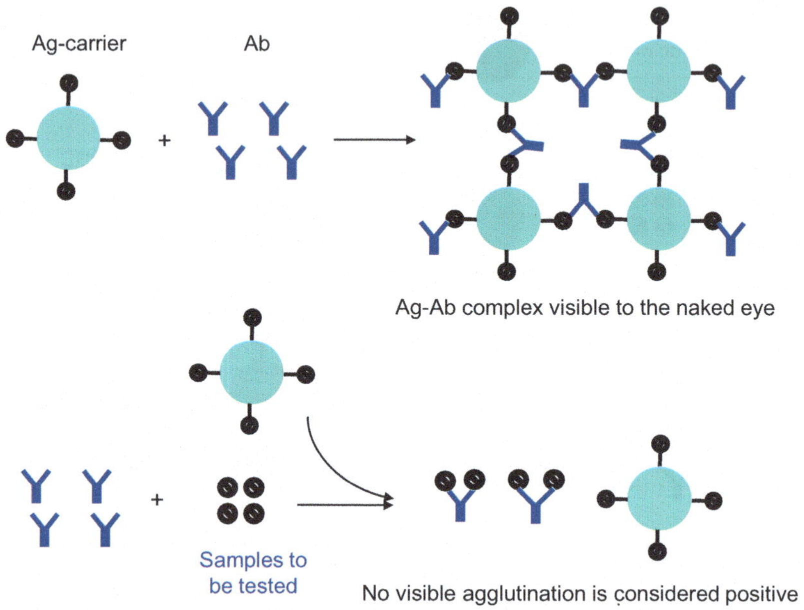

Fig. 6.22 Schematic diagram of the principle of the agglutination inhibition assay. Antigens are known to bind to particulate carriers to form Ag-carrier particles, which when combined with antibodies form an antigen–antibody complex visible to the naked eye. If this agglutination test is used as an indicator system and the sample to be tested is mixed with antibodies and Ag-carrier particles are added, there will be no naked-eye visible agglutination, i.e., a positive result. For example, the HCG test kit is used to detect early pregnancy

6.3.6 Fluorescent Antibody Assay

Fluorescein is a class of organic dye compounds that can produce obvious fluorescence, in the ultraviolet and blue ultraviolet irradiation, it can absorb light quantum caused by the outer electron leap and is in the unstable high energy state, in which electron after 10^{-8} s can jump back to the original position of the electronic layer, at the same time, release energy in the form of light quantum, and this excited visible light is fluorescence. In the process of excitation of fluorescence there is a partial loss of energy, so the wavelength of fluorescence is greater than that of the excitation light. Fluorescent antibody assay is a special immunoassay technique that uses fluorescein-labeled antibodies to detect the corresponding antigen in a sample and observes the fluorescence with a fluorescence microscope (Fig. 6.24) and is widely used in the diagnosis of infectious diseases.

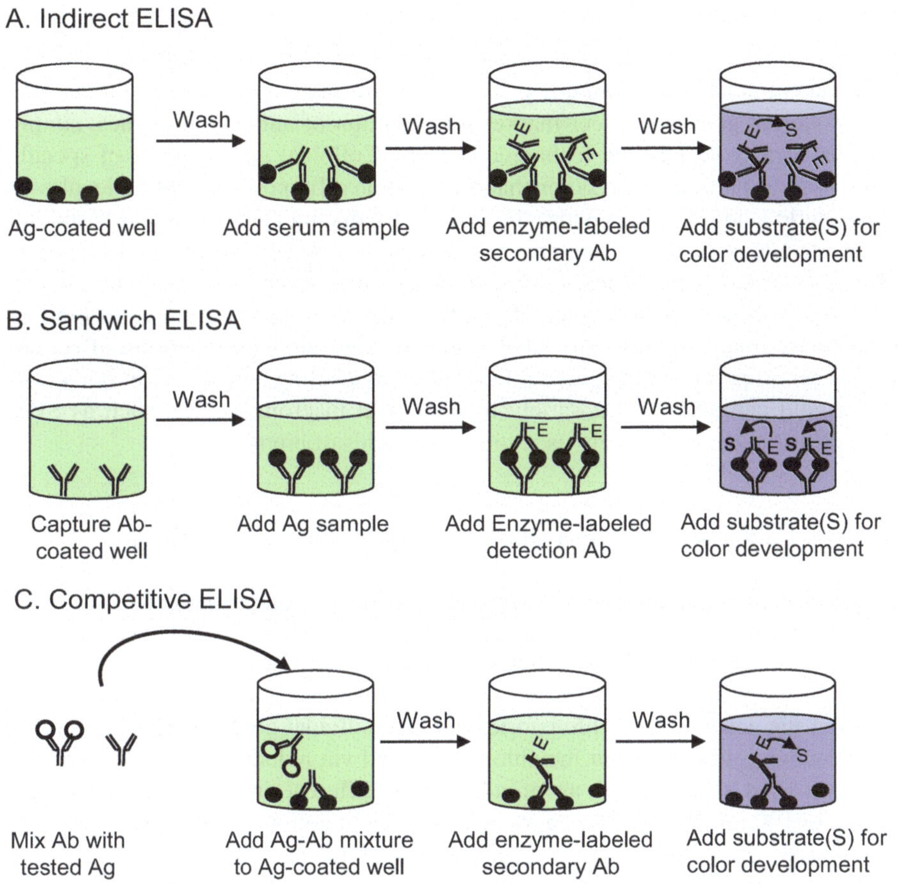

Fig. 6.23 Schematic diagram of several common ELISAs. (**a**) Indirect ELISA: The ELISA microplate is coated with the antigen, washed and blocked by adding a blocking solution (e.g., PBS containing 0.5–2% BSA), washed again and added with antibody for Ag–Ab reaction, washed and added with the enzyme-labeled secondary antibody (HRP–Ab2), washed and added with the substrate (H_2O_2 + OPD) for color development, added with stopping solution for termination, the OD450 or OD490 value would be measured by a microplate reader, and the OD value is proportional to the amount of the sample to be tested. This method can be used to detect both antigens and antibodies. (**b**) Sandwich ELISA: The ELISA microplate is coated with purified capture antibodies, washed and blocked by adding the blocking solution, washed again and added with the antigen to be tested for Ag–Ab reaction, washed and added with the enzyme-labeled detection Ab for reaction, and washed and added with the substrate (H_2O_2 + OPD) for color development. This method is used for antigen detection. (**c**) Competitive ELISA: The antibody is mixed with the antigen to be detected, the antigen–antibody mixture is added to the antigen-coated microplate wells (which have been blocked), the enzyme-labeled secondary antibody is added after washing, and the substrate is added after washing for color development. The concentration of the antigen to be detected is inversely proportional to the OD value, this method is mainly used for antigen detection. The cut-off value set-up: The mean OD value of the negative sample + 2 SD (standard deviation). Thus, the ELISA result would be considered positive if the OD measurement of the sample to be tested is $\geq$ the cut-off value

6.3.7 *Immune Colloidal Gold Technique*

Immune colloidal gold technique is a new type of immune labeling technique that uses colloidal gold as a tracer marker for detection of antigen-antibodies, simply referred to as gold labeling technique (GICT). Colloidal gold is made of specific sizes of gold particles with color that are generated by polymerized molecules of chloroauric acid (HAuCl4) under the action of reducing agents (such as trisodium citrate, tannic acid, etc.) and remains a stable colloidal state due to the electrostatic effect. Colloidal gold carries a negative charge in a weak base environment, can combine with protein molecules of positive charge groups to form solid binding (that is, labeling), because this label is electrostatic binding, so do not affect the biological properties of the protein. In addition to labeling antigen or antibody, colloidal gold can also label a variety of biological macromolecules, such as SPA, PHA, ConA, etc., thus widely used in various fields of biomedicine.

6.3.7.1 Colloidal Gold Preparation, Principles, and Basic Methods

Preparation of chloroauric acid (HAuCl4): Chemical reaction equation.

$$\mathrm{Au} + \mathrm{HNO_3} + 4\mathrm{HCl} = \mathrm{HAuCl4} + 2\mathrm{H_2O} + \mathrm{NO}?$$

After the generation of chloroauric acid, the colloidal gold particles are formed by reducing gold element in the chloroauric acid via reducing agent. If trisodium citrate is used as a reducing agent, addition of different amount of 1% trisodium citrate to 100 mL 0.01% chloroauric acid will yield colloidal gold particles of different sizes and colors. For example, addition of 1.5 mL 1% trisodium citrate to 100 mL 0.01% chloroauric acid will produce orange-red colloidal gold particles with a diameter of 24.5 nm, while addition of 1 mL of trisodium citrate will produce red colloidal gold particles with a diameter of 41 nm, and addition of 0.7 mL of trisodium citrate will produce purple-red colloidal gold particles with a diameter of 71.5 nm.

6.3.7.2 Application of Immune Colloidal Gold Techniques

As a tracer technique, colloidal gold labeling can be used in a variety of assays, such as liquid phase immunoassays (combining colloidal gold with antibodies to establish a microagglutination test to detect the corresponding antigen, similar to an indirect hemagglutination assay, where the agglutinated particles can be directly observed with the naked eye), gold-labeled flow cytometry (using colloidal gold-labeled sheep anti-mouse Ig antibodies for flow cytometry), and colloidal gold solid-phase immunoassay (dot immunogold examination method, colloidal gold-labeled test strip assay, etc.). The colloidal gold-labeled test strip assay is the most

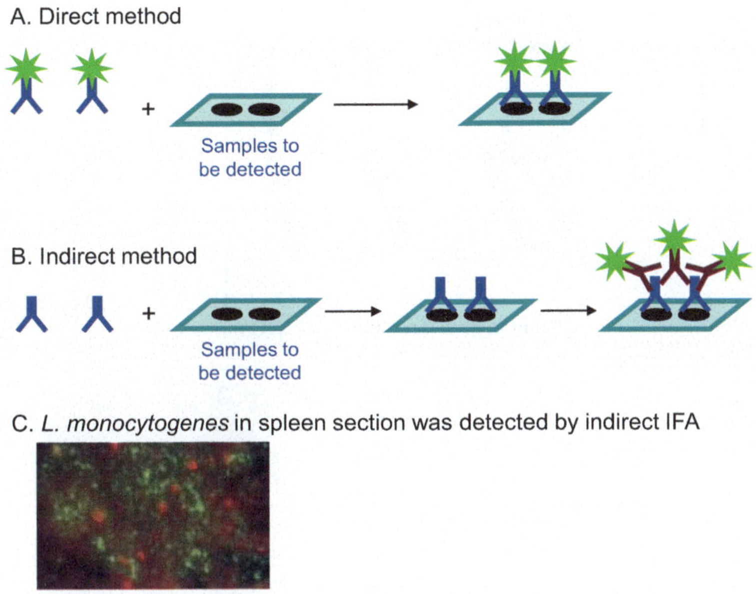

Fig. 6.24 Schematic diagram of the principle for the technique of fluorescent antibody assay. The common ways for the detection of antigen by fluorescent antibodies are direct and indirect methods. (**a**) Direct method: The sample to be examined (e.g., blood smear, tissue section, etc.) is fixed, non-specifically blocked, then a fluorescent antibody is added to the sample dropwise and incubated for half an hour. After washes, the sample is covered by coverslip and examined under a fluorescence microscope. (**b**) Indirect method: The sample to be examined is fixed, non-specifically blocked, added with detection antibody, and then incubated for half an hour, followed by sufficient washing and addition of fluorescently labeled secondary antibodies, washed, covered by coverslip, and examined under a fluorescence microscope. (**c**) An indirect method of fluorescent antibody assay for the detection of *Listeria monocytogenes* in spleen sections. The *Listeria monocytogenes* in the spleen sections were probed by anti-*Listeria* antibodies as primary antibodies, followed by incubation with Alexa-488-labeled secondary antibodies, Listeria monocytogenes in green and apoptotic cells (examined by TUNEL-staining) in red were observed under a fluorescence microscope

widely used method, such as drug residue testing, antigen/antibody testing, and testing for small molecules (e.g., hormones, drugs) (Fig. 6.25).

6.3.8 Radioimmunoassay Techniques

Radioimmunoassay is the first immunolabeling technique used to detect small-molecule antigens. The basic principle is that isotope-labeled antigens compete with unlabeled antigens to bind high-affinity antibodies. Usually the isotope-labeled antigen is bound to the antibody to saturation, and the unlabeled sample antigen is added to compete with it. The amount of antigen in the sample to be tested is

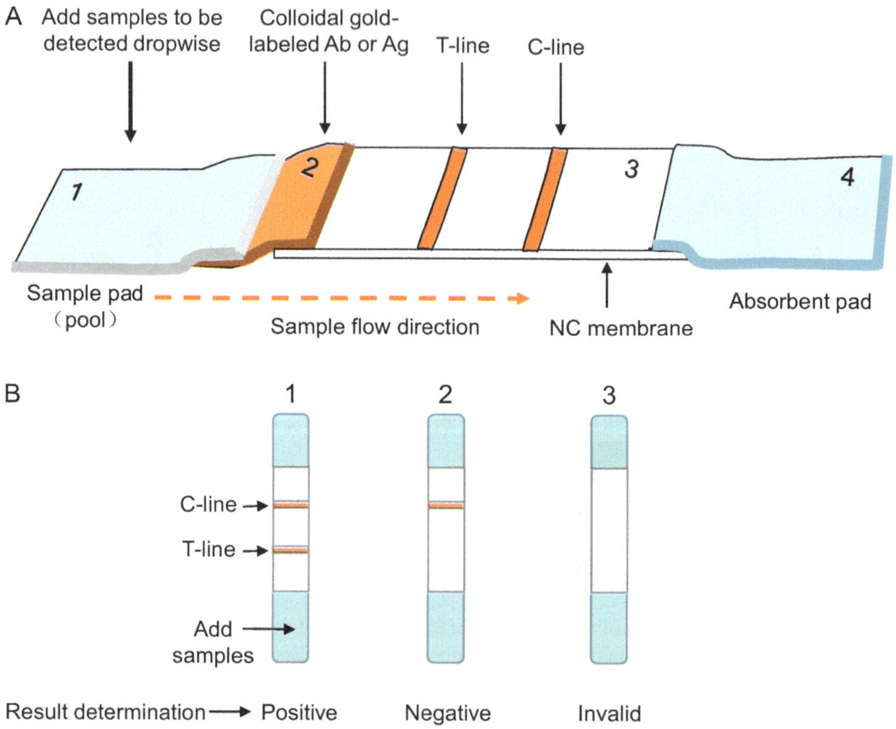

Fig. 6.25 Schematic diagram of the principle for colloidal gold-labeled test strip examination. The basic principle of colloidal gold-labeled test strips is that the antigen or antibody to be tested binds to the corresponding colloidal gold-labeled antibody or antigen, and the resulting colloidal gold-labeled Ag–Ab complex is captured in the liquid-phase flow on a nitrocellulose membrane to form a positive reaction line of gold particles visible to the naked eye, which is a simple immunochromatographic technique. (**a**) The components of colloidal gold-labeled test strip mainly include, taking HCG (human chorionic gonadotropin) colloidal gold-labeled test strip for example, sample pad (pool), NC (nitrocellulose) membrane and absorbent pad. A1 is the sample pool; A2 contains a colloidal gold-labeled anti-HCG monoclonal antibody (mouse Ab); A3 is the NC membrane, previously coated with anti-HCG capture antibody (T-line) and anti-mouse secondary antibody (e.g., rabbit anti-mouse or sheep anti-mouse antibody) as quality control (C-line); A4 is the absorbent pad. (**b**) Check up the results. The results will be available in a few minutes after dropping urine samples on to the sample pool. Judgment of results: B1. Test strip with two lines (detection line and quality control line) indicates a positive result, i.e., HCG is present in the urine; B2. Test strip with one line (quality control line) indicates a negative result, i.e., HCG is not present in the urine; B3. Test strip with no reaction line indicates an invalid test result

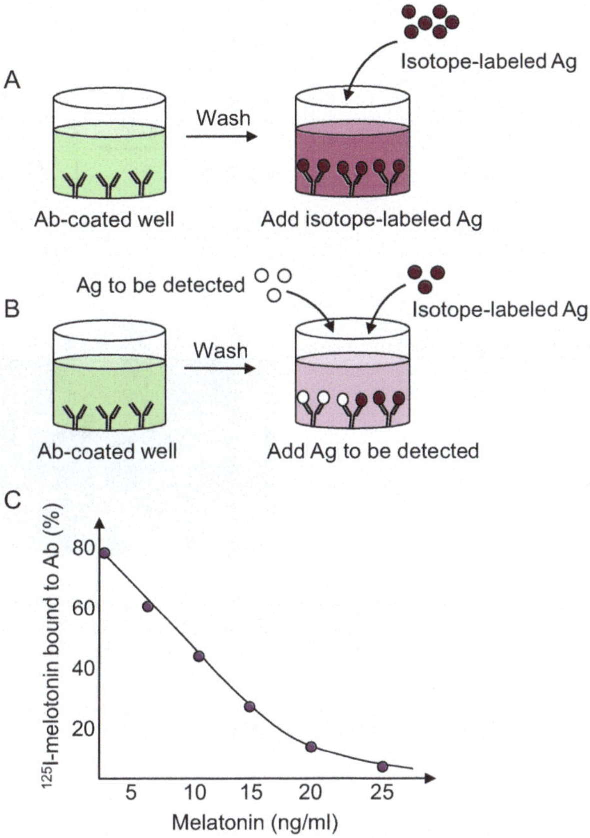

Fig. 6.26 Schematic diagram of the principle for radioimmunoassay technique. (**a**) The reaction well of plate is coated with antibodies, the isotope-labeled antigen is added, and after washes the radioactive intensity is detected by equipment after adding scintillation solution. (**b**) The antigen to be tested is added to the isotope-labeled antigen–antibody reaction system to participate in competitive binding, and the concentration of antigen in the sample is inversely proportional to the radioactive intensity. (**c**) The standard curve is prepared using different concentrations of standard antigen, which can be used to convert and calculate the content of antigen in the sample

determined by detecting the amount of isotope in the antigen–antibody complex (Fig. 6.26). This method was first established by S. A. Berson and Rosalyn Yalow in 1960 for the detection of insulin and has since been extended. The method proved so useful in detecting small-molecule traces that Yalow was awarded the 1977 Nobel Prize in Medicine.

6.3.9 Immuno-Polymerase Chain Reaction

Immuno-polymerase chain reaction (immuno-PCR) is a new technique developed in the 1990s that combines the specificity of antigen–antibody reactions with the high sensitivity of polymerase chain reaction (PCR) to form a highly sensitive and specific assay technique. In contrast to ELISA, immuno-PCR detects antigens with DNA-labeled antibodies by PCR amplification of labeled DNA on the antigen–antibody complex (Fig. 6.27), and ELISA, which uses labeled-enzymes to catalyze the substrate for color development to reflect the presence of antigens or antibodies.

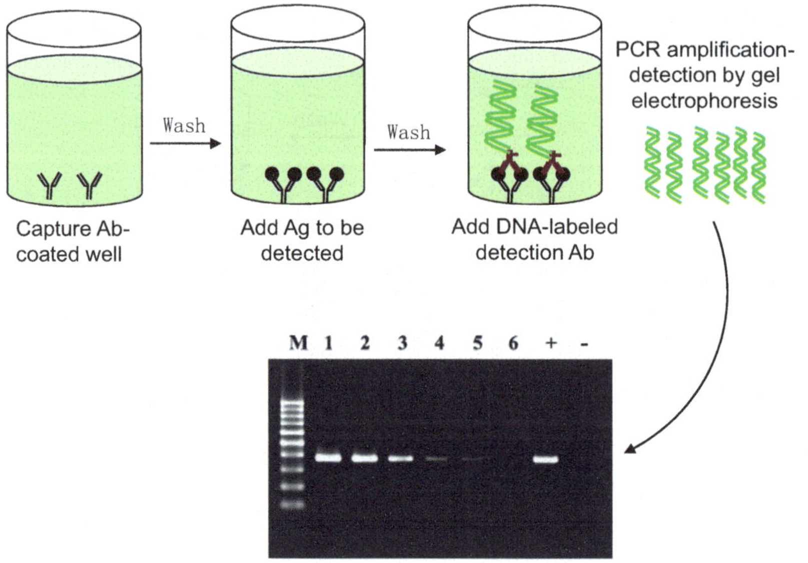

Fig. 6.27 Schematic diagram of the principle for immuno-PCR. The reaction plate is coated with a capture antibody, followed by addition with the antigen to be tested. After washes, the detection antibody labeled with DNA is added, and after sufficient washes, various components (bases, polymerase, etc.) for PCR are added for PCR amplification, and the reaction products are detected by agarose gel electrophoresis. The result in the figure is the detection of neurotoxin A of Clostridium botulinum by immuno-PCR

Immuno-PCR usually uses agarose gel electrophoresis to detect the amplification product of the labeled DNA to show the results of the assay.

6.3.10 Enzyme-Linked Immunospot Assay

The enzyme-linked immunospot assay (ELISPOT) is essentially a new technique developed using the basic principles of ELISA combined with cell culture techniques, and is used to detect immune cells that secrete cytokines (Fig. 6.28). Thus, assessment of the direction, type, and intensity of the immune response at the cellular level is an important improvement in immunoassay technology. Commercial ELISPOT assay kits are already commercially available.

The reaction wells are coated with antibodies against the cytokine to be detected, and the cell suspension is added to the reaction wells for incubation. The number of colored spots represents the number of cytokine-secreting cells, thus ELISPOT represents the number of cells secreting a particular cytokine.

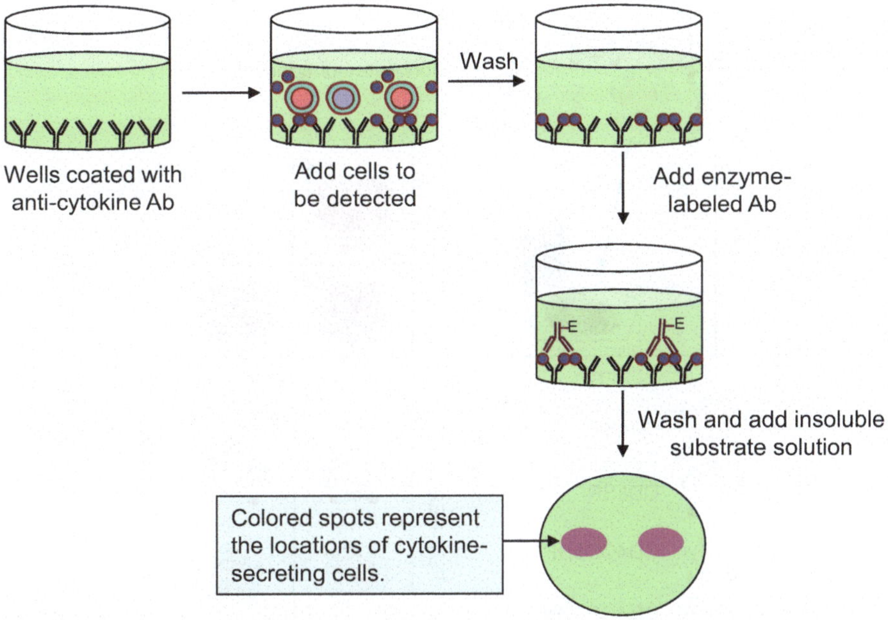

Fig. 6.28 Schematic diagram of the principle for ELISPOT

6.3.11 Flow Cytometry

Flow cytometry is a new immunoassay technique developed by combination of immunofluorescence technology with computer technology and is an important milestone in the development of immunoassay technology. The technique detects antigens on the surface or inside cells by means of fluorescent antibodies. When cells bound with fluorescent antibodies pass through a single-cell channel, a laser beam excites the fluorescence on the cells, and a receiver collects the excited fluorescence signal to analyze qualitatively or quantitatively multiple antigens expressed by the same cell via powerful computer analysis (Fig. 6.29). Thus, if antibodies are labeled with fluorescence of different colors, different antigens will be detected simultaneously, and the intensity of cellular fluorescence luminescence is proportional to the expression of antigen molecules on the cell surface. Flow cytometers that simultaneously use 16 or even more color fluorescent antibodies for antigen detection are now available and have greatly accelerated the research and diagnostic process.

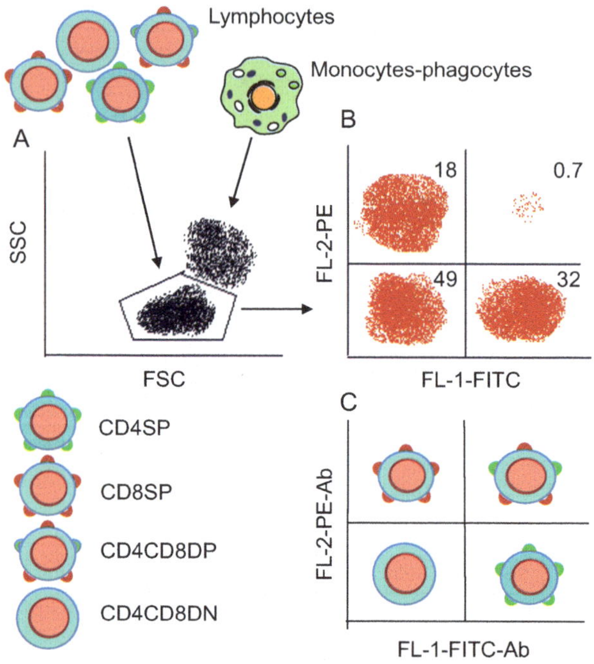

Fig. 6.29 Schematic diagram of detecting cell surface antigen by flow cytometry. Cell suspensions (e.g., spleen cell suspensions) contain monocytes-phagocytes, lymphocytes, etc. If the proportion of $CD4^+$ and $CD8^+$T cells is examined, the cell suspensions can be stained directly with FITC-CD4 and PE-CD8 fluorescent antibodies, followed by analysis with flow cytometry. (**a**) FSC represents the size of the cells, and SSC indicates the smoothness of the cell surface. When the cell suspension is examined by the instrument, two groups of cells will appear more clearly on the computer screen. Based on the size and smoothness, the cells can be divided into two major groups, lymphoid cells and mononuclear phagocytes. The lymphoid cell group includes B cells, T cells, NK cells, and NKT cells, while the mononuclear phagocyte group mainly includes macrophages, neutrophils, and a small number of dendritic cells. (**b**) If T cells are analyzed, the group of lymphoid cells is gated (also called set gate or gating) and this group of cells in (**a**) can be further analyzed in another viewport as indicated in (**b**). All cells in (**b**) are gated ones in (**a**). The *X* and *Y* axes represent the fluorescence luminescence intensity of the fluorescent antibody, respectively, and the numbers in (**b**) represent the percentages of the four subpopulations of cells that sum up to 100%. If the total number of cells in each sample is known, the absolute number of cells in the different subpopulations can be calculated from the cell percentages, providing an important reference for objective measurement of the immune response. (**c**) The enlarged cell represents the cell subpopulations in (**b**), with 32% CD4SPT cells, 18% CD8SP T cells, 49% CD4CD8DN T cells (containing mainly B cells, with a small number of NK and NKT cells), and 0.7% CD4CD8DPT cells

6.3.12 Gene Microarray Technology for Detecting Gene Expression

Gene microarray technology is an emerging technique for detecting differential expressions of target genes. Although this technology is not a detection technique based on the principle of antigen–antibody interaction, it plays an important role in molecular immunology research, so the basic principles of the technique are described here. This technique performs large-scale screening and analysis of known phenotype-associated genes at an mRNA level, providing information on upregulated or downregulated expressions of certain genes and laying the foundation in further in-depth studies of related gene functions (Fig. 6.30).

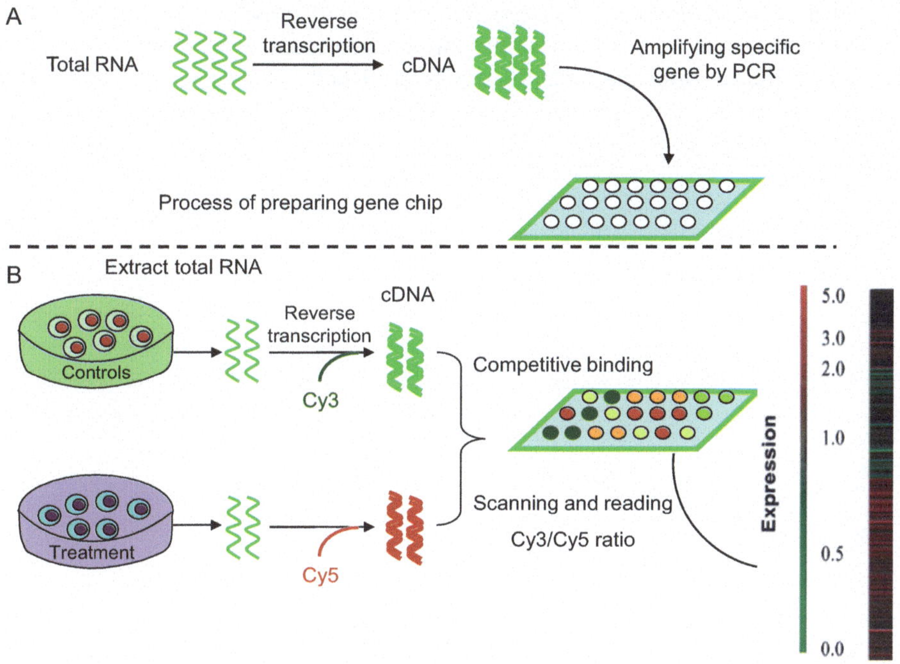

Fig. 6.30 Schematic diagram of the principle of gene chip technology. (**a**) Preparation of gene chips: Total RNA is extracted from tissues or cell samples, cDNA is obtained by reverse transcription, and the chips are prepared by amplifying the target gene with specific primers, all of which have been commercialized for this process. (**b**) An example of relevant gene expression during T-cell activation: the splenocyte treated with ConA is the treatment group, and the normal splenocyte culture as controls. Total RNA was extracted from the treatment and control groups, and fluorescein-labeled nucleotides were added during reverse transcription. Cy5-labeled cDNA (red fluorescence) was obtained from the treatment group and Cy3-labeled cDNA (green fluorescence) from the control group. The two cDNAs were mixed in a 1:1 ratio to hybridize with the genes on the microarray, and scanned and read with a scanner. If the fluorescence is yellow or orange, it indicates no change in the expression of the target gene post treatment, if red, it indicates upregulation of the target gene expression in the treatment group, and if green, it indicates downregulation of the target gene expression. Using the software, the relative folds of the target gene expression changes can be analyzed for up- or downregulation, laying the foundation in further study of the gene's function

Study Questions

1. What is the definition of antigen?
2. What are the definitions of antigenic epitopes, haptens, and complete antigens?
3. What are the basic structures of antibodies?
4. What role does each Fab/Fc fragment play?
5. What is the role of ADCC?
6. What is the basic principle for monoclonal antibody preparation?
7. What are the definitions of isotype, allotype, and idiotype?
8. What are the basic principles for the various serological assays?
9. How is the cut-off value for an ELISA determined?

Chapter 7
Major Histocompatibility Complexes and Antigen Presentation

Overview:
Antigens that are present on the cell surface and involved in immune rejection are called histocompatibility antigens. A highly polymorphic group of genes encoding major histocompatibility antigens are called the major histocompatibility complex (MHC). The proteins encoded by MHC are associated with immune recognition, rejection and immune response, thus MHC determines the susceptibility of an individual to infectious diseases and the tendency to autoimmune diseases, and in particular plays a key role in the adaptive immune response.

7.1 Composition and Genetic Characteristics of MHC

7.1.1 Discovery of MHC

In the mid-1930s, Peter Gorer studied blood group antigens in inbreed mice and found that there were four groups of genes, I–IV, that encoded blood group antigens on blood cells. In 1940–1950, Peter Gorer and George Snell studied these four groups of antigens and confirmed that group II antigens were involved in tissue and tumor graft rejection. Snell called these group II genes "histocompatibility genes", thus called mouse MHC "H-2". For his outstanding contributions to the discoveries of histocompatibility antigens and their inheritance significance, Snell was awarded the Nobel Prize in Medicine in 1980.

S. J. Zheng, *Veterinary Molecular Immunology*,
https://doi.org/10.1007/978-981-99-8929-4_7

Table 7.1 Nomenclature of MHC in humans and other animals

Type	MHC name	Chromosome number where the MHC is located
human	HLA	6
Mouse	H-2	17
Rat	RT1	20
Porcine	SLA	7
Cattle	BoLA	23
Rhesus	RhLA	19, 20
Dog	DLA	12, 2, 4, 14
Guinea pig	GPLA	UN
Rabbit	RLA	12
Horse	ELA	20, 16, 14, 23, 5
Poultry	B	16, 1
Sheep/goat	OLA, GLA	23
Feline	FLA	B2

7.1.2 Composition of the MHC

The MHC is a genomically linked gene cluster located on chromosome 6 in humans and on chromosome 17 in mice. Different animals have different names for MHC, e.g., human MHC is called HLA, porcine MHC is called SLA, etc. (Table 7.1).

Human and mouse MHC are now better understood, and their gene classification and encoded proteins are shown in Fig. 7.1.

7.1.3 Location of MHC Genes in the Genome

MHCs are groups of genes that are closely-linked on the same chromosome, such as HLA on chromosome 6 and H-2 on chromosome 17. This definition originated from research results on human and mouse MHC, however, not all animal MHCs conform to this observation, e.g. avian have 2 MHC regions on the same chromosome, while some fish have MHCs located on 3 different chromosomes, and there are certain MHC molecules (e.g. CD1) that are not in the traditional MHC regions. There is growing evidence that MHCs in different animals may be distributed on different chromosomes (Table 7.1).

MHC-III class genes are relatively conserved in all animals and mainly encode complement components (C2, C4, factor B), hydroxylases (21-OHA, 21-OHB), tumor necrosis factor (TNF-α, TNF-β), and heat-shock protein (HSP).

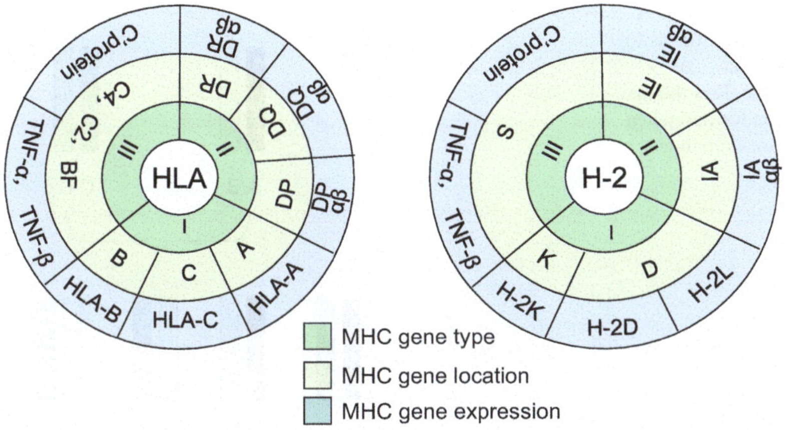

Fig. 7.1 Schematic representation of the composition of human and mouse MHC. HLA class I (HLA-I) loci contain B, C, and A, encoding HLA-B, HLA-C, and HLA-A molecules, respectively, HLA class II (HLA-II) loci contain DP, DQ, and DR, encoding DP-αβ, DQ-αβ, and DR-αβ molecules, respectively, and HLA class III (HLA-III) loci contain C4, C2, and BF, encoding C′-protein, TNF-α, TNF-β, etc. H-2 class I loci contain K and D, encoding H-2K, H-2D, and H-2L, respectively, H-2 class II loci contains IA and IE, encoding IA-αβ and IE-αβ, respectively, and H-2 class III loci contain S, encoding C'-protein, TNF-α, TNF-β, etc

7.1.4 The Genetic Characteristics of HLA

HLA inheritance is characterized by haplotype inheritance, high polymorphism, co-dominant expression, and linkage disequilibrium. HLA inheritance is also known as haplotype inheritance, where the MHC genes located on a chromosome in a tightly-linked set of specific combinations of allelic loci are called a haplotype, and the traits reflected in their expression products are called phenotypes, and analysis of the traits is called phenotyping. HLA inheritance conforms to Mendelian inheritance laws, such that each pair of chromosome has two haplotypes, one from paternal side and the other from maternal side (Fig. 7.2). Each offspring has one half MHC from father and the other half from mother, and the probability of sibling with the same MHC is 1/4. If there are more than four siblings from the same parent, the probability of finding an exact HLA match increases. HLA matches are required for tissue or organ transplantation. An individual with two identical HLA haplotypes is called homozygotes, and the one with two different haplotypes are called heterozygotes.

Human HLA inheritance is of polymorphism, and HLA polymorphism refers to a high degree of heterogeneity in the composition of HLA molecules between populations, where more than two genotypes can exist for the same locus. If each human haplotype has alleles of seven loci each, then a homozygote encodes 7 HLA molecules and a heterozygote encodes 14.

There are 27 known A loci, 59 B loci, 10 C loci, 26 D loci, 24 DR loci, 9 DQ loci, and 6 DP loci. Thus there are 161 loci in total, and the combinations of 161 loci account for 5.4×10^8 haplotypes. In addition, there are subtypes and variants, and

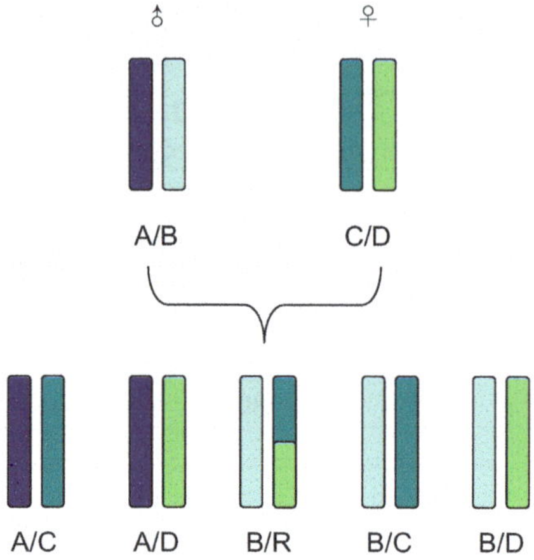

Fig. 7.2 Human HLA inheritance conforms to Mendelian laws of inheritance. R is the new haplotype formed when alleles are exchanged

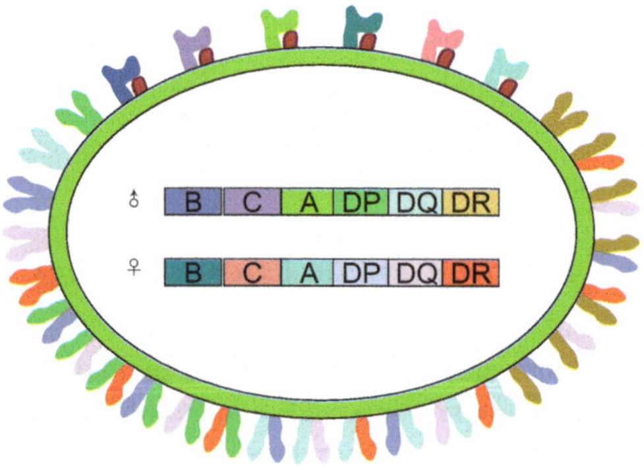

Fig. 7.3 Schematic diagram of inheritance of co-dominant expression of human HLA-I and HLA-II molecule haplotypes

the genotypes and expressions of HLA are numerous and exceed the total number of human population. HLA genes are co-dominantly expressed, i.e., each cell surface expresses both paternal and maternal HLA molecules (Fig. 7.3).

There is a linkage disequilibrium in HLA expression. Although HLA loci are tightly linked, some genes appear in a haplotype with high frequency, while others appear with low frequency, a phenomenon called linkage disequilibrium in which genes are not randomly distributed on haplotypes. For example, the frequencies of HLA-B8 and HLA-DR3 alleles in North American whites are 0.09 and 0.12, respectively, and the probability of appearance in the same haplotype should be $0.09 \times 0.12 = 0.0108$, but in reality it is 0.07, which is the result of linkage disequilibrium, which helps the accumulation of excellent genes to adapt to different geographical and living environments.

Co-dominant expression of the mouse H-2 gene is important in molecular immunology and genetic studies (Fig. 7.4), and the F1 generation carries and expresses MHC from both parents.

The mouse H-2 gene is derived from both parents and expresses both the paternal and maternal H-2 genes on the cell surface.

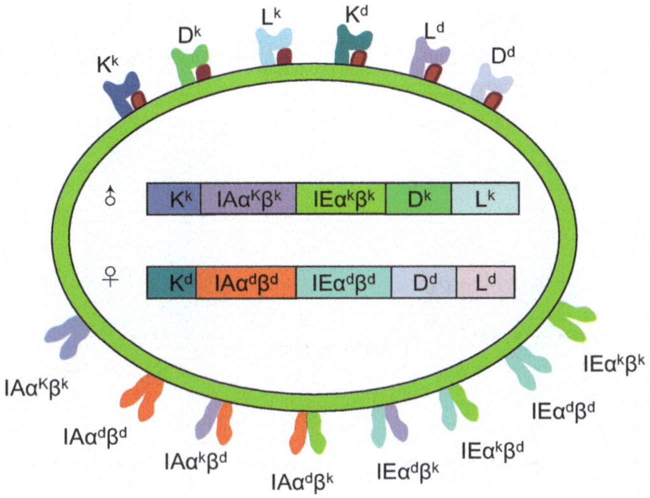

Fig. 7.4 Schematic diagram of co-dominant expression of H-2 gene in mice

7.2 Regulation of MHC Distribution and Expression

MHC class I molecules are distributed on the surface of all nucleated cells, whereas MHC class II molecules are mainly distributed on the surface of specialized antigen-presenting cells, such as B cells, dendritic cells, macrophages, etc. The expression of MHC is regulated by various factors, such as certain stimulants (bacteria, viruses), cytokines, etc., which can upregulate or suppress MHC expression.

7.2.1 Structure and Distribution of MHC-I Molecules

MHC class I molecules consist of α-chain transmembrane proteins and β2-microglobulin (Fig. 7.5).

MHC class I molecules are membrane-bound proteins, mainly consisting of a 45kD α-chain and a 12kD β2-microglobulin, of which the α-chain contains three regions, α1, α2 and α3, each containing about 90 amino acids, and the α3-linked transmembrane region contains 25 amino acids in the transmembrane region and 30 amino acids in the intracellular region. α3 region are highly conservative and has the binding site for CD8 molecules of CD8T cells. MHC-II class molecules belong to membrane-bound glycoproteins and consist of α and β chains. α chain molecules are 33 kDa in size and contain two regions, α1 and α2, while β chains are 28 kDa in size and contain two regions, β1 and β2. α2 and β2 are both connected to the transmembrane region and further extended to the intracellular region. α2 contains superantigen-binding site, and β2 contains sequences recognized by T cell CD4 molecules.

The α1 and α 2 regions of MHC class I molecules form a peptide-binding cleft that accommodates short peptides presenting 8–11 amino acids. β2-microglobulin,

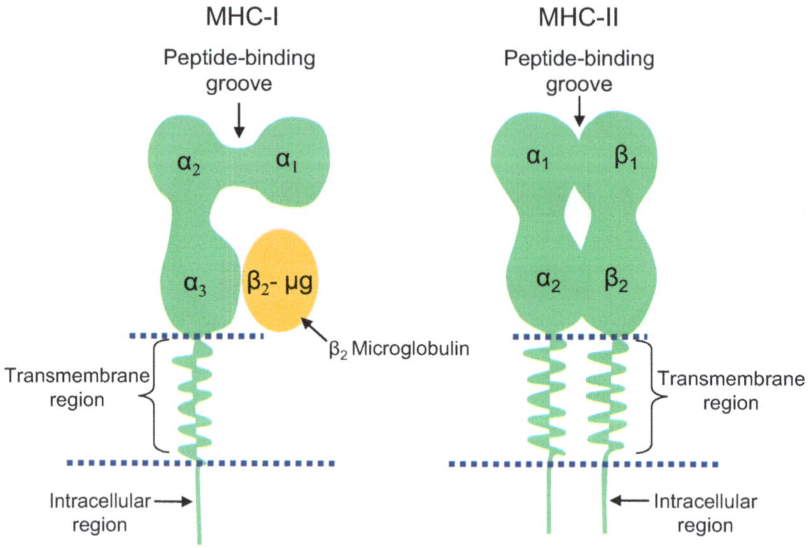

Fig. 7.5 Schematic diagram of the molecular composition of MHC-I and MHC-II molecules

which is encoded by a gene on a different chromosome, is noncovalently bound to the α chain and is required for the expression of MHC class I molecules. Without β2-microglobulin, MHC class I molecules are not expressed on the cell membrane. Experimental evidence that elucidates the critical role of β2-microglobulin in the expression of MHC class I molecules is that Daudi tumor cells cannot synthesize β2-microglobulin, and although tumor cells contain MHC class I genes that can be transcribed into mRNA and also translated into proteins, there is no MHC class I molecule alpha chain on the cell membrane. If Daudi tumor cells were transfected with the gene encoding functional β2-microglobulin, MHC class I molecules could appear on their cell membranes. This evidence strongly suggests that β2-microglobulin is important for the presence of MHC class I molecules on cell surface.

MHC class I molecules are expressed in all nucleated cells, with the most abundant on the surface of lymphocytes (1000–10,000 molecules per cell), with relatively few in liver, lung, and kidney, and even fewer in brain and muscle cells.

The functions of MHC class I molecules are mainly: (1) involved in the development and maturation of thymic CD8T cells; (2) involved in endogenous antigen presentation; (3) involved in transplant rejection, MHC class I molecules are the major antigens that induce immune response; (4) involved in cell-mediated immunity, MHC class I molecules are target molecules recognized by CTL (CTL killing target cells are MHC-restricted, i.e., they recognize their own MHC class I molecules by TCR to kill target cells).

7.2.2 Structure and Distribution of MHC-II Class Molecules

The α1 and β1 regions of MHC class II molecules form peptide binding grooves (regions) (Fig. 7.5) that bind and present short peptides of 13–18 amino acids. Analysis of the protein crystallographic structure indicates that MHC class II molecules are dimers of heterodimers composed of α/β chains. Two peptide binding clefts facing each other in opposite directions promote the binding of two T cell receptors (TCRs) and two CD4 molecules, which are required for signaling.

MHC-class II molecules are expressed only in antigen-presenting cells (B cells, dendritic cells, macrophages, etc.), thymic epithelial cells, vascular endothelial cells, etc. All cells expressing MHC class II molecules also express class I molecules.

The functions of MHC-II class molecules include: (1) participation in the development and maturation of CD4T cells in the thymus; (2) exogenous antigen presentation; (3) acting as a molecule recognized by CD4 T cells, and as a target molecule for CD4 recognition; (4) participation in immune response and immune regulation; and (5) participation in allograft rejection.

MHC-II molecules are also important target antigens responsible for causing transplant rejection and act as stimulatory antigens in graft versus host reaction (GVHR) and mixed leukocyte reaction (MLR) cultures, causing proliferation and differentiation of immune cells.

7.2.3 Gene Arrangement of MHC-I and MHC-II Molecules

Regardless of the number of introns in the genomic DNA encoding MHC-I or MHC-II molecules, the transcribed mRNA forms two complete strands (Fig. 7.6), whereas the β2-microglobulin, which constitutes the MHC-I molecule, is not on the same chromosome as the α strand.

7.2.4 Regulation of the Expression of MHC Molecules

The expression of MHC molecules is influenced by several factors, such as cytokines, pathogenic infections, and certain compounds. IFN-γ and TNF can enhance the expression of MHC class I molecules. IL-4 increases the expression of MHC class II molecules, while glucocorticoids and prostaglandins decrease the expression of MHC class II molecules. Viral infection can affect MHC expression (Table 7.2), but whether these phenomena are related to pathogenic mechanisms remains to be further investigated.

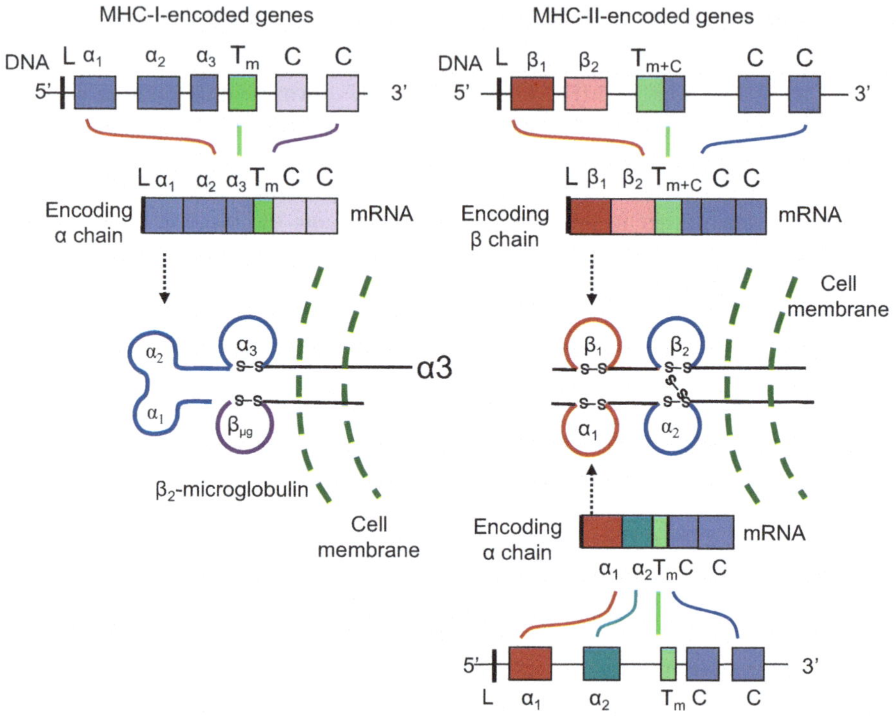

Fig. 7.6 Schematic diagram of transcription and translation of genomic DNA of MHC-I and MHC-II molecules

7.3 The Relevance of MHC to Medicine

Human organ transplantation requires histocompatibility testing to determine donor and recipient HLA. The key to transplant success is the degree of compatibility of the major histocompatibility antigens between the donor and recipient. Since HLA plays a role in antigen-presention, it determines the capability of adaptive immune response in individuals. Tumor cells escape immune surveillance by reducing HLA expression, and IFN-γ enhances the specific killing effect of CD8+ CTL cells by promoting the expression of HLA class I molecules in tumor cells.

Since HLA inheritance follows Mendelian law and is haplotyped, it is possible to conclude kinship in forensic medicine by testing for HLA. In addition, HLA typing is widely used in anthropological studies.

7.3.1 The Relationship Between MHC and Disease

Some diseases are highly correlated with the location of HLA alleles, such as certain autoimmune diseases, allergies, viral infectious diseases, and neurological disorders that may be genetically associated with HLA. By analyzing the association

Table 7.2 Effect of viral infection on MHC expression

Virus	MHC-I	MHC-II
Adenovirus (Adv)	↓	? (unknown)
Cytomegalovirus (CMV)	↓	↓
Infectious demodicosis virus (ectromedia virus, EV) in mice	↓	?
Hepatitis B virus (HBV)	↓	?
Herpes simplex virus (HSV)	↓	?
Human immunodeficiency virus (HIV)	↓	↑
Human papilloma virus 16 (HPV16)		↑
Measles virus (MeV)	↑	?
Moloney leukemia virus (MoMLV)	↑	?
Rous sarcoma virus (RSV)	↓	↓
Simian immumodeficiency virus (SIV)	↓	↑
Vaccinia virus (VACV)	↓	?
Vesicular stomatitis virus (VSV)	↓	?
West Nile virus (WNV)	↑	Virus MHC-I MHC-II Adenovirus (Adv) Cytomegalovirus (CMV) Infectious demodicosis virus (ectromedia virus, EV) in mice West Nile virus (WNV) ↑

between the frequency of a disease at an HLA allele and the incidence of the disease in the general population, the Relative Risk (RR) value for the disease at that allele is calculated and can be used as an indicator of disease inheritance (Table 7.3). If the RR value is equal to 1, the disease is not associated with the heritability of the HLA allele, and the larger the RR value, the greater the genetic association of the disease.

7.3.2 Relationship Between MHC and Tumors

Since CTL recognizes and kills target cells of its own MHC-presenting antigen, changes in MHC expression directly affect immune recognition and immune clearance. Since tumor cells have decreased or completely lost MHC expression, thereby escaping immune surveillance and clearance, a clinical lack of MHC expression in tumor cells is indicative of a possible poor prognosis for the condition. If the expression of MHC in tumor cells can be increased, the anti-tumor effect of the immune system can be enhanced, e.g. IFN-γ can increase the expression of HLA-I molecules in human tumor cells, and therefore enhance the specific killing effect of CTL on tumor cells. In addition to circumventing immune clearance by reducing MHC expression, tumor cells can also inhibit the killing effect of CTL through immune negative regulation of costimulatory molecules (immune checkpoint) (see Chap. 4), and how tumor cells evade immune surveillance and are able to develop has always been a key question to be addressed in medical research.

7.3.3 MHC and Organ Transplantation

In 1954, Drs. Merril and Murray of Harvard University performed the first successful kidney transplant between twin brothers, and the transplanted kidney survived long term without the use of any immunosuppressive drugs in the recipient. This was a historic breakthrough in organ transplantation and pioneered clinical organ transplantation, for which Drs. Merril and Murray were awarded the 1990 Nobel Prize in Medicine. In the late 1980s, with the improvement of surgical techniques, the development of transplantation immunology and the application of immunosuppressive drugs, the survival rate of clinical organ transplantation has been increasing, and now kidney transplantation has become a routine operation in transplantation centers of major hospitals in China. The survival rate of kidney transplant patients has reached the international advanced level.

Regardless of the type of organ transplant (kidney, liver, heart, lung, bone marrow), tissue matching is required prior to surgical operation. The most important test is HLA matching, requiring that the HLA between donor and recipient matches as closely as possible. The main clinical test is whether the HLA-I and HLA-II molecules of the organ transplant recipient match the donor. the HLA-I gene has

Table 7.3 Association between HLA alleles and genetic disorders

Name of disease	HLA alleles	RR	Disease characteristics and clinical manifestations
Ankylosing spondylitis	B27	90	The disease is an injurious inflammatory disease of the joints and periarticular tissues of the spine. It usually first affects the sacroiliac joints and then gradually involves the lumbar, thoracic and cervical vertebrae as the lesion develops, resulting in blurring of the small joint spaces, loss of fusion and destruction of vertebral osteophytes and ossification of ligaments, leading to spinal ankylosis or humpback status. The main symptoms are pain at the site of onset and limitation of movement, and it is a common disease of unknown etiology
Goodpasture's syndrome (pulmonary hemorrhage-nephritis syndrome)	DR2	16	The disease was first described by Ernest Goodpasture (1918) and is characterized by simultaneous pulmonary and renal involvement, displaying hemoptysis, pulmonary infiltrates, glomerulonephritis, and anti-glomerular basement membrane antibodies in the blood and tissues involved. (1) recurrent sudden hemoptysis and dyspnea with a self-limiting course, migrating infiltrative foci on lung X-ray and iron-containing macrophages in sputum; (2) long-term pulmonary bleeding
Gluten-sensitive enteropathy	DR3	12	This disease, also known as celiac disease, non-tropical stomatitis diarrhea and maltoprotein enteropathy, is a genetic disorder in which an allergy to maltoprotein that is not tolerated causes alterations in the small intestine, resulting in a digestive malabsorption syndrome. This genetic disorder is due to abnormal sensitivity of the patient to gluten. Gluten is found in wheat, rye, barley and oats. Gluten molecules bind to antibodies in the small intestine, flattening the brush-like surface layer of the normal small intestine and greatly diminishing the digestion and absorption of food. After avoiding gluten-containing foods, intestinal function is gradually restored
Hereditary hemochromatosis (hereditary hemochromatosis)	A3 B14 A3/ B14	9.3 2.3 90	The disease is a disorder of iron regulation caused by excessive iron absorption. Over time, iron overload and deposition in the tissues leads to various chronic diseases and premature death. It is the most common genetic disorder among Caucasians (Caucasians) in the United States. Early signs and symptoms of the disease: elevated liver enzymes (AST, ALT), reduced physical strength, stress, fatigue, joint pain, impotence, absence of menstruation. Late stage: grey or coppery skin, cirrhosis, liver cancer, heart disease/heart failure, diabetes, hypopituitarism, hypogonadism, chronic abdominal pain, severe fatigue, increased rate of bacterial infections, and splenomegaly

(continued)

Table 7.3 (continued)

Name of disease	HLA alleles	RR	Disease characteristics and clinical manifestations
Narcolepsy	DR2	130	The disease is a disorder characterized by irresistible short-term episodes of sleep. It tends to start in childhood or young adulthood and has a similar incidence in both sexes. Some patients may have a history of encephalitis or craniocerebral trauma. The pathogenesis is not yet clear, but may be related to reduced function of the brainstem reticular superior activating system or hyperfunction of the pontocaudal reticular nucleus. Most patients have other symptoms such as sudden collapse disorder, sleep paralysis, and sleep hallucinations, collectively referred to as the tetralogy of sleep attacks
Reiter's syndrome	B27	37	The disease is multi-systemic and was first reported in a male with urethritis, conjunctivitis, dermatitis and arthritis following dysentery. Since then, the disease has been reported following epidemics of intestinal infections such as dysentery, while in countries such as the United Kingdom and the United States it is thought to be associated with gonorrhea and non-gonococcal urethritis, with the most prominent manifestations being urethritis, conjunctivitis, arthritis and skin lesions, which can occur either simultaneously or sequentially
Sjogren's syndrome	DW3	6	It is a systemic autoimmune disease related to the exocrine glands, mainly invading the lacrimal and salivary glands, manifesting dry keratoconjunctivitis, oral dryness and other rheumatic diseases such as rheumatoid arthritis, and causing damage to the respiratory tract, digestive tract, urinary tract, blood, nerves, muscles, joints, etc.
Type I- diabetes (insulin-dependent) (diabetes mellitus)	DR4/ DR3	20	The disease is an autoimmune disease caused by self-reactive T cells attacking insulin-producing beta cells
Multiple sclerosis	DR2	5	Multiple sclerosis (MS) is a disease of the brain and spinal cord in which the self-immune system attacks the myelin sheath of nerve fibers, resulting in interruption or disruption of nerve conduction. The clinical signs of the disease include multiple sclerosis, muscular dystrophy or poliomyelitis, which can lead to disability
Myasthenia gravis	DR3	10	The disease is an autoimmune disorder in which transmission at the neuromuscular junction is impaired due to a decrease in acetylcholine receptors. Antibodies to the acetylcholine receptor (AChR) are the main autoantibodies responsible for its development. The anti-ACHR antibody bind to ACHR and block transmission at the neuromuscular junction, resulting in weakness of the eye muscles, swallowing muscles, respiratory muscles and skeletal muscles of the limbs. This means that the nerves that innervate muscle contraction are unable to transmit "signal commands" to the muscles properly due to a variety of causes, causing the muscles to lose contractile function. The etiology of the disease is systemic, but the affected muscles have different clinical manifestations depending on the focus

Rheumatoid arthritis	DR4	10	Rheumatoid arthritis (RA) is a commonly-seen autoimmune disorder characterized by acute or chronic connective tissue inflammation that can recur and involve the heart. It manifests clinically as wandering joint and muscle aches and pains
Systemic lupus erythematosus	DR3	5	The disease is an autoimmune disease in which the patient's immune system attacks his or her own chromosomal DNA or histone proteins, involving multiple systems and organs of the body, and multiple autoantibodies can be found in the patient's blood and organs. In some patients with lupus erythematosus due to genetic factors, the immune response is triggered by certain external or internal factors, such as viral infection, sun exposure, psychological trauma, drugs, pregnancy, etc., resulting in the production of multiple antibodies and immune complexes against their own tissues, causing damage to the body's own tissues, such as skin, blood vessels, heart, liver, kidney, brain and other organs and tissues

three loci, A, B, and C, which encode the antigenic molecules HLA-A, HLA-B, and HLA-C, respectively (Fig. 7.1). Whether these three loci match or not is associated with long-term survival of the transplanted organ. HLA-II genes include loci such as DR, DP, and DQ, among which HLA-DR is closely related to the survival time of the transplanted organ. In renal transplantation, the order of importance of histocompatibility matching is: HLA-DR > HLA-B > HLA-A and HLA-C loci, the more loci are identical, the better. In organ transplantation, in addition to HLA matching, ABO blood type matching, complement dependent cytotoxicity (CDC), and panel reactive antibody are also examined, and results are required before organ transplantation.

The ABO blood type matching for organ transplantation should follow the principle of blood transfusion, and it is best to transplant organs of the same blood type. If the ABO blood type does not match, rejection reactions of varying degrees are likely to occur and such transplantation should be avoided.

The complement-dependent lymphocyte cytotoxicity test, also known as the cross-match or cross-cytotoxicity test, detects the specific antibodies against donor lymphocyte HLA in recipient blood by mixing the recipient's serum with the donor's lymphocytes and observing for cell lysis in the presence of complement. If the lymphocytes of the donor are killed, the reaction is judged to be positive, indicating that the recipient serum has antibodies against the HLA of the donor lymphocytes and that the mortality of the lymphocytes represents the antibody strength. If the cross-cytotoxic reaction is positive, acute rejection will occur after transplantation, resulting in graft failure.

Panel reactive antibody (PRA) test is used to determine whether a recipient is in a hypersensitive immune state as a predictor of acceptance of HLA antigens from the transplanted organ. Based on experiences of clinical doctor (Qian and Yuan 2012), the sensitivity of recipient immune reaction to donor organ is: $PRA \leq 10\%$ considered as insensitive; PRA 11–50% moderately sensitive; $PRA \geq 50\%$ highly sensitive; $PRA \geq 80\%$ extremely sensitive. If the panel reactive antibody test result is moderately sensitive or above, it is clinically prone to acute rejection, and especially extremely-sensitive is contraindicated in transplantation.

The general procedure for organ transplantation is that the recipient awaiting transplantation is first HLA-typed and the data entered into a database. When a donor is available, the donor's HLA is then HLA-typed, and the data is then matched manually or by computer software using database information to screen a matching donor and recipient for transplant surgery.

7.4 Antigen Processing and Presentation

T cells recognize antigenic peptide-MHC complexes via TCR. The process by which antigenic proteins is degraded into peptides that bind to MHC and form complexes is called antigen processing. The process of transporting the antigenic peptide-MHC complex to the surface of the cell membrane for presentation is called antigen presentation.

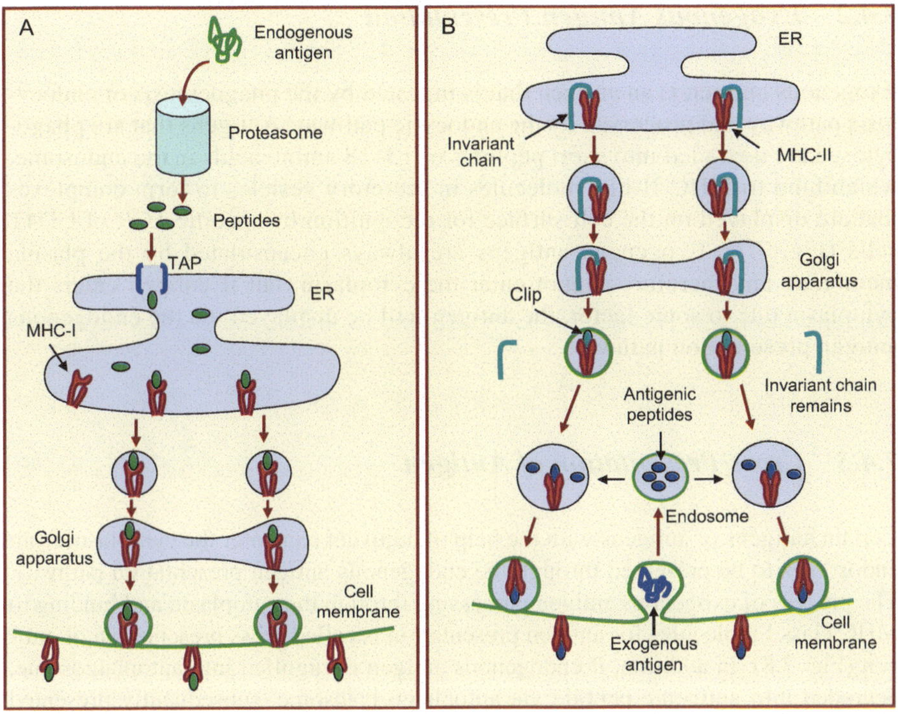

Fig. 7.7 Schematic diagram of endogenous and exogenous antigen presentation. (**a**) Endogenous antigen presentation pathway, also known as the cytoplasmic pathway, presents antigens in the cytoplasm, which are ubiquitinated, processed by the proteasome, and degraded into short peptides of 8-11 amino acids, which enter the endoplasmic reticulum (ER) through TAP and bind to MHC class I molecules to form antigenic peptide-MHC I complex and are displayed on the cell membrane. (**b**) Exogenous antigen presentation pathway, also called the endocytic pathway. MHC-II molecule in the ER binds to an invariant chain for the protection of antigen-binding groove. When the secretory vesicle fuses with the endosome, the antigenic short peptide replaces the protective clip of the invariant chain and enters the antigenic binding groove, and the secretory vesicle fuses with the cell membrane to present the antigenic peptide-MHC-II complex. *TAP* transporter associated with antigen processing

7.4.1 Endogenous Antigen Presentation

Endogenous antigens are antigens that are produced and processed (degraded by the proteasome pathway) within the cell, including proteins produced by normal cells, tumor proteins, and proteins produced by viruses or bacteria within infected cells. The endogenous antigen presentation pathway, also known as the cytosolic pathway, must present antigens within the cytosol. After ubiquitination, intracellular antigens are processed by the proteasome and degraded into 8–11 amino acid antigenic peptides, which enter the endoplasmic reticulum through TAP (transporter associated with antigen processing) and bind to MHC-I molecules to form antigenic peptide-MHC-I complex, which is displayed on the cell membrane surface to be recognized by the specific TCR of CD8+T cells (Fig. 7.7).

7.4.2 Exogenous Antigen Presentation

Exogenous antigen is an antigen that is ingested by the phagocytosis or endocytosis pathway and processed by the endocytic pathway. Antigens that are phagocytosed are degraded into short peptides of 13–18 amino acids in the endosome, which bind to MHC-II-like molecules in secretory vesicles to form complexes that are displayed on the cell surface for recognition by specific TCR of CD4T cells (Fig. 7.7). Exogenous antigens are always encapsulated by the plasma membrane and therefore cannot enter the cytoplasm, but if antigen enters the cytoplasm due to some factor, the antigen will be displayed via the endogenous antigen presentation pathway.

7.4.3 Cross Presentation of Antigen

Certain antigens or antigens with the help of adjuvant can enter the cytoplasm from endosomes to be presented through the endogenous antigen presentation pathway. The process of exogenous antigen processing through the cytoplasm and binding to MHC class I molecules for antigen presentation is called cross presentation of antigen (Fig. 7.8). In addition, if endogenous antigen is engulfed into autophagosome, degraded into antigenic peptide via autophago-lysosome, subsequently presented by MHC-II molecules for recognition by CD4T cells, this process is also called cross presentation of antigen.

7.4.4 CD1, MRI and Antigen Presentation

CD1 and MRI molecules have a similar structure to MHC class I molecules and also have antigen-presenting functions.

CD1 exclusively presents lipid or glycolipid antigens, which are recognized by the NKT or γδ T-cell receptor (TCR), and there are different subclasses of CD1 molecules, CD1a, CD1b, CD1c, CD1d molecules, etc., responsible for the presentation of Mycobacterium tuberculosis cell wall antigens, gangliosides, and certain phospholipids, synthetic glycolipids (α galactosylceramide, α-GalCer), etc.

MRI (MHC-I like related molecule) is expressed in all mammalian cells, is highly conserved across species, and has a highly similar molecular structure to MHC-I and CD1 molecules, but has a distinctly different antigen-binding groove. MRI has been found to bind and present mainly vitamin metabolites or derivatives, such as metabolites of vitamin B9 (folic acid) and bacterial derivatives in the synthesis of vitamin B2 (riboflavin). MRI presents antigen molecules to mucosal-associated invariant T cells (MAIT). The TCR of MAIT contains an invariant alpha chain (Vα7.2Jα33-) Vβ2 or Vβ13 to form TCR

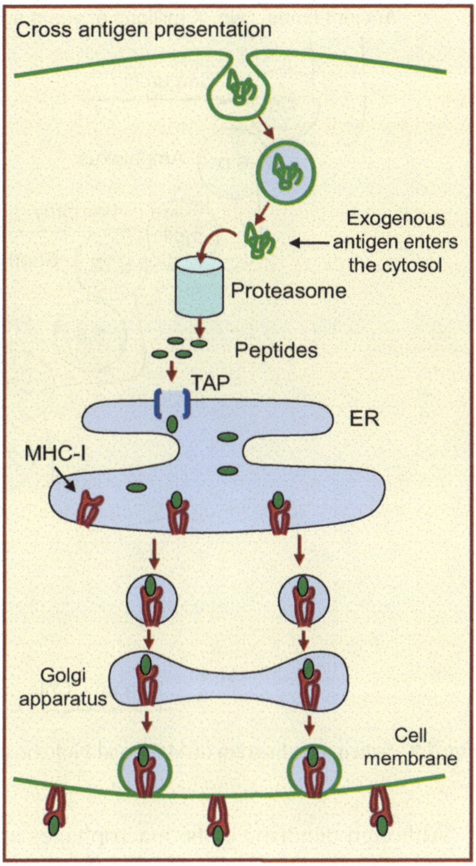

Fig. 7.8 Schematic diagram of cross presentation of antigen

antigen recognition region that recognizes the vitamin metabolite or derivative presented by MRI.

7.4.5 Classification of Antigen-Presenting Cells

Since all nucleated cells express MHC class I molecules, they can all present endogenous antigens. There are cells that express not only MHC class I and II molecules but also the costimulatory molecule B7, thus these cells are called professional antigen-presenting cells (APCs), including B cells, dendritic cells and macrophages. Other cells that transiently express MHC class II and co-stimulatory molecule B7 besides MHC class I molecules at certain times are called non-professional antigen-presenting cells, including fibroblasts in the skin, microglia in brain tissue, beta cells in the pancreas, thymic epithelial cells, thyroid epithelial cells, and vascular endothelial cells.

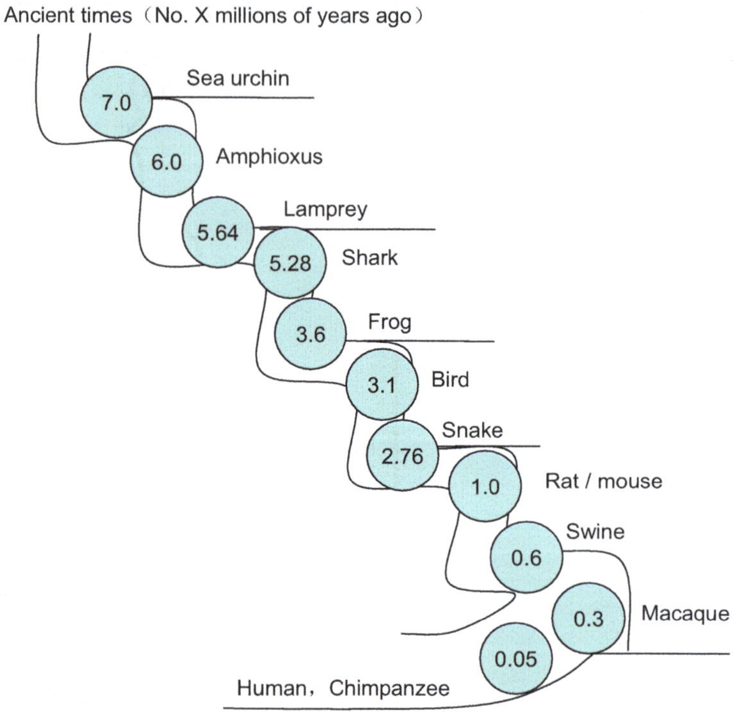

Fig. 7.9 Schematic diagram of MHC and biological evolution

Although dendritic cells, macrophages and B cells are all professional antigen-presenting cells, each of them have their own characteristics. For example, dendritic cells are the most efficient antigen-presenting cells, and these cells express high levels of MHC-II molecules and the costimulatory molecule B7 in the normal state, so they can activate the naive Th cells; macrophages express MHC-II molecules and the costimulatory molecule B7 only after they have phagocytosed the antigen; B cells, although expressing MHC-II molecules in the normal state, they must be activated before expressing the costimulatory molecule B7.

7.5 MHC and Evolution

Biological evolution can be analyzed based on the haplotype inheritance and polymorphism of MHC molecules. MHC gene sequence analysis indicates that the earliest organism found to have an MHC molecule is the sea urchin, which is approximately 7 million years old. Primate MHCs appeared about 50,000 years ago (Fig. 7.9).

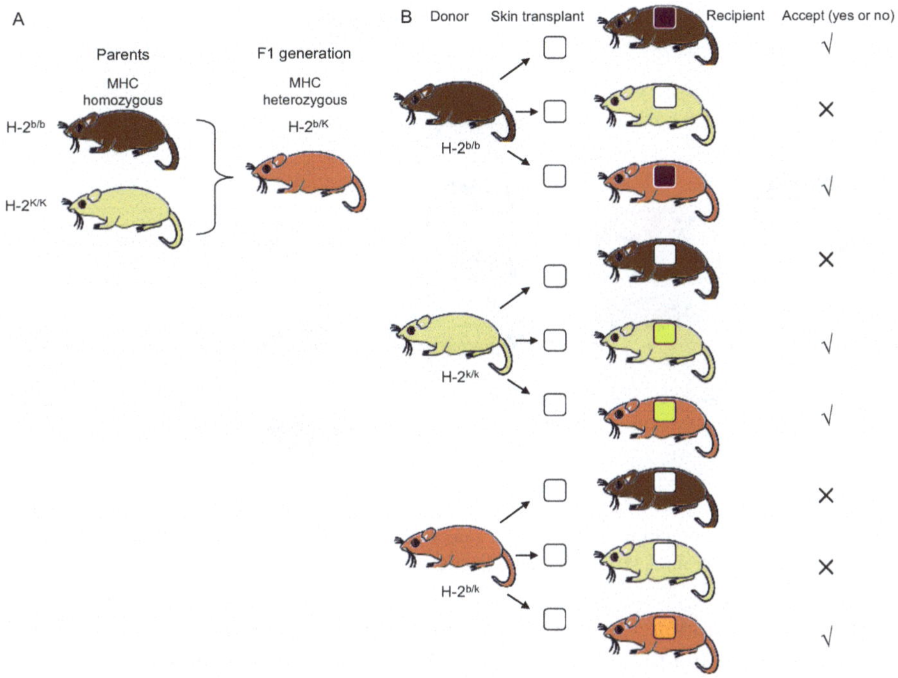

Fig. 7.10 Schematic diagram of skin graft experiments in inbreed mice. (**a**) Mouse parents H-$2^{b/b}$ and H-$2^{k/k}$ homozygotes cross to produce F1 generation, heterozygotes H-$2^{b/k}$; (**b**) Because MHC is co-dominantly expressed, the F1 generation expresses both H-2^b and H-2^k, and can therefore accept transplanted organs from both parents, but both parents cannot accept any organs donated by F1 offspring, resulting in rejection because the donor organ has half of the non-self genes expressed. Rejection of grafted skin leads to an inflammatory response in the form of a stiff skin flap with inadequate blood supply, ulceration, or even detachment

7.6 MHC and Experimental Findings of Antigen Presentation

7.6.1 Tissue Transplant and Rejection Experiments

The skin grafting experiments in inbreed mice are shown in Fig. 7.10.

7.6.2 The Experiment for Genetic Selection of Mouse Strain B (H-$2^{k/k}$) to Obtain the Genetic Background of Strain A (H-$2^{b/b}$) but Retaining the H-$2^{k/k}$ Inbreed Characteristics

This experiment used the principle of skin graft rejection in different strains of mice for genetic selection (Fig. 7.11).

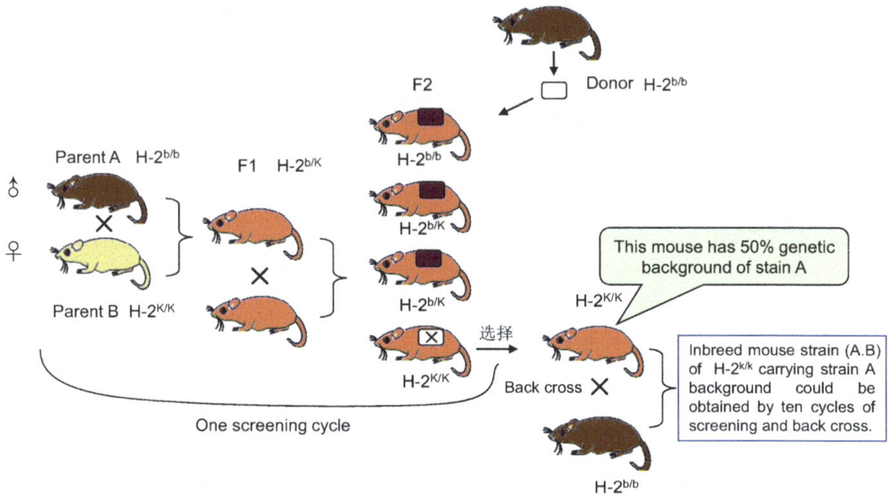

Fig. 7.11 Schematic diagram of the genetic selection of strain B mice retaining H-$2^{k/k}$ homozygotes and obtaining strain A (H-$2^{b/b}$) genetic background

In the F2 generation, the mice that could accept the skin graft of A inbreed mice were eliminated, and the mice that rejected the skin graft of A inbreed mice were selected to be backcrossed with A inbreed mice.

7.6.3 Experimental Findings that T cell Activation Requires Antigen Processing and Presentation

As early as 1959, immunologists realized that T and B cells might recognize antigens by different mechanisms, but at that time it was always believed that the immune system recognized only the molecular structure of intact natural antigens, so when P. G. H. Genll and B. Benacerraf discovered in their experiments that antigens, whether denatured or not could elicit a secondary cell-mediated immune response, whereas denatured antigens could not elicit a secondary antibody immune response, the antigen molecule could elicit a secondary antibody immune response only when it retained its natural structure. This discovery was not taken seriously, only considered as an interesting phenomenon at that time, and was completely ignored by the academic community until 1980, when its importance was recognized.

At that time, it was thought that T and B cells recognized antigens by similar mechanisms, both recognizing natural antigenic structures, but K. Ziegler and E. R. Unanue found in their experiments that if APC was inactivated with paraformaldehyde beforehand, the antigen could not activate CD4T cells, but if the antigen was mixed with APC for 1 h and then inactivated, CD4T cells could be activated (Fig. 7.12a, b). This result confirms that APC is required in processing antigen to

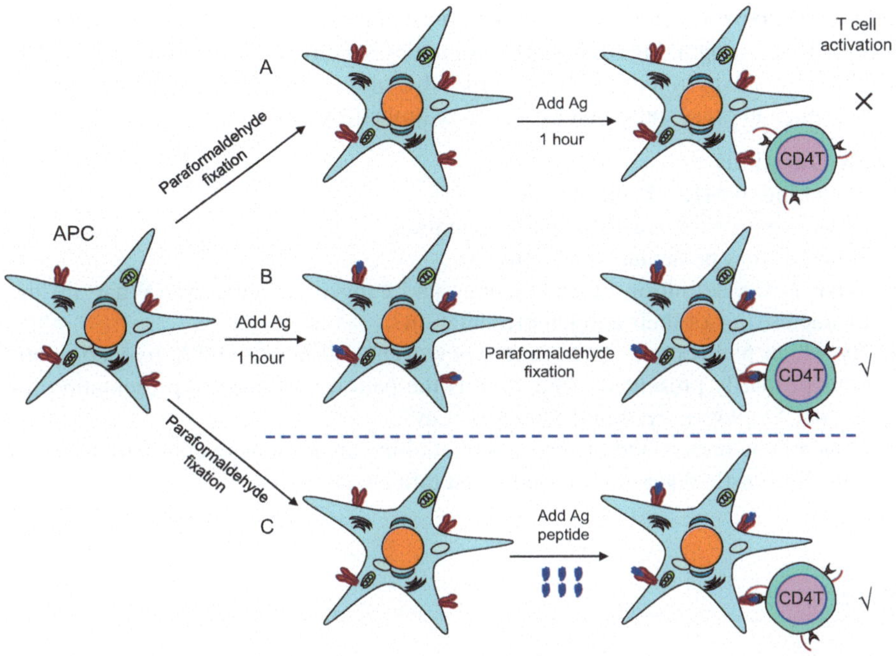

Fig. 7.12 Schematic diagram of the role of antigen processing and presentation by APC in activating CD4 T cells. (**a**) Paraformaldehyde-fixed APC lost the function of antigen processing and presentation, and CD4T cells could not be activated. (**b**) Antigen added to APC culture was phagocytosed by APC and degraded into short peptides displayed on the cell surface within 1h, and the antigen presented by MHC after paraformaldehyde-fixation could still activate CD4T. (**c**) Antigenic peptide could be directly loaded on APC MHC-II molecules to form a complex to activate CD4T cells

activate CD4T cells. Subsequently R. P. Shimonkevitz confirmed in his experiments that antigenic short peptides could be loaded directly onto APC without the need for phagocytosis and processing (Fig. 7.12c).

7.6.4 Experimental Findings of Two Pathways of Antigen Processing and Presentation

In their study, L. A. Morrison and T. J. Braciale found that specific CTL recognizing MHC-I-HA killed only influenza virus-infected target cells, whereas specific CTL recognizing MHC-II-HA killed both influenza virus-infected target cells and target cells treated with inactivated influenza viruses. If emetine is added to influenza virus-infected target cell cultures to inhibit viral protein synthesis, target cells can be recognized and killed by MHC-II-HA-specific CTL, but not by MHC-I-HA-specific CTL. If chloroquine is added to influenza virus-infected target cell cultures to block endocytotic pathway of antigen presentation, target cells can be killed by specific CTL recognizing MHC-I-HA but not by specific CTL of MHC-II-HA. The results of this experiment established that the endogenous and exogenous antigen processing and presentation pathways were quite different.

Study Questions

1. Sketching of MHC-I/II molecules.
2. What is antigen processing and presentation?
3. What is cross presentation of antigen?
4. What is the definition of endogenous and exogenous antigens? What are the characteristics of their antigen presentation?
5. By which pathway are *Listeria monocytogenes*, *Brucella* and *Salmonella* antigens primarily presented? And what is the pathway of antigen presentation for *E. coli*, *Staphylococcus* and *Streptococcus*?
6. How do you understand the process of eliciting an immune response to infection with Newcastle disease virus and avian influenza virus?

Reference

Qian Y, Yuan M (2012). Practical complete book of kidney transplantation. Beijing: People's Military Medical Publishing House

Chapter 8
Adaptive Immunity and Immune Control

Overview:
Adaptive immunity, also known as acquired immunity, is the specific immunity formed by antibodies and effector T cells produced by the immune system in response to antigenic stimulation. In the process of adaptive immune response, the immune system exhibits the characteristics of specific immune recognition and memory. Involved in this process are T cells, B cells, and antigen-presenting cells as well as various types of cytokines. Immune control is a preventive or therapeutic measure (e.g. vaccination or use of immunological preparations) against an epidemic using immunological methods.

8.1 The Adaptive Immune Response

The adaptive immune response is a series of immune responses of the immune system in response to stimulation by antigenic substances (e.g., vaccines), including antigenic recognition, antigen processing and presentation, activation of T and B lymphocytes, production of cytokines, antibodies and effector T cells, and antigen clearance. The adaptive immune response of the body includes both humoral and cell-mediated immune responses. The animal establishes resistance to a certain pathogen through the immune response, for example, vaccination is to make the animal produce an immune response and establish immunity to resist a specific pathogenic infection, to achieve the purpose of immune control.

Higher organisms have both innate and adaptive immunity, whereas lower organisms have only innate immunity, and the adaptive immunity possessed by higher organisms is the result of "survival of the fittest" evolution. Thus, adaptive immunity is more efficient and specific, exhibiting orientation of immune response toward humoral immune or cell-mediated immune response depending on the mature of invading pathogens. For example, cell-mediated immunity is the main immune

S. J. Zheng, *Veterinary Molecular Immunology*,
https://doi.org/10.1007/978-981-99-8929-4_8

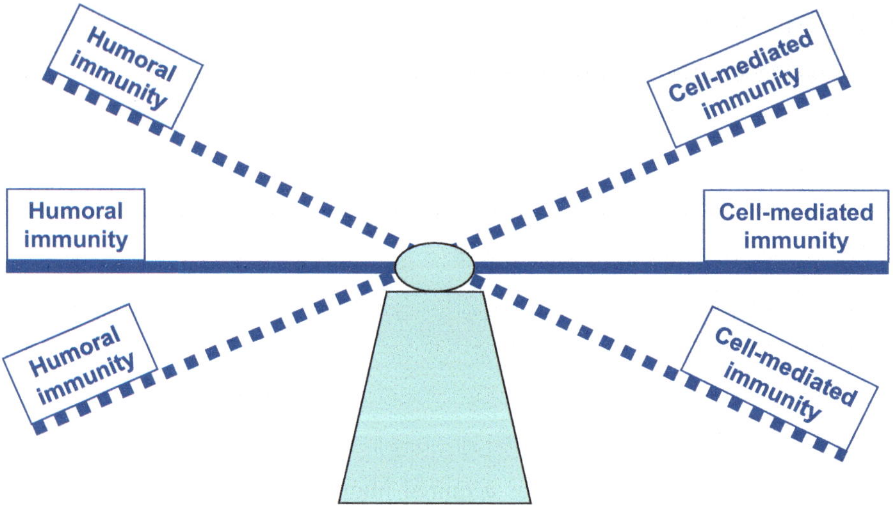

Fig. 8.1 Balanced relationship between humoral and cell-mediated immunity

response to intracellular pathogens (viruses, intracellular bacteria), while humoral immunity is the main immune response to extracellular pathogens. Cell-mediated immunity and humoral immunity are often in balance, just like a seesaw, regulating each other, with one side being high and the other side being low (Fig. 8.1).

8.1.1 Humoral Immunity

Humoral immunity is the entire physiological response process in which B cells recognize antigens through BCR receptors for activation, proliferate and differentiate into plasma cells to produce antibodies, and neutralize and remove antigens through antibodies. Humoral immunity is also known as antibody immunity because the primary mediator substance for antigen clearance is antibody. The primary step in this process is antigen recognition by B cells for activation (i.e., clonal selection; see Chap. 5 for details).

8.1.1.1 Primary Immune Response

Primary immune response refers to the immune response caused by the first entry of an antigen into the body to stimulate the immune system. The entire process of primary immune response includes: antigen recognition, antigen presentation, specific lymphocyte activation (clonal selection and expansion), antibody production, and antigen clearance, etc. Therefore, a certain latency period is required. Three to 4 days after exposure to antigen, IgM could be detected in the blood, subsequently isotype class switching occurs to produce IgG. Thus the total concentration of antibodies gradually increases and the antigen is gradually cleared. After the antigen is cleared, the antibody-producing plasma cells enter the apoptotic phase, and the concentration of antibodies in the blood gradually decreases over time, eventually approaching normal levels, with the formation of memory cells persisting in the body.

8.1.1.2 Secondary Immune Response

When an animal that has undergone a primary immune response is stimulated again by the same antigen, an unusually strong and long-lasting immune response is produced, a process called the secondary immune response, also known as the anamnestic response (Fig. 8.2). Because the animal contains a certain number of memory cells after the primary immune response, the same antigen, when re-entering the body, is immediately recognized by memory cells that proliferate and produce antibodies. This phase has a short induction latency, a rapid rise in antibody concentration in the blood, a long duration of antibody peaks, and the predominant isotype of antibody is IgG.

The antigen stimulates the body (e.g. vaccination with vaccine A) to induce a primary immune response, and after clearing the antigen, the body would be recovered. If the animal is vaccinated with a combined vaccine (vaccine A + vaccine B), the animal produces secondary immune response (anamnestic response, also called recall response) to vaccine A, while still producing a primary immune response to vaccine B.

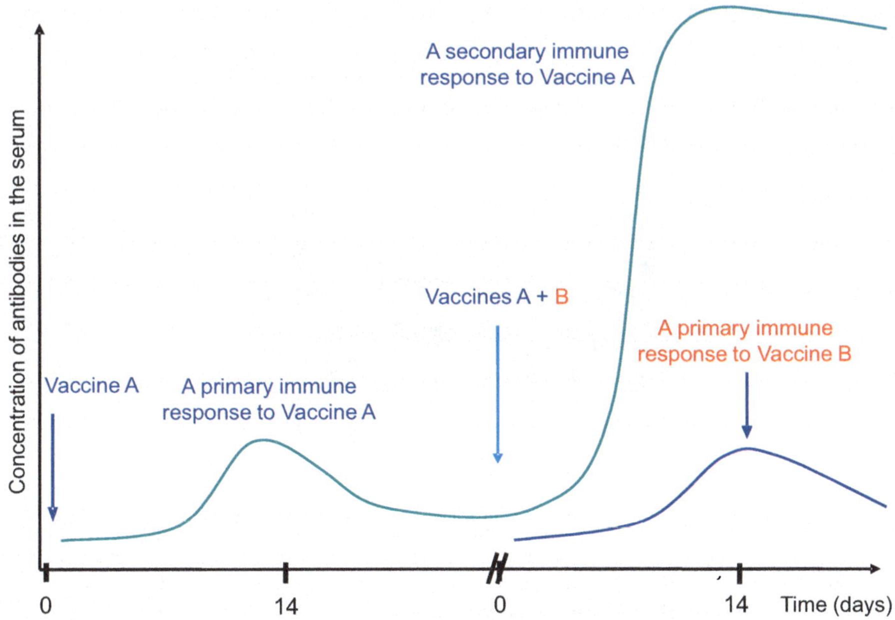

Fig. 8.2 Schematic diagram of the primary and the secondary immune responses

8.1.2 Cellular Immunity

Cellular immunity, also known as cell-mediated immune response, is a physiological response process in which T lymphocytes recognize MHC-antigen complexes for activation, proliferate and differentiate into effector T lymphocytes that perform killing of target cells, and this process is mainly carried out by CD4T and CD8T cells.

When endogenous antigens (e.g., viruses, intracellular parasites, or protozoa) are presented on the cell surface via MHC class I molecules, they are recognized by the T-cell receptor (TCR) of CD8T cells, which, with the stimulation by costimulatory factors, is activated (Fig. 8.3) and proliferate to generate cytotoxic T cells (CTL) that have a powerful ability to kill target cells. CTL kill target cells by killing mediators such as granzyme, perforin, FASL, TRAIL, TNF, IFN-γ. The initiating signal for CD8T cell activation is the recognition by the TCR of antigenic peptides presented by MHC class I molecules, while its CD28 molecule acts as a co-stimulatory molecule and binds to B7 of the APC to form a co-stimulatory signal as the second activating signal, and only when these two signals exist working together are CD8T cells completely activated. Activated CD8T cells (CTL) further proliferate and perform killing tasks, however, CTL cannot proliferate indefinitely. Activated CTL are induced to express the negative costimulatory factor CTLA4, which bind to B7 with a much greater affinity than that of CD28 to B7, so the negative regulatory signal formed by CTLA4-B7 binding acts as "a brake" to slow down CTL proliferation, avoiding over-reaction and immune damages. The activation of CD4T cells is similar to that of CD8T cells (see Chap. 4).

CTL killing of target cells is MHC-restricted, i.e., CTL only kills target cells of its own MHC-presenting antigens (see Chap. 4). Because antibodies can only directly bind antigens in body fluids (act by neutralizing, opsonization, ADCCs, etc.) and cannot enter the cell through the cell membrane to kill pathogens, cellular immunity is important for the clearance of intracellular bugs (viruses, intracellular bacteria, parasites or protozoa), and CTL releases pathogens by killing target cells

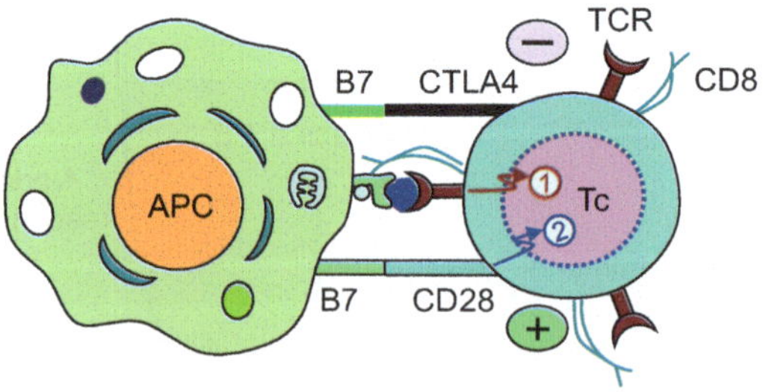

Fig. 8.3 Schematic diagram of CD8T cell activation. CD8T cells obtain the first signal for activation through the recognition of antigenic peptides presented by MHC-I class I molecules by the TCR, and the second signal is formed by the interaction of co-stimulatory molecule CD28 with B7, and both signals act together to completely activate CD8T cells. Activated CTL are induced to express CTLA4, and the B7-CTLA4 interaction forms a negative regulatory signal that suppresses CTL action

to destroy the environment on which intracellular pathogens depend for survival and proliferation, thereby facilitating antibody neutralization of pathogens, promoting phagocytosis of phagocytes (antibody-mediated opsonophagocytosis) and clearance of pathogens (Fig. 8.4).

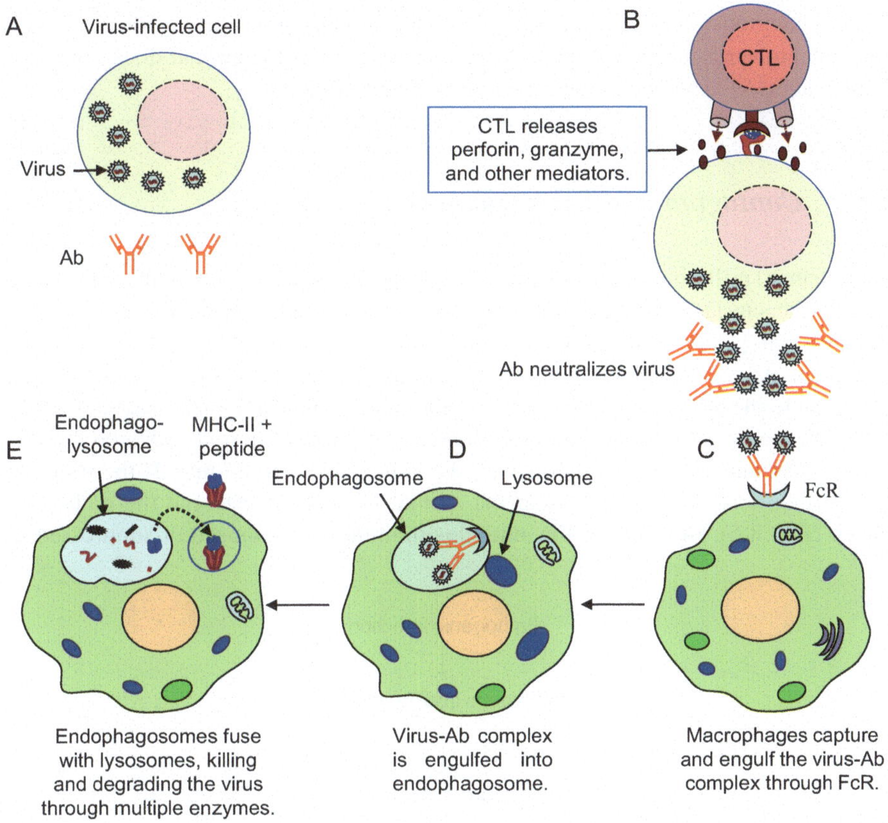

Fig. 8.4 Schematic illustration of the basic process by which intracellular pathogens (in the case of viruses) are killed and cleared. (**a**) Antibodies are not able to enter the cell to neutralize the virus; (**b**) CTL recognize the antigen presented by MHC-I and release perforin, granzyme, and other mediators, killing virus-infected target cells of self-MHC to release the virus, and antibodies bind the released and free virus for neutralization; (**c**-**e**) Macrophages capture and engulf the virus-antibody complex through FcR, and endophagosomes fuse with lysosomes to form endophagosomes-lysosomes, which kill and degrade the virus through multiple enzymes. The antigenic epitopes of viral proteins after degradation are presented to the cell surface via MHC-II molecules

8.2 Pathogenic Infections and Herd Immunity

8.2.1 The Concept of Contagion

The process by which microorganisms invade the animal and cause a series of pathological reactions of the animal is called contagion. If there are clinical signs, it is called contagious disease. If it is infectious, the contagion is taken as infection, and the contagious disease is called an infectious disease. If the animal does not show clinical signs, it is a subclinical infection or latent infection.

8.2.2 Contagion and Consequences

More than 30 million species of various organisms have been found to be infectious, including bacteria, viruses, protozoans (such as coccidia), arthropods, worms, etc. However, one of the reasons we do not feel the occurrence of so many infectious diseases in daily lives is that many pathogens invade the organism without posing a life threat to the animal population, the animal may be in a subclinical state of infection. The pathogen causing the subclinical infection is quickly cleared by the immune system, and as a consequence, the animal gains immunity. If the immune system fails to clear the pathogen, the animal is likely to become the pathogen-carrier (Fig. 8.5). However, if the animal is immunocompromised or suppressed, it

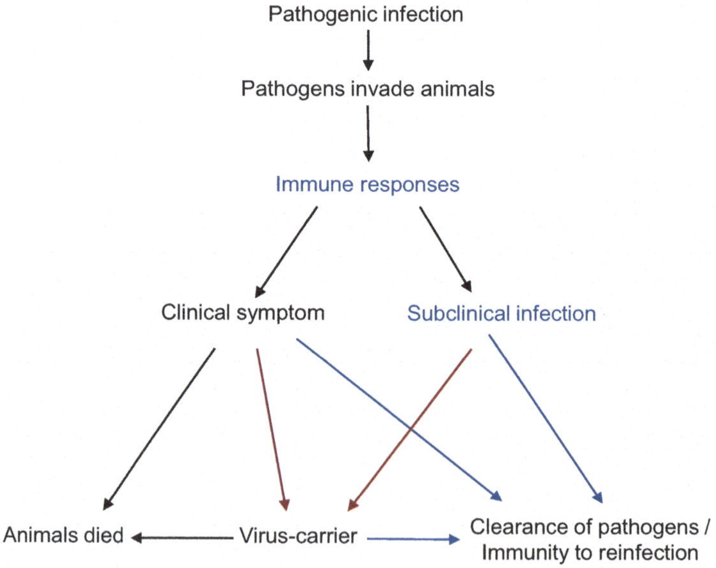

Fig. 8.5 Schematic diagram of the consequences of pathogenic infection

is likely that subclinical infection will display clinical signs and cause a new outbreak of clinical disease. If the animals display clinical signs after pathogenic infection, the animals that recover by relying on the immune system to clear the pathogen will gain immunity to reinfection by the same pathogen. If the immune system fails to clear the pathogen, the animal dies and, even if alive, becomes a carrier of the pathogen and continues to shed pathogens to the environment.

8.2.3 The Relationship of Immunity with Diseases

Immunity is the resistance of the animal to pathogenic microorganisms. Animals have developed natural anti-defense ability over a long period of evolution. Furthermore, animals acquire specific resistance to certain diseases as a result of the exposure of individual animal to external factors (pathogens and their products). Herd immunity is the resistance of a herd to a particular epidemic and determines whether a disease breaks out and spreads among animals (Fig. 8.6).

Although improving animal herd immunity is beneficial for the prevention and control of animal diseases, experimental evidence indicates that production performance will not be fully excerted when animals are in a high immune state (Fig. 8.7), so development of proper immunization program in control of environmental pathogens is fundamental and prerequisite for disease prevention and control to ensure optimal animal production performance (Fig. 8.8).

The genetic, nutritional and immune status of the animal directly affects the resistance of the organism to infectious diseases, however, the virulence and quantity of pathogens (bacteria, viruses, fungi, etc.) in the environment struggle with the animal's resistance and pose a constant threat to animal health, so effective control of pathogens in the environment is fundamental and prerequisite for successful prevention and control of infectious diseases.

8.3 Vaccines and Immunization Control of Infectious Diseases

8.3.1 Types of Immunization

Immunization of animals is divided into active and passive immunization, the later referring to the administration of antibodies (such as antisera, yolk antibodies, etc.) and the former referring to the vaccination or natural pathogenic infection. Vaccination is one of the most important means of preventing and controlling infectious diseases of animals and can improve the specific resistance of animals to diseases.

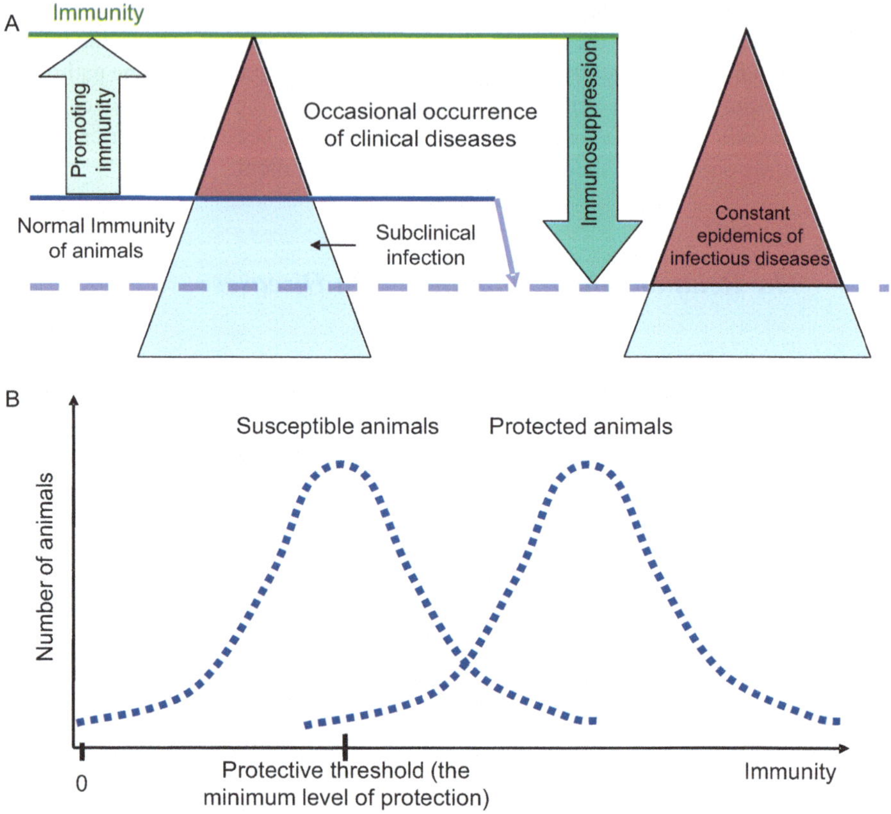

Fig. 8.6 Schematic illustration of the relationship of herd immunity with diseases. (**a**) Normally, the resistance of animals to a disease is shown in normal distribution (bell curve), and there are always a small number of animals with low resistance to the disease. If a pathogen is introduced, animals with low resistance to diseases will be susceptible to diseases. Herd-specific immunity can be improved by immunization, and if most animals have specific immunity, the animal population remains stable. However, if immunosuppression occurs in the herd, it leads to a decreased function of the immune system, making the animals susceptible to more pathogens, and the herd is constantly afflicted with infectious diseases, allowing epidemics to occur. (**b**) As to a certain virulent pathogen, the population of susceptible animals has low herd immunity, and is prone to disease epidemics. When herd immunity is improved by vaccination, most animals acquire specific immunity, and only a few individual animals have a resistance below the threshold (the protective level), so that no large-scale epidemic occurs even if individual animals had diseases

8.3.2 Types of Vaccines and Immune Responses

There are many types of vaccines, mainly including live attenuated live vaccines, inactivated vaccines, genetically engineered vaccines, subunit vaccines, synthetic peptide vaccines, toxoid vaccines, DNA vaccines, mRNA vaccines, etc. The type of immune responses induced by different types of vaccines varies (Table 8.1), and the

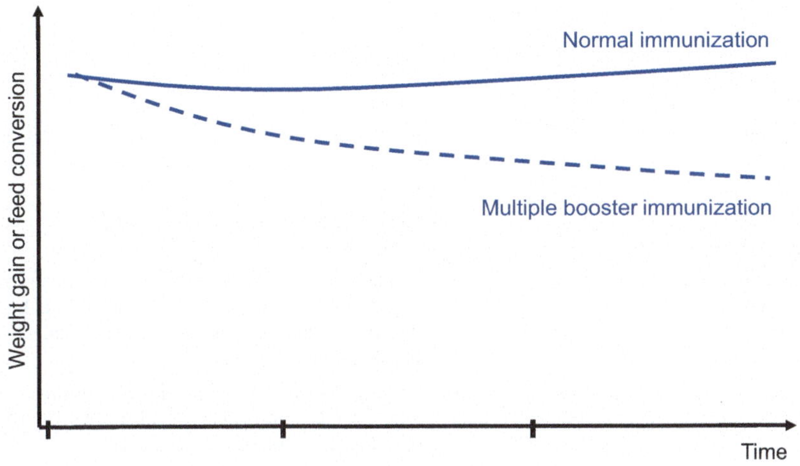

Fig. 8.7 Schematic diagram of the relationship of enhanced immunization with pig feed conversion and weight gain

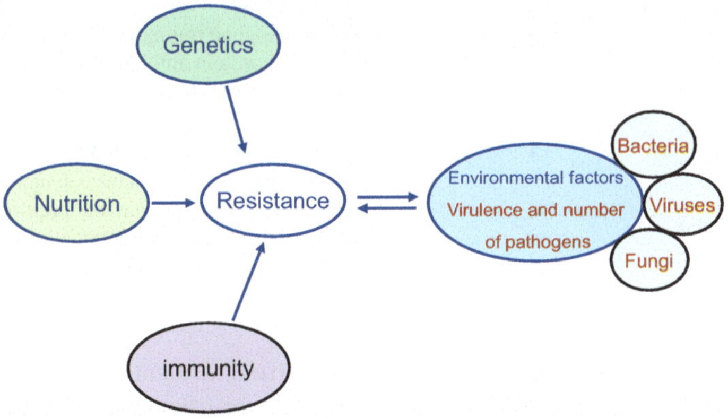

Fig. 8.8 Relationship of animal disease resistance with various factors

appropriate vaccine should be selected based on the animal's age, disease characteristics, epidemic status, and risk factors.

1. **Attenuated live vaccines** are vaccines developed by artificially attenuating virulence of the pathogen or a naturally avirulent or mild virulent strain isolated from an animal. Attenuated live vaccine strains maintain the main antigenic characteristics of the virulent pathogen and have a decreased pathogenicity. Vaccination with an attenuated live vaccine gives the animal a strong specific resistance and induces both humoral and cellular immunity. Examples include hog cholera virus-lapinized vaccine, attenuated anthrax spore vaccine, attenuated Newcastle disease vaccine, and attenuated infectious bursal disease vaccine.

Table 8.1 Types of vaccines and immune responses

Type of vaccines	Humoral immunity (Ab)	Cellular immunity (CTL)	Remark
Attenuated live vaccine	+	+	Effective but risky (revert to virulence or have the risk of generating mutant or recombinant new strains)
Inactivated vaccine	+	−	Protective effects are strain-dependent
Genetically engineered vaccine	+	+	Recombinant vaccine protection is related to antigen sequence; gene-deficient vaccines achieve better results but have the risk of generating recombinant new strains
Subunit vaccine	+	−/+	The protective effect is related to the antigen sequence; addition of specific adjuvants (liposomes) can induce a cellular immune response
Synthetic peptide vaccine	+	−/+	The protective effect is related to the synthetic peptide sequence; addition of specific adjuvants (liposomes) can induce a cellular immune response
Toxoid	+	−	The protective effect is related to the type and action of the toxin
DNA vaccine	+	+	The protective gene fragment of the vaccine and the gene encoded protein antigen should be used alternatively for immunization to achieve better immunity
mRNA vaccine	+	+	Immunization efficacy is related to the encoded protective antigen, and this class of vaccines is safe, efficient, and easy to produce, but requires avoidance of recognition by intracellular pattern recognition receptors to avoid inflammatory responses

2. **Inactivated vaccines are** vaccines prepared from virulent strains of pathogens inactivated by inactivating agents. Adjuvants are often added to vaccines to improve immunization efficacy and mainly cause humoral immunity, such as inactivated avian influenza vaccine, inactivated Newcastle disease vaccine, inactivated porcine reproductive and respiratory syndrome vaccine, etc.
3. **Genetically engineered vaccines** are vaccines including recombinant vaccines and gene deficiency vaccines. Genetic recombinant vaccines are live vector vaccines made by inserting protective gene fragments of pathogens into live vectors using genetic engineering techniques, which can cause both humoral and cellular immunity, such as live pox virus vaccine, live adenovirus vector vaccine, and live turkey herpes virus (HVT) vector vaccine. Gene deficiency vaccines are vaccines made by knocking out the virulence gene of a pathogen to make it less or none pathogenic, such as porcine pseudorabies gene deficiency vaccine. In recent years, virus rescued by reverse genetic techniques has been developed into an attenuated vaccine, which is characterized by the maximum modification

of the virus using molecular biology techniques, and this approach to the development of vaccine represents the future trend of research.

4. **Subunit vaccines** are vaccines that are divided into conventional subunit vaccines and genetically engineered subunit vaccines. Traditional subunit vaccines are those made by physically or chemically inactivating a pathogen and then extracting its effective protective antigen component, thus they are also known as component vaccines, such as polyvalent pneumococcal polysaccharide vaccines. Genetically engineered subunit vaccines are developed by cloning protective antigen gene, inserting the gene into eukaryotic or prokaryotic expression vectors by molecular biological techniques, expressing the antigen in eukaryotic or prokaryotic expression system, and mixing the purified antigens with adjuvants, such as foot-and-mouth disease VP1 subunit vaccine and infectious bursal virus VP2 subunit vaccine. Subunit vaccines primarily elicit humoral immunity, but can also elicit cellular immune responses when specific adjuvants (e.g., liposomal adjuvants) are included in the vaccine.
5. **Synthetic peptide vaccines** are vaccines prepared by chemical synthesis of protective antigenic peptides or by molecular biology techniques, and subsequently by mixing the peptides with appropriate adjuvants. These vaccines mainly induce humoral immunity, but can also induce a cellular immune response if special adjuvants (e.g. liposomal adjuvants) are included in the vaccine, such as the foot-and-mouth disease peptide vaccine.
6. **Toxoid vaccines** are vaccines prepared by inactivating bacterial toxin (produced by the metabolism of a pathogen) with formaldehyde and mixing with an appropriate adjuvant. This type of vaccine mainly causes humoral immunity, e.g. tetanus toxoid vaccine.
7. **DNA vaccines** are vaccines prepared by inserting a protective antigen gene fragment into a eukaryotic expression vector. They are usually vaccinated alternatively with the encoding protein vaccine to achieve better immunity. These vaccines induce both humoral and cellular immunity.
8. **mRNA vaccines** are vaccines prepared by modifying mRNA encoding protective antigens, which are translated directly into antigenic proteins in host cells after vaccination to induce an immune response for the purpose of disease prevention and control. These vaccines are safe, efficient and easy to produce, and can induce both humoral and cellular immunity.

8.3.3 Polyvalent and Combined Vaccines

If a pathogen has multiple serotypes (e.g., *E. coli*, *Streptococcus suis*, etc.), a vaccine prepared by mixing vaccine strains from multiple serotypes together is called a polyvalent vaccine. Vaccination of animals with polyvalent vaccines protects against attack by multiple serotypes of pathogens. A combined vaccine is a vaccine prepared by mixing vaccine strains from multiple pathogens. Vaccination with a combined vaccine can prevent several diseases at the same time, such as Newcastle disease + infectious bronchitis + infectious bursal disease (NII) triple vaccine, Newcastle disease + infectious bronchitis + egg reduction syndrome (NIE) triple vaccine.

8.3.4 Immunization Routes

The common immunization routes are divided into four categories as follows.

1. **Injection** vaccination by intramuscular, subcutaneous, intradermal, intrathoracic, intraperitoneal, and intravenous injections.
2. **Oral intake** vaccination mainly by the drinking route.
3. **Aerosols** vaccination by inhalation, primarily by spraying, forming droplets of aerosol and mist.
4. **Nose or eye drops** vaccination by dripping vaccine into the nose or eye of the animals

8.3.5 Immunization Programs

The immunization program is a vaccination plan based on the age of the animal, the type of vaccine, and the prevalence of the disease, including the timing, dose, route, and frequency of vaccination (see Appendices VI and VII).

8.3.6 Monitoring the Effectiveness of Immunization

The effectiveness of immunization is evaluated by monitoring specific antibodies after vaccination, and the main serum antibody detection techniques commonly used are the hemagglutination inhibition (HI) test, agar diffusion test, indirect hemagglutination (IHA) test, enzyme-linked immunosorbent assay (ELISA), and immune colloidal gold technique (GICT). The main techniques for monitoring cellular immunity are solid-phase enzyme-linked immunospot assay (ELISPOT) and flow cytometry (see Chap. 6).

8.3.7 Factors of Immunization Failure

The outcome of failure to produce protective immunity after vaccination, as demonstrated by monitoring, is called immunization failure. The causes of immunization failure may be related to the following factors.

8.3.7.1 Vaccine Quality

The viability of attenuated live vaccine determines the efficacy of vaccination. The production and transportation of vaccines should meet quality control standards, especially the live vaccines that require cold chain transportation, and they should be guaranteed to be sufficiently viable before use.

8.3.7.2 Vaccine Serotypes

For pathogens with multi-serotypes, such as foot-and-mouth disease virus (FMDV), the vaccine strain should be serotypically identical to the endemic strain of FMDV, otherwise immunization failure will occur because cross-protection cannot be formed between FMDV serotypes.

8.3.7.3 Too Late Timing for Immunization

The timing of vaccination is chosen according to the age of disease occurrence. and immunization failure will occur if the time for immunizing animals is too late, especially later than the time of wild virus infection.

8.3.7.4 Animals Fail to Produce an Immune Response

1. **Immunosuppression**: If an animal is infected with certain immunosuppressive pathogens (e.g. PRRSV, HCV, IBDV, ALV, etc.), the function of the immune system is disrupted and a normal immune response to the antigen cannot be induced, resulting in immunization failure.
2. **Immune tolerance**: If an animal receives a large amount of antigen or a very small amount of antigen at once, this often results in anergy, a phenomenon known as immune tolerance. Thus, if the vaccine is used with an excessive dose, immunization failure can occur.
3. **Genetic factors**: Certain animals have a congenital immune deficiency and do not develop a normal immune response to vaccination, which is determined by genetic factors.
4. **Interference of antibody** (maternal or serum antibodies): Newborn animals acquire maternal antibodies through colostrum or yolk, and as maternal antibodies gradually diminish or even disappear over time, animals need to be vaccinated to obtain active immunity (Fig. 8.9). If vaccination is done too early, maternal antibodies can severely affect the effectiveness of immunization and even cause immunization failure. Therefore, the interference of maternal antibodies should be considered when developing immunization programs.

8.3.8 Relationship Between Immunization Dose and Immunization Effect

In general, immunization doses are determined through rigorous testing, and a single dose of vaccine, administered according to a scheduled immunization procedure, can protect animals against challenge by 10 lethal doses of virulent pathogen. However, in clinical practice, the disease cannot be controlled even after vaccination, so some farmers tend to solve the problem by increasing the vaccine dose and the number of immunizations. As a consequence, the animals still develop disease although a high concentration of antibodies in the blood could be detected. So, in the case of high antibody titers in the blood, why do animals still get sick? What is the best dose for immunization? and how many times should animals be immunized? These questions should be considered from the following perspectives.

Although antibodies cannot enter the cell, they can neutralize the virus to prevent it from infecting the target cell. Virus infected target cells by recognizing and binding to specific receptors on the target cell membrane, and the affinity of the virus

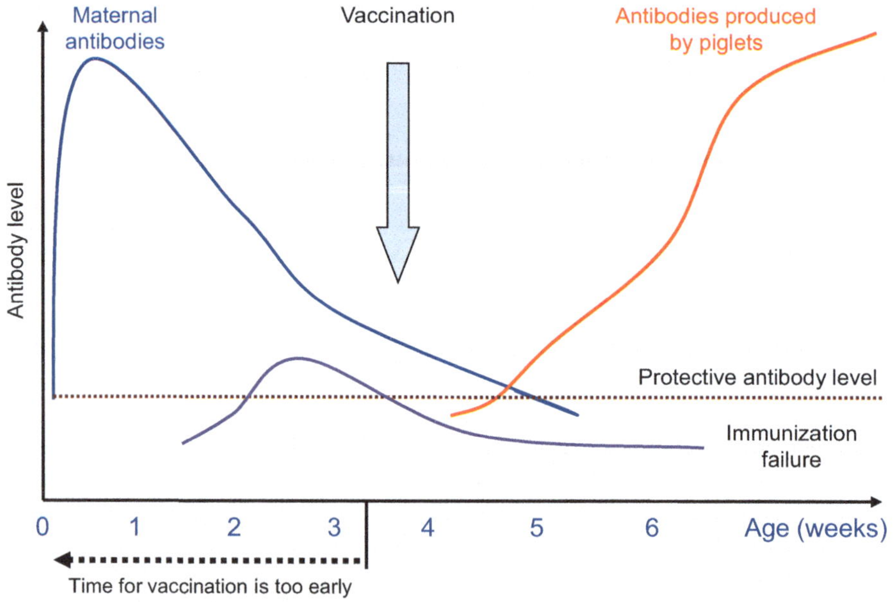

Fig. 8.9 Schematic illustration of the effect of maternal antibodies on vaccine efficacy. The concentration of maternal antibodies in the blood of animals reaches its peak within 1 week of birth. As the age of animals increases, the intestine of animals gradually develops to completion a week after birth, and the large molecules such as antibodies cannot be absorbed directly into the blood, so the concentration of maternal antibodies in the blood gradually decreases and eventually disappears. If vaccination is given too early, the maternal antibodies will neutralize the vaccine, which results in immunization failure. Piglets immunized at 3.5–4 weeks of age can obtain better immunization effect

binding to the membrane receptors is the key to successful virus infection. The binding of the antibody to the virus is reversible, so the ability of the antibody to prevent viral infection depends on the neutralizing ability (affinity) of the antibody (Fig. 8.10).

Each B cell has a certain number of antigen receptors on its surface (300–10,000 BCRs), and all BCRs on the surface of the same B cell are identical, but the BCRs of different cells are different, which allows for the existence of B cells that recognize and bind to the same antigen with different affinities, and B cells containing high-affinity BCRs produce high-affinity antibodies after activation (Fig. 8.11).

B cells with high-affinity BCRs have a strong ability to recognize and bind to antigens, can be activated by low concentrations of antigen, and produce antibodies that are also high-affinity antibodies, which are characterized by a strong ability to neutralize antigen. In contrast, B cells with low-affinity receptors require higher concentrations of antigen to be activated, and the antibodies produced by these B cells are also low-affinity antibodies, which have a weak neutralizing ability against pathogens (Fig. 8.12). In addition, if the concentration of antigen is too high, the B cell with high affinity B cell receptors will bind to excessive numbers of antigen molecules at the same time, which are more than required for B cell activation, and immune tolerance (anergy) will occur. Under this situation, low affinity B cells can be activated to produce antibodies, and the high concentrations of such low-affinity antibodies in the blood are often detected, and because of the low affinity, they cannot block virus infecting cells (Fig. 8.10). Even though a high antibody titers in the serum of animals can be detected, there is still the occurrence of diseases in the animals.

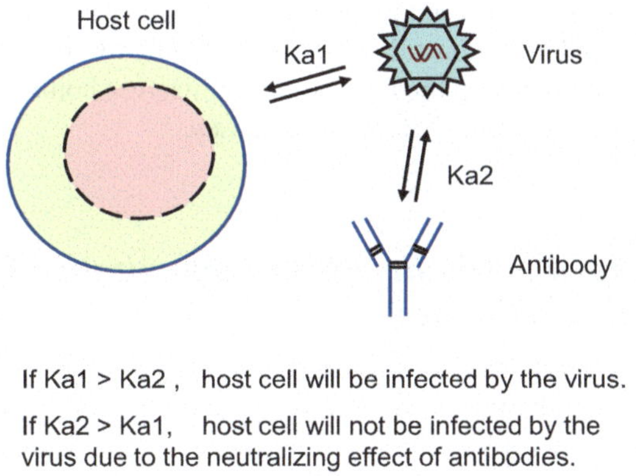

Fig. 8.10 Schematic illustration of antibody-virus interaction for the blockade of virus infecting target cells. The affinity constant for virus binding to the receptor of target cell is Ka1, and the affinity constant for antibody binding to the virus is Ka2. If Ka2 > Ka1, the antibody effectively neutralizes the virus and prevents the virus from infecting the target cell; if Ka1 > Ka2, the virus cannot be prevented from infecting the cell, even if the antibody concentration is comparatively high

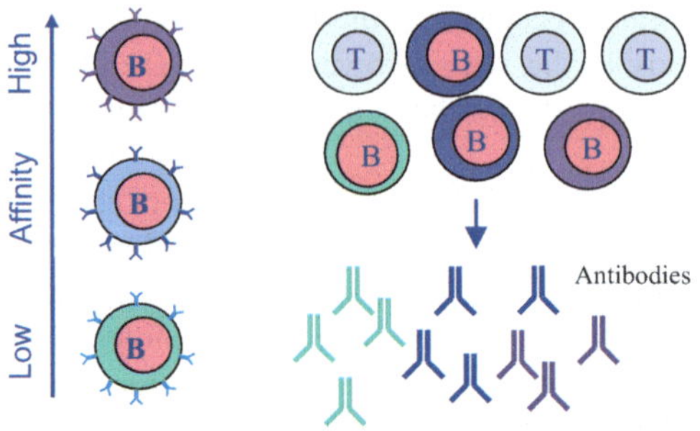

Fig. 8.11 Schematic illustration of the B cells with different affinity receptors in recognition of the same antigen. There are B cells in the body that recognize the same antigen, but they have a high or low affinity for binding to the same antigen, and B cells with high affinity BCRs produce high affinity antibodies after activation

According to the theory of clonal selection, the lower the concentration of antigen, the higher the affinity of B cells will be activated for antibody production, which is the rationale for the preparation of high-affinity monoclonal antibodies by low-dose and long-time course immunization. Although animal immunity can be established through vaccination, the ability of the vaccinated animals to defend against pathogenic infection is also limited, and disease can still occur in the presence of a large number of circulating pathogens. Thus, the high dose of vaccine and multiple booster immunizations may not necessarily resist the attack by virulent pathogens. An important factor threatening the safety of herds is the number of pathogens in the environment, so environmental control and reduction of the number of pathogens with biosecurity measures are the prerequisites for successful prevention and control of infectious diseases, and only on this basis can vaccination be implemented to achieve the desired prevention and control results. In practice, only by adhering to the epidemic control principle of "maintaining a good biosecurity environment plus implementing appropriate immunization control" can we achieve the desired results of controlling infectious diseases.

8.3.9 Difficulties and Challenges in Immunization Control of Infectious Diseases

Vaccination is undoubtedly the most effective method of controlling infectious diseases, and there are numerous examples of its success. The theoretical basis of controlling epidemics by immunization is that vaccination induces humoral and cellular immune responses that produce specific antibodies and effector T lymphocytes and, most importantly, memory cells, so that when a pathogen containing the same antigen invades again, the host produces a strong recall immune response to clear the pathogen more effectively and maintain its own homeostasis. The specific antibodies produced by the humoral immune response play a role in neutralizing pathogens by binding to and preventing pathogens from infecting target cells, and the

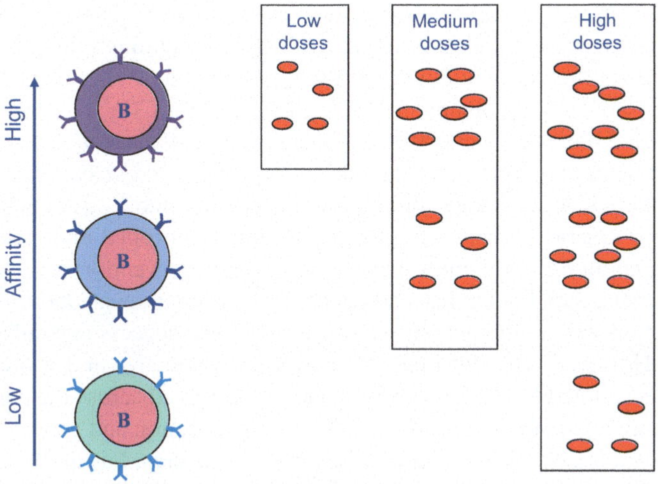

Fig. 8.12 Schematic illustration of the relationship between the dose of antigen and the activation of B cells with different affinities. All three kinds of B cells recognize the same antigen, but their BCRs recognize and bind to the antigen with different affinities. The binding of BCRs to antigen is competitive; low-dose antigen can be recognized by high-affinity B cells, but as the amount of antigen increases, low-affinity B cells are also able to come into contact with the antigen and be activated, while high-dose antigen puts high-affinity B cells in a tolerant state. Thus, high-dose antigen activates low-affinity B cells, producing low-affinity antibodies that have a weak ability to neutralize pathogens or even unable to prevent pathogenic infection

CTL of cellular immune response kills target cells infected by pathogens to expose and release pathogens, facilitating the role of antibodies in neutralizing pathogens, with the ultimate task of phagocytosis and clearance of pathogens being performed by macrophages. Phagocytosis and clearance of antigens by macrophages through Fc receptor-bound antibody-antigen complexes (opsonin-mediated phagocytosis) is one of the major ways in which the immune system combat pathogenic infection. However, the target cells of some pathogens are macrophages, and such pathogens not only infect the target cells by directly binding to cell membrane receptors, but also infect macrophages via taking advantage of antibody-mediated opsono-phagocytosis of pathogens by macrophage, resulting in a situation in which the higher the antibody titer the more severe the viral infection. This phenomenon is called antibody-dependent enhancement of diseases (ADE). The commonly seen viruses causing ADE are porcine reproductive and respiratory syndrome virus (PRRSV, also known as blue ear disease virus), African swine fever virus (ASFV), etc. In clinical practice, it has been found that pigs with high antibody levels and good nutritional status tend to have severe morbidity and high mortality, which is attributed to the strong inflammatory response caused by the pathogenic infection, resulting in a cytokine storm and septic shock, ultimately leading to the death of animals. Therefore, it is a challenge to develop vaccines against immunosuppressive pathogens that cause immunosuppression in host by infecting immune cells (e.g. macrophages, B cells, T cells).

In terms of prevention and control of epidemics, an epidemic will be controlled if any of the three links of an infectious disease epidemic (the source of infection, transmission route, and susceptible animals) are controlled, such as eliminating the pathogen or isolating the source of infection, cutting off the transmission route, and protecting susceptible animals (or populations). The main measures to prevent and control the epidemic of SARS in 2003 and the COVID-19 in 2019–2022 were traditional biosecurity control measures such as isolation/quarantine, segregation/containment, and pathogen elimination. As long as these measures are taken properly, they will have a substantial preventive and control effect, which is well proven in our practice, where we have successfully controlled and eradicated two major airborne human epidemics through biosecurity prevention and control measures at all costs, with human life and health as the primary consideration, and with unified national action, thus achieved a world-renowned success in combating major epidemics. However, why did PRRSV (Porcine reproductive and respiratory syndrome virus) and ASFV (African swine fever virus) spread across the country causing significant economic losses without effective prevention and control? The main reason is not the lack of prevention and control measures, but the lack of proper implementation of these measures, and the key issue is the cost of implementation of these measures. For example, ASFV is a typical virus infecting hosts by direct contact with animals rather than through the air or aerosol. As long as the route of virus transmission is cutoff to prevent the pathogen into the pig herd, the disease will not occur. However, in practice, this measure is not done in place. As a consequence, this virulent virus quickly spread into the pig farm across the country through a variety of ways, causing huge economic losses to stakeholders. If the disease prevention and control is focused on improving animal herd immunity by vaccination, it is tantamount to endorsing the long-term presence of the pathogen, and instead, the priority is to safeguard animals from disease. It should be clear that for most pathogens, vaccination for disease control can only improve the specific herd immunity of animals (refer to Fig. 8.6) and reduce the incidence of disease in susceptible animals or even prevent disease, but it cannot eradicate the pathogen and stop the transmission of pathogens in animals. For example, PRRSV, Avian Influenza Virus (AIV), Infectious Bursal Virus (IBDV), etc. all have the control problem because of strain mutation, and every time a new highly pathogenic strain emerges, it will cause a major outbreak, giving immunization and control a lot of pressures. How to carry out effective measures for the prevention and control of infectious diseases in animals is an urgent issue to be solved.

8.4 Immune Adjuvants and Stimulants

8.4.1 Immune Adjuvants

1. **Concept of immune adjuvants**: Those substances that are mixed with antigens to immunize animals and enhance the immunogenicity of the antigen are called immune adjuvants. Adjuvants are generally used when the immunogenicity of antigen is weak or the amount of antigen is small.

2. **Mechanisms of immune adjuvant action**: **There are five mechanisms of immune adjuvant action** (a) prolongation of antigen duration in vivo; (b) enhancement of co-stimulatory signals; (c) enhancement of local inflammatory responses; (d) nonspecific stimulation of lymphocyte proliferation; and (e) prolongation of antigen-presenting cell (APC) activity.
3. **Main types of immune adjuvants.**

   (a) **Insoluble aluminium salt adjuvants** include aluminium hydroxide, alum, tricalcium phosphate, etc.
   (b) **Oil-water emulsion adjuvants**: Water-in-oil (W/O), water-in-oil-in-water (W/O/W) biphasic emulsions, Freund's complete adjuvants and Freund's incomplete adjuvants, etc.
   (c) **Microbial and metabolite adjuvants**: Certain killed bacteria, G-bacterial outer membrane lipopolysaccharide (LPS), mycobacteria (muramyl dipeptide, MDP), small short rod-shaped bacilli, lactobacillus components, lipophosphatidic acid (LTA) of G+ bacteria, protein toxins of bacteria (e.g. cholera toxin, tetanus toxin, pertussis toxin), etc.
   (d) **Nucleic acids and analog adjuvants** include nucleic acids, CPGDNA sequences, etc.
   (e) **Cytokine adjuvants** include IL-1, IL-2, IFN-γ or other cytokines, etc.
   (f) **Liposomes** are lipid vesicles consisting of a single or contiguous multilayer of phospholipids containing an aqueous phase space, which serve both as carriers of antigens and as immune adjuvants. The main types of liposomes are compound liposomes, large monolayer liposomes, and small monolayer liposomes.
   (g) **Immunostimulatory complex adjuvant** Immunostimulatory complex (ISCOM) is a mixture of bifid antigen with saponin (QuilA) and cholesterol in a 1:1:1 molecular ratio, and is a novel antigen delivery system with high immunological value, whose mechanisms of action include: (1) activation of CD4T cells, CD8T cells and B cells; and (2) long duration of immune response.
   (h) **Propolis adjuvant** is a natural substance made from the resin secreted by bees collecting plant buds, which is formed by the mixture with secretions from the bees' palate glands, beeswax, pollen, and other organic and inorganic substances, and is a by-product of beekeeping industry.

8.4.2 Immune Stimulants

1. **Concept of immune stimulants**: Substances given alone that enhance the immunity of animals are collectively referred to as immune stimulants.

   The difference between immune stimulant and adjuvant is that an immune stimulant is given "independently", whereas an adjuvant is used in mixture with an antigen.

2. **Types of immune stimulants**: There are four main categories of immune stimulants.

    (a) **Biological immune stimulants** include Transfer factors, immune ribonucleic acid, thymic hormones, interferons, plant and herbal polysaccharide components, etc.
    (b) **Cellular immune stimulants**: Corynebacterium parvum, BCG vaccine, bacterial lipopolysaccharide, etc.
    (c) **Chemical immune stimulants** include Levamisole, pyran, tilorone, polynucleotides, etc.
    (d) **Nutritional immune stimulants** include Vitamins, trace elements, etc.

Study Questions

1. What is the relationship between immunization dose and immunization effectiveness?
2. What is the theoretical basis for control of epidemics by vaccination?
3. What is the role of vaccination in epidemic prevention and control?
4. What is the mechanism of action of immune adjuvants?
5. What is the concept of immune stimulants? How do they differ from adjuvants?

Chapter 9
Immune Regulation

Overview:
Immune regulation is the regulation and control of the immune response by the immune system in order to maintain homeostasis of the organism. Once started, the immune response is like an accelerating train that must be braked to keep it running smoothly and eventually come to a stop. In the whole process of immune response, immune regulation excerts a braking effect, which is realized through immune regulatory cells, so the main body of immune regulation is immune regulatory cells. By regulating the direction and intensity of the immune response, immune regulatory cells orient the immune response towards cellular immunity or humoral immunity and keep it in the most appropriate state of physiological response, thus maintaining the physiological balance and stability of the body. However, if the immune regulation is out of control, it can cause the body to have abnormally excessive immune response, leading to immune damage or even death by septic shock. Therefore, immune regulation determines to some extent the physiological state and immune tolerance of animals, especially in the adaptive immune response where immune regulation plays a key role.

9.1 Important Immune Regulatory Cells

Both innate and adaptive immune responses are accomplished by involvement of multiple cells. These cells involved in the immune response can be broadly divided into immune response dominant cells (e.g., phagocytes in the innate immune response, and T and B lymphocytes that undertake the adaptive immune response) and helper immunoregulatory cells. Immunoregulatory cells regulate the course and intensity of the immune response by secreting cytokines in the early stages of the immune response. For example, in the early innate immune response to pathogenic infection, phagocytes further mature under the action of IFN-γ. Mature phagocytes

S. J. Zheng, *Veterinary Molecular Immunology*,
https://doi.org/10.1007/978-981-99-8929-4_9

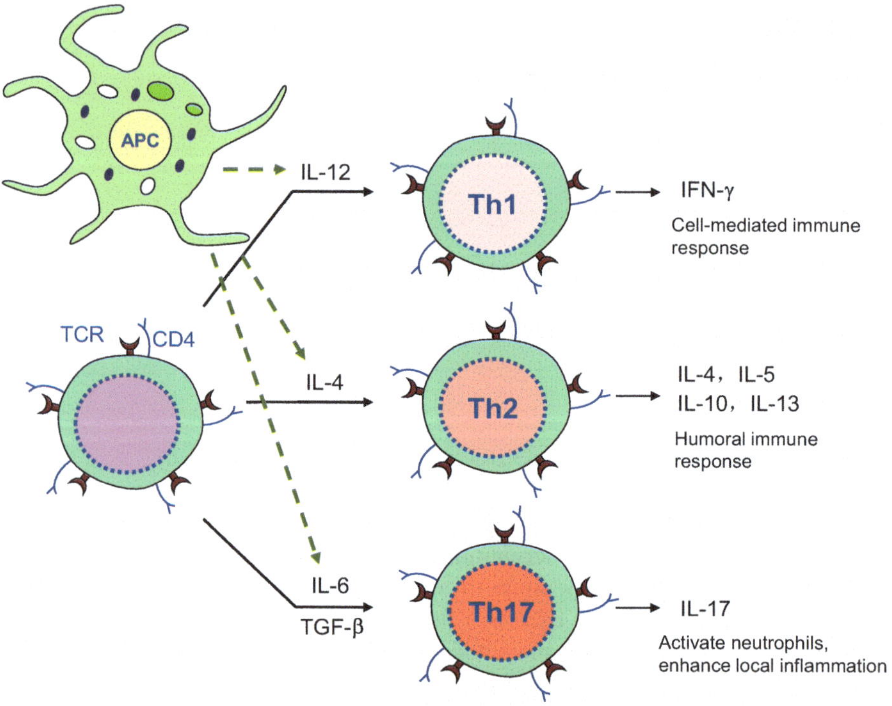

Fig. 9.1 Schematic illustration of the differentiation of naive CD4T cells into different subsets of Th cells after activation in the presence of indicated cytokines produced by dendritic cell

acquire a powerful ability to phagocytose, kill and clear foreign pathogens, and IFN-γ, which plays an important role in this process, is secreted by NK and NKT cells, thus NK and NKT cells play an important supporting role in the process of phagocytosis.

Based on their roles in the immune response, immunoregulatory cells can be broadly divided into two categories: (1) immunoregulatory cells that promote the immune response, such as NK cells, NKT cells, Th1CD4T cells, Th2CD4T cells, and Th17CD4T cells; (2) immunoregulatory cells that suppress the immune response, also known as regulatory T cells, which are more diverse, including some subsets of CD4T cells and certain subpopulations of CD8T or $CD4^{-}CD8^{-}$T cells. Upon activation, dendritic cells contribute to the immune response by secreting specific cytokines (IL-4, IL-12, IL-6/TGF-β, etc.) to drive further differentiation of naive CD4T cells into Th1, Th2, or Th17 cells that produce the corresponding cytokines to drive the immune response in different directions (Fig. 9.1), and these immunoregulatory cells have a promoting effect on the immune response. For example, Th1 cytokines (mainly including IFN-γ and TNF) are important for activating macrophages, promoting cellular immune responses and clearing intracellular pathogens, whereas Th2 cytokines (mainly including IL-4, IL-5, and IL-13) are important for promoting humoral immune responses and clearing helminth infections.Th1 and Th2 cytokines are regulated by each other, such as Th2 cytokine (IL-4) suppresses Th1 immune response and promotes Th2 immune response, while Th1 cytokine (IFN-γ) or certain cellular immunity-promoting cytokines (IL-12) suppress Th2 immune response. Some CD4T cell subpopulations produce both Th1 cytokine (IFN-γ) and Th2 cytokine (IL-4) upon activation, and such cells are classified as Th0 cells in immunology. CD4T cells differentiate into Th17 T cells in response to IL-6/TGF-β and secrete IL-17, a cytokine associated with activation of neutrophils, eliciting local inflammatory responses, and clearance of pathogens. In addition, innate lymphocytes ILC1, ILC2, and ILC3 secrete cytokines similar to Th1, Th2, and Th17, for example, ILC1 secretes IFN-γ and TNF-α; ILC2 produces cytokines IL-4, IL-5, IL-9, and IL-13; ILC3 produces cytokines IL-17 and IL-22, and through these cytokines ILC regulates the immune response (refer to Chap. 2, Fig. 2.4).

Regulatory T cells have a suppressive effect on the immune response and are associated with immune tolerance, suppression of the immune inflammatory response, suppression of autoimmune diseases, and tumor immunity. A comprehensive description of regulatory T cells is the focus of this chapter.

9.2 Regulatory T Cells

9.2.1 Overview of Regulatory T cells

Regulatory T cells (Treg) are a specical subset of T cells whose main function is to suppress the immune response, including immune tolerance, suppression of immune inflammatory response, suppression of autoimmune diseases and tumor immunity, etc. The target cells of Treg include effector T cells, B cells, $CD4^{+}$T cells, NK cells, dendritic cells (DCs), and so on. Through the inhibition of target cells, the immune response is regulated. Thus, Treg is an important cell in the regulation of immune responses.

9.2.1.1 Discovery and Classification of Treg

Since Beredka discovered that guinea pigs were immunotolerant to antigens in food more than 100 years ago, the mechanisms of immune tolerance have been continuously explored, and the role of $CD8^{+}$T cells in suppressing other immune cells was first suggested in the 1970s. Since then, the immunosuppressive effect of $CD8^{+}$T cells and the relationship with immune tolerance have become a hot topic of immunological research. However, due to the lack of effective experimental methods and techniques, as well as the lack of substantial research progress, people gradually lost interest in the study of $CD8^{+}$T cell immunosuppression. In the 1980s, with the development of organ and tissue transplantation technology, people paid more and more attention to the study of immune tolerance. in the late 1980s, two types of cells were found to have immunosuppressive effects; one type is the IL-10-producing $CD4^{+}$T-cell subset, which plays a suppressive role in immune response, and was named Tr1 (type I regulatory T cell); the other is T cells that do not express CD4 and CD8, called DNT cells (double negative T cells), which also have immunosuppressive effects. In 1995, immunologists discovered that a group of $CD4^{+}$T cells expressed IL-2 receptor alpha chains in large numbers, and that the IL-2 receptor alpha chains were classified as CD25 on CD antigens. It was found that $CD4^{+}CD25^{+}$T cells had immunosuppressive functions and were called regulatory T cells (Treg). This class of cells played an important role in the mechanism of maintaining autoimmune tolerance, and if autoimmune tolerance was broken, a variety of autoimmune diseases would occur. Although $CD4^{+}CD25^{+}$T cells had immunosuppressive characteristics, they belonged to a subet of $CD4^{+}$T cells, which was not in line with people's imagination and expectation, so the discovery did not attract sufficient attention. At that time, it was widely believed that T cells should have at least three subpopulations, namely: effector T cells, helper T cells, and suppressor T cells. Effector T cells, or Teff for abbreviation, also known as cytotoxic T (CTL) cells, whose primary surface marker is $CD8^{+}$, perform the role of killing target cells; helper T cells, or Th for abbreviation, whose surface marker is $CD4^{+}$ and whose primary role is to assist and promote the immune response; and suppressor T cells, or Ts for abbreviation, whose role is to suppress and regulate the immune response. Both CTL and Th cells had clear characteristic surface markers. However, after two or three decades of intensive research with great enthusiasm, the imagined subpopulation of Ts cells was still not found, and the idea that T cells have a subpopulation of Ts cells had to be finally abandoned.

Since no subpopulation of Ts cells has been identified, which cells play the immunosuppressive and regulatory roles? There are two main categories of cells with immunosuppressive and regulatory roles that have been reported: (1) Treg (natural Treg, nTreg) that develop and mature naturally in the thymus; and (2) Treg (induced/adaptive Treg, iTreg) that are induced by peripheral environmental factors (e.g., antigenic stimulation, inflammatory response, etc.).

Important Types of Treg Cells and Their Characteristics

1. $Foxp3^{+}$Treg This type of T cells is characterized by $CD4^{+}CD25^{+}$, and expresses the transcriptional regulator Foxp3 (Forkhead box P3), which is subdivided into nTreg and iTreg depending on the origin of $Foxp3^{+}$Treg.
2. DNTreg ($CD4^{-}CD8^{-}$Treg) This type of Treg is characterized by the expression of α/βTCR but not CD4, CD8, Foxp3 and NK cell surface markers, and accounts for approximately 1-5% of the α/βT cell population.
3. Tr1 (Type I Treg) This type of Treg is characterized by the production of IL-10 without expression of Foxp3. Another characteristic of this subpopulation of T cells is the co-expression of CD49b (integrin) and lymphocyte activation gene 3 (LAG-3)
4. Th3 (Type 3 helper T cell) This type of Treg belongs to $CD4^{+}$ T cells and is characterized by production of TGF-β.

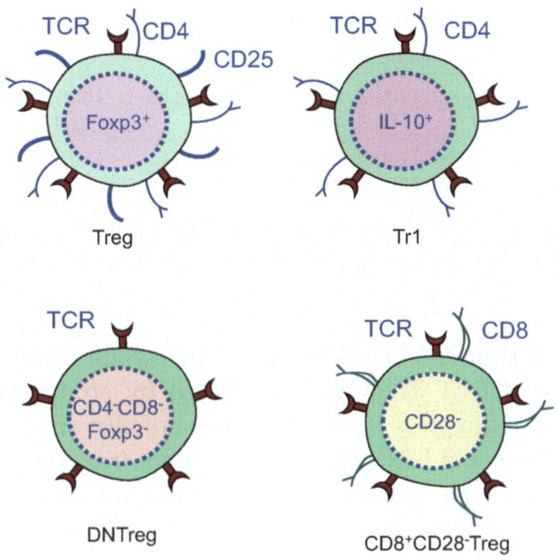

Fig. 9.2 Schematic illustration of T cell subsets with regulatory roles

5. $CD8^+$ Treg There is no strictly uniform surface marker for this type of Treg, but some markers have been reported for identification, such as Qa-1 restricted $CD8^+$, $CD8^+CD28^-$, $CD8^+CD122^+$ (CD122 is the beta chain of IL-2 receptor), $CD8^+CD103^+$, $CD8^+Foxp3^+$, $CD8^+CD122^+PD\text{-}1^+$, $CD8^+CD25^+$, $CD8^+CD45RO$, $CD8^{++}CD11c^+$, $CD8^+LAP^+$, etc. The identification markers of $CD8^+$Treg are numerous and there is no way to unify them yet, whether $CD8^+$Treg has such diverse subpopulations or just different manifestations of a certain cell subpopulation at different stages remains to be studied in depth. Since $CD8^+CD28^-$ Treg also expresses FoxP3, GITR, CTLA-4, CD25, CD103, CD62L, and 4-1BB with a phenotype very similar to $CD4^+CD25^+Foxp3^+$Treg, some literature refers to them as Ts cells.

As more cell types with immunosuppressive effects have been identified, this chapter is primarily focused on $CD4^+D25^+Foxp3^+$Treg, DNTreg, Tr1, Th3, and CD8Treg cell subsets (Fig. 9.2).

9.2.1.2 Generation of Tregs and their Characteristic Markers

1. **$CD4^+CD25^+Foxp3^+$**Treg: T cells are born and differentiated in the thymus. As animals grow, nTreg originate naturally, develop and differentiate in the thymus, whereas iTreg are generated from CD4T cells in the peripheral due to certain environmental factors. Depending on the origin, $CD4^+CD25^+Foxp3^+$Treg cells are also divided into nTreg and iTreg, and a comparison of the production and surface markers of these two types of Treg is shown in Table 9.1.

Table 9.1 Generation and surface markers of nTreg vs. iTreg

Treg	Generation site	Common markers	Special markers
nTreg	Thymus	Foxp3, CD4, CD25, GITR, CTLA-4, LAG-3, PD-1	Helios, Nrp1, PD-1, Swap70
iTreg	Peripheral lymphoid organs and tissues or sites of inflammation	Foxp3, CD4, CD25, GITR, CTLA-4, LAG-3, PD-1	Dapl1, Igfbp4

Foxp3 forkhead box P3, *GITR* glucocorticoid-induced tumor necrosis factor receptor family related gene, *CTLA-4* cytotoxic T lymphocyte-associated antigen-4, *LAG-3* lymphocyte activation gene-3, *PD-1* programmed death-1

2. **Tr1**: Tr1 cells are generated from CD4T cells induced by certain factors in the peripheral, and this type of CD4T Treg is characterized by producing large amounts of IL-10, but does not express Foxp3. Tr1 cells are also characterized by the co-expression of CD49b and LAG-3, which is distinct from other CD4T subpopulations (Th1, Th2, Th3, Th17, Foxp3^{+}Treg). In vitro cell culture, naive CD4T cells in response to antigenic stimulation in the presence of IL-10 or IL-10^{+} immunosuppressants (dexamethasone or rapamycin) in the culture medium would be induced to differentiate into Tr1 cells. It was found that IL-27 expression by immune tolerant dendritic cells induced the generation of Tr1 cells. In addition, IL-6 was also found to have a role in inducing the generation of Tr1 cells, and since Tr1 inhibits the inflammatory response by producing IL-10, which helps to reduce inflammation to restore homeostasis.
3. **DNTreg**: The origin of DNTreg is not well understood, some suggest thymic origin, while others propose that it is generated in the peripheral and that under certain conditions CD8T or CD4T cells can be converted to DNTreg. Its surface marker is TCR^{+}CD4^{-}CD8^{-}NK1.1^{-}, and this type of cells does not express Foxp3.
4. **Th3**: Th3 cells are TGF-β-secreting CD4 T cells formed primarily by immune tolerance due to GI food stimulation, and there are currently no specific surface markers for this subpopulation of T cells, characterized only by high expression of TGF-β.
5. **CD8 Treg**: CD8 Treg is generated and differentiated in the thymus. However, after CD8T cells enter the peripheral tissues, they can also differentiate into CD8 Treg due to certain factors, acting as a negative regulator for the immune response.

9.2.2 Modes of Action of Regulatory T cells

The main function of Treg is to suppress the immune response. Different subgroups of Treg may have different modes of action, which are summarized as follows: (1) direct contact with target cells to inhibit them; (2) secreting inhibitory cytokines for anti-inflammatory response; (3) keeping APC in an unresponsive state (anergy); (4) affecting metabolic processes of target cells; (5) killing target cells (exerting cytolytic effect): inducing apoptosis in target cells.

1. **Direct contact inhibition**: Treg inhibits target cells by direct contact with them, and Treg regulate T cell function by expression of the inhibitory receptors CTLA-4 and PD-1 (Fig. 9.3).
2. **Secretion of inflammatory suppressor cytokines**: Treg secretes suppressor cytokines such as IL-10 and TGF-β. IL-10 inhibits T-cell expression of IL-2, IFN-γ, and GM-CSF and suppresses T-cell proliferation (Fig. 9.4).

   Similarly, TGF-β has a role in suppressing the T-cell immune response, and the Th cells that secrete TGF-β are called Th3, which plays an important role in mucosal immune tolerance. In addition, IL-10 has the ability to inhibit IL-12 expression by APCs, inhibiting IFN-γ expression by activating SOCS1, and playing a role in inducing the conversion of T cells to Treg.
3. **Treg puts APC in an unresponsive state (anergy)**: Treg secret IL-10 cytokines inhibiting APC expression of costimulatory molecules and inflammatory response cytokines, while activating the expression of multiple immune toler-

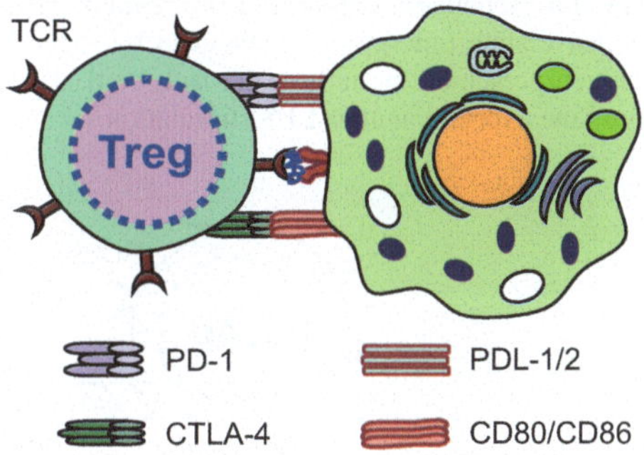

Fig. 9.3 Schematic diagram of direct contact inhibition of Treg cells. *CTLA-4* cytotoxic T-Lymphocyte antigen 4, *PD-1*: programmed death 1, *PD1L* PD1 ligand. CTLA-4 and PD-1 are immune response inhibitory receptors

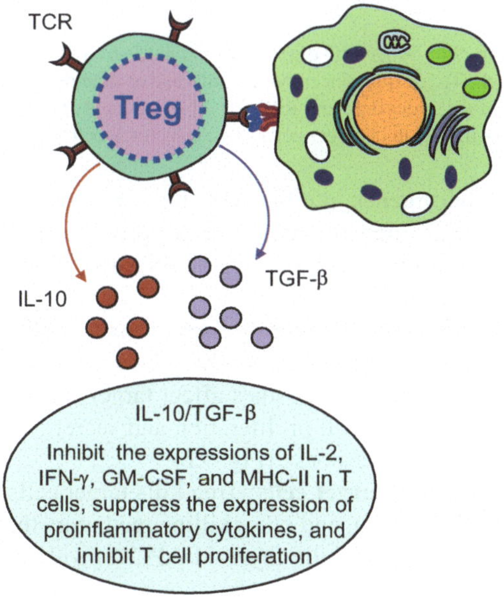

Fig. 9.4 Schematic diagram of the mode of action of Treg secreting inflammatory inhibitory cytokines

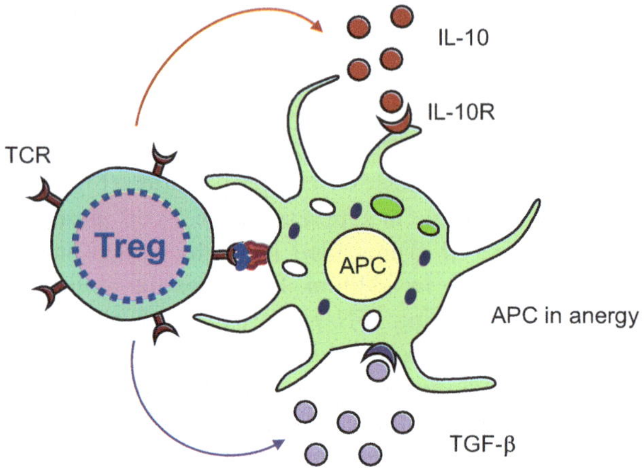

Fig. 9.5 Schematic diagram of the regulation of APC by Treg. Treg induces APC into an unresponsive state (anergy), as evidenced by reduced expression of MHC molecules, which affects antigen presentation, and by the lack of IL-12 production and inhibition of other inflammatory cytokine expression after LPS stimulation

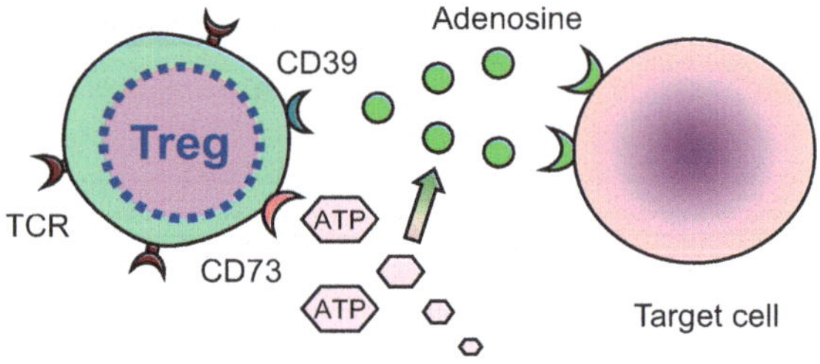

Fig. 9.6 Schematic diagram of the mode of action of Treg affecting cellular metabolism of target cell. Treg expresses enzymatically active membrane molecules, CD39 and CD73, which have the ability to degrade extracellular ATP and catalyze the production of adenosine, which inhibits the proliferation of target cells (e.g., effector T cells) and suppresses cytokine expression

ance genes, putting APC into an unresponsive state (anergy). This APC in the nonresponsive state does not produce IL-12 in response to pathogen-associated molecular patterns (e.g., LPS), and affects the host Th1 cell immune response (Fig. 9.5).

4. **Affecting the metabolism of target cell**: Treg express proteins with enzymatic activity such as CD39 and CD73 molecules, which degrade ATP to yield adenosines. The adenosines affect target cell metabolism by binding to the receptor, inhibiting cell proliferation and secretion of cytokines, thereby suppressing the immune response (Fig. 9.6).
5. **Killing target cells**: Treg kills target cells by secreting granzyme A, granzyme B, perforin, and other killing mediators that cause target cell lysis. For example, human Tr1 kills myeloid APCs (myeloid APCs) by recognizing antigens pre-

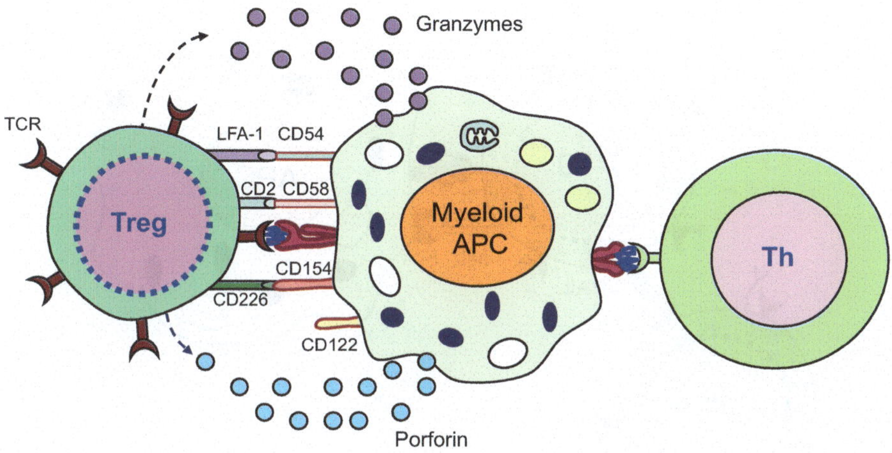

Figure 9.7 Schematic diagram of cytolytic effect of Treg. CD122 is the beta chain of IL-2 receptor and interacts with IL-2 to generate immune signaling. cd54/LFA-1, CD58/CD2 and CD154/CD226 interact to prompt Treg to release granzyme A (GzA) and granzyme B (GzB) as well as porforin, killing target cells (APCs) and preventing Th cell activation

sented by HLA-I molecules. This killing differs from NK cells, which kill target cells independent of MHC-I expression (Fig. 9.7).

6. **Inducing apoptosis in target cells**: Treg induces apoptosis in target cells via the membrane receptor (FasL-FasR or TRIAL-DR5) mediated apoptotic pathway. For example, DNTreg directly contacts target cells (effector T cells or DCs) to induce apoptosis via the FasL-Fas or TRIAL-DR5 apoptotic pathway (Fig. 9.8)

9.2.3 Molecular Mechanisms of Regulatory T-cell Action

1. **CD4$^+$CD25$^+$Foxp3$^+$Treg** This type of regulatory cells can exert inhibitory effects in at least two ways: (1) by inhibiting target cells through direct contact; and (2) by secreting anti-inflammatory response cytokines such as IL-10 and TGF-β.

   Foxp3 is not only expressed in CD4$^+$CD25$^+$Foxp3$^+$Treg, but also in CD8$^+$CD28$^-$Treg. It has been found that all Foxp3$^+$T cells have immunosuppressive effects. Foxp3$^+$T cells exhibit the following characteristics: (1) they cannot help producing IL-2 after receiving TCR stimulation signal; (2) IL-7Rα expression was very low; (3) with high expression of CD25, CTLA-4 and GITR.

   Foxp3 is both a transcriptional regulatory repressor and an activator. Foxp3 interacts with the transcriptional regulator NF-AT (see Fig. 4.13 in Chap. 4 for function) to inhibit NF-AT entry into the nucleus, thereby suppressing the expression of immune response genes. In addition, Foxp3 also interacts directly with the activated AP-1 to suppress immune response. Foxp3 upregulates the

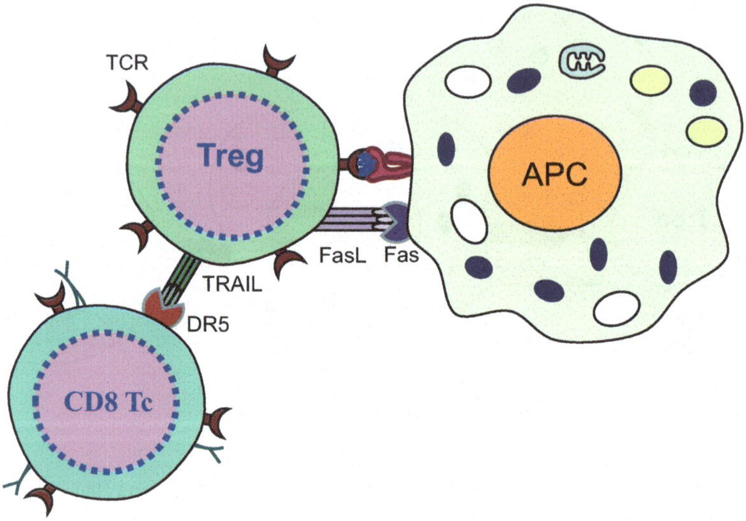

Fig. 9.8 Schematic diagram of Treg-induced apoptosis in target cells. Treg induces apoptosis in APC cells via the FasL-FasR mediated pathway and inhibits the immune response by inducing apoptosis in effector T cells in the TRIAL-DR5-mediated (DR5 in mice, DR4 in humans) pathway. Both FasL and TRIAL are members of the TNF superfamily and have apoptosis-inducing effects. *FasL* fatty acid synthase ligand, *FasR* fatty acid synthase receptor, *TRAIL* TNF-related apoptosis-inducing ligand, *DR4* death receptor-4, present in humans, *DR5* death receptor-5, present in rodents

expression of genes such as CTLA-4, IL-10, CD5, and FasL, while suppressing the expression of IL-2, TNF-α, IFN-γ, IL-17, and IL-4.

Foxp3 regulates immune response not only at the protein level, but also at the level of microRNAs (miRNAs). miRNAs are short-stranded RNAs (20-24 nucleotides in length) transcribed from non-coding regions of the genome, and these miRNAs regulate gene expression by acting on mRNAs transcribed from target genes to degrade or inhibit their translation. Therefore, miRNAs play an important regulatory role in tissue and organ development, cell differentiation and their own stability. The miRNAs that have been found to play important roles in Treg include miR-155, miR-146a and miR-142-3p. miR-155 negatively regulates IL-2R signaling by acting on SOCS1 (which is a transcriptional repressor), miR-146a regulates STAT1 expression that in turn regulates Th1 immune response, and miR 142-3p regulates the expression of cAMP.

All in all, Foxp3 is consistently expressed in Treg, serves as an important transcriptional regulator of Treg, is a key gene in the maintenance of Treg suppressor function, and plays an important role in the regulation of the immune response.

2. **Tr1** Although Tr1 transiently expresses Foxp3 upon activation, its inhibitory effect is not dependent on Foxp3 expression. Human immune dysregulation polyendocrinopathy enteropathy X-linked (IPEX) is caused by mutations in the Foxp3 gene, but the body can still produce large amounts of Tr1Treg in this disease, so the immunosuppressive mechanism of Tr1 is different from that of

Foxp3⁺Treg. Tr1 is induced after antigen stimulation in the peripheral and exerts its regulatory role mainly in the late stage of life, thus cooperating with early thymus-produced Foxp3⁺Treg to regulate the immune tolerance of the organism. Tr1 activation is characterized by producing large amounts of IL-10 and TGF-β, expressing small amounts of IL-2, IL-5 and IFN-γ, and no expression of IL-4. The mechanisms by which Tr1 inhibits the immune response include: (1) Secretion of IL-10, TGF-β and other inhibitory cytokines. The immunosuppressive effect of IL-10 is manifested by inhibiting IL-12 expression by APC, promoting Treg production, and activating SOCS1 to inhibit IFN-γ expression. Tr1 inhibits the activation of APC through suppressive cytokines. (2) Tr1 mediates T cell function through expression of inhibitory receptors CTLA4, PD-1 and ICOS.

3. **DN Treg** DNTreg does not express CD4, CD8, Foxp3 and NK cell surface markers, and suppresses target cells mainly through direct contact, including recognition of APC-presented antigens (mainly self-antigens) by TCR to highly express CTLA-4, inhibiting APC B7 expression, or inducing apoptosis in APC cells through the FasL-FasR pathway, or secreting perforins, etc. to kill target cells (e.g., B cells, etc.). Normally, DNTreg exerts suppressive effects without producing IL-10. The experimental results from animal tissue transplantation demonstrated that transfer of DNTreg to recipient animals prolongs graft survival, suggesting that DNTreg plays an important role in anti-transplantation rejection.

IL-10 is an important cytokine secreted by Treg that plays an important role in suppressing immune response, and the molecular mechanism of IL-10 action is complex and is now partially elucidated (Fig. 9.9).

9.3 The Relationship Between Regulatory T cells and Diseases and their Use in Therapy

9.3.1 The Relationship Between Regulatory T cells and Diseases

$CD4^+CD25^+Foxp3^+$Treg was found to affect multiple diseases or pathogenic infections by suppressing the immune response (Table 9.2).

9.3.2 Application of Regulatory T cells to Immunotherapy

Because of the strong affinity of Treg's TCR for the self-antigenic peptide-MHC complex, Treg has a wide range of promising applications in the prevention and control of autoimmune diseases as well as anti-transplant rejection. However, there are challenges of using Treg for anti-infection and anti-tumor immunotherapy in the clinical setting.

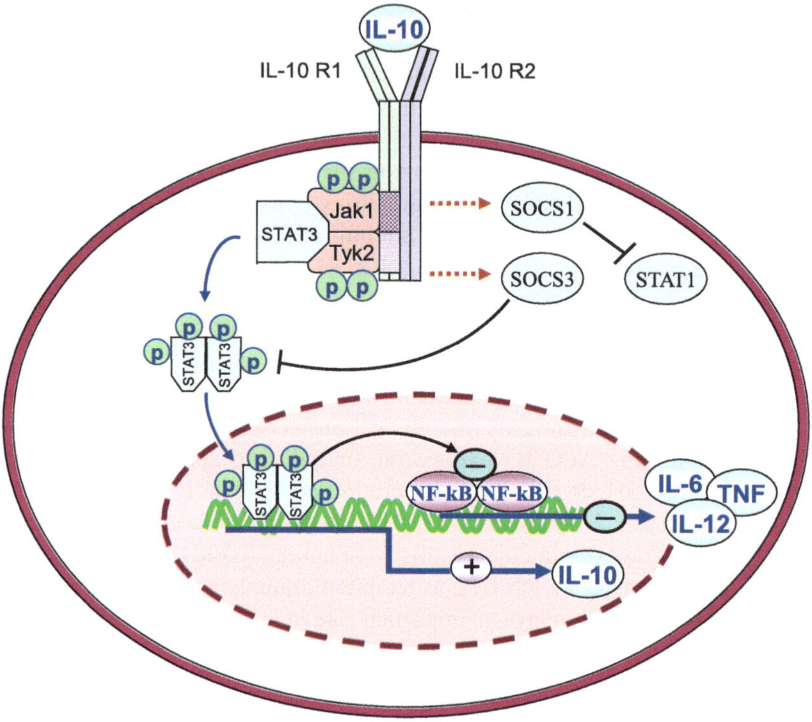

Fig. 9.9 Schematic illustration of the molecular mechanism of IL-10 action. The IL-10 receptor is a tetrameric molecule composed of IL-10R1 and IL-10R2 dimers, and is a transmembrane receptor. When IL-10 binds to the receptor, the intracellular region of the IL-10R1 molecule will be activated, recruiting and phosphorylating Jak1 and Tyk2 adaptor molecules, and the phosphorylated Jak1 and Tyk2 recruit and phosphorylate STAT3. The phosphorylated STAT3 enters the nucleus to bind to the IL-10 promoter and initiate IL-10 expression. STAT3 inhibits the activity of transcriptional regulator NF-κB and suppresses the expression of pro-inflammatory cytokines (IL-6, IL-12, TNF, etc.), thereby dampening the extent of the immune response.IL-10 also activates SOCS1 and SOCS3, and SOCS1 inhibits STAT1, thereby suppressing the expression of IFN-γ and exerting an inhibitory effect on the immune response

Table 9.2 The relation of $CD4^+CD25^+Foxp3^+$Treg to diseases

Disease	mechanism of Treg action	Effects of Treg on diseases
Rheumatoid arthritis	Inhibition of pro-inflammatory cytokine TNF-α expression	Amelioration
Type I diabetes	Inhibiting the expression of pro-inflammatory cytokines TNF-α and IFN-γ	Has preventive and control effects
Inflammatory bowel disease (IBD)	Inhibiting the expression of the pro-inflammatory cytokine IFN-γ and cell proliferation, and promoting the expression of IL-10 and TGF-β	Has preventive and curative effects
Neoplasms	Inhibiting tumor-specific CTL activity by expression of CTLA-4, PD-1L, LAG3 and other immune checkpoints	Inhibiting the antitumor activity of CTL
Viral infection	Inhibiting the expression of pro-inflammatory cytokine IFN-γ as well as specific CTL activity	unfavorable to virus clearance and easily helping virus establish persistent infections
Parasitic infection	Inhibiting expression of the pro-inflammatory cytokine IFN-γ and cell proliferation, and promoting IL-10 expression	Causing persistent infection and allowing host maintaining a carrier state
Allergic reaction	Inhibiting IgE production, inhibiting the expression of IL-4 and IL-13, and inhibiting eosinophil activity	Suppressing allergic reactions
Organ transplant rejection	Inhibiting cell proliferation, and promoting IL-10 expression	Suppressing organ transplant rejection

Study Questions

1. How many types of Treg are there? What are their main functions?
2. What are the main mode of action of Treg?
3. What are the promising clinical applications of Treg?

Chapter 10
Cytokines

Overview:
Cytokines are a class of active protein molecules secreted by cells that perform signaling and whose main chemical components are proteins or glycoproteins. Both immune and non-immune cells can secrete cytokines in response to certain factors, but different cells or subsets of cells secrete different cytokines, which exert regulatory effects by binding to membrane receptors and acting on target cells for signal transduction. This chapter focuses on the biological properties of cytokines, cytokine receptors, important cytokines secreted by different CD4T cell subsets, immunomodulatory effects of cytokines, and the relationship between cytokines and diseases.

10.1 Biological Properties of Cytokines

10.1.1 Modes of Cytokine Action

As cytokines act by binding to specific cell membrane receptors, the intracellular portion of receptors engaged by cytokines recruits adaptor molecules to transduce signals that initiate activation and expression of target genes (Fig. 10.1). Thus, the presence of a corresponding cytokine receptor in the cell membrane determines whether that cell responds to the cytokine. Cytokines bind to receptors with high affinity (10^{-12} to 10^{-10} kDa), and minute amounts (picomolar) of cytokines are able to bind to membrane receptors to exert biological effects.

Although cytokines act by binding to membrane receptors, the mode of cytokine action can be classified as autocrine, paracrine, and endocrine actions, depending on the mode of cytokine action on target cells. In autocrine action, the cell secretes cytokines for its own use only. In paracrine action, the cytokines act on nearby target cells. In endocrine action, the cytokines are transported through the peripheral circulation system and act on distant target cells (Fig. 10.2).

S. J. Zheng, *Veterinary Molecular Immunology*,
https://doi.org/10.1007/978-981-99-8929-4_10

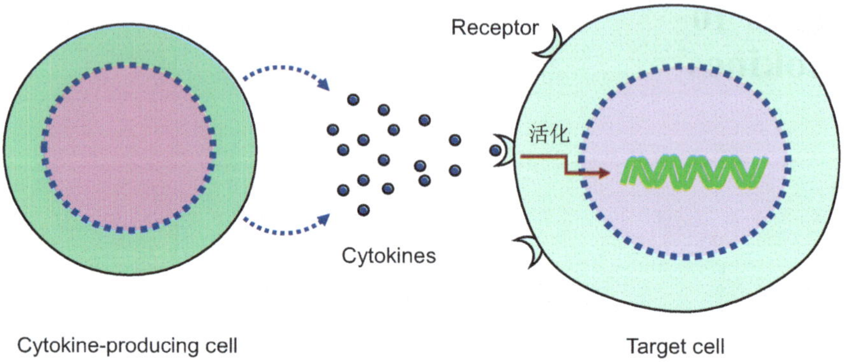

Fig. 10.1 Schematic illustration of mode of cytokine action

The mode of cytokine action is closely related to their half-life, and cytokines with a short half-life are hardly endocrine, but more autocrine or paracrine, such as IL-2. However, IL-1, IL-4, IL-6, IL-10, IL-12, IFN-γ, TNF, etc. play an endocrine role in regulation of an immune response. Therefore, when detecting cytokines in serum, all that can be detected are cytokines that play an endocrine role.

10.1.2 The Activity of Cytokines

The activities of cytokines are summarized in five categories: pleiotropy, redundancy, synergy, antagonism, and cascade induction.

1. **Pleiotropy** refers to the action of a cytokine on different target cells to exert different effects. For example, IL-4 acts on B cell for its activation, proliferation and differentiation, whereas it acts on thymocytes or mast cells to promote their proliferation.
2. **Redundancy** refers to the fact that two or more cytokines perform similar functions. For example, both IFN-α and IFN-β have antiviral effects, while IL-2, IL-4, and IL-5 all have proliferative effects on B cells.
3. **Synergy** refers to the additive effects of two cytokines acting together that is greater than the sum of their individual effects. For example, IL-4 and IL-5 acting together induce IgE production in B cells (isotype switching).
4. **Antagonism** refers to the inhibition or counteracting effect of one cytokine by another. For example, IFN-γ inhibits IL-4-induced antibody isotype switching in B cells.
5. **Cascade induction** refers to the fact that the action of cytokines on target cells causes the target cells to secrete more cytokines, a phenomenon known as cascade induction. **For** example, IFN-γ acts on macrophages to make them secrete IL-12, and IL-12 acts on activated Th cells to make them secrete Th1 cytokines such as IFN-γ, TNF-α, IL-2, and others.

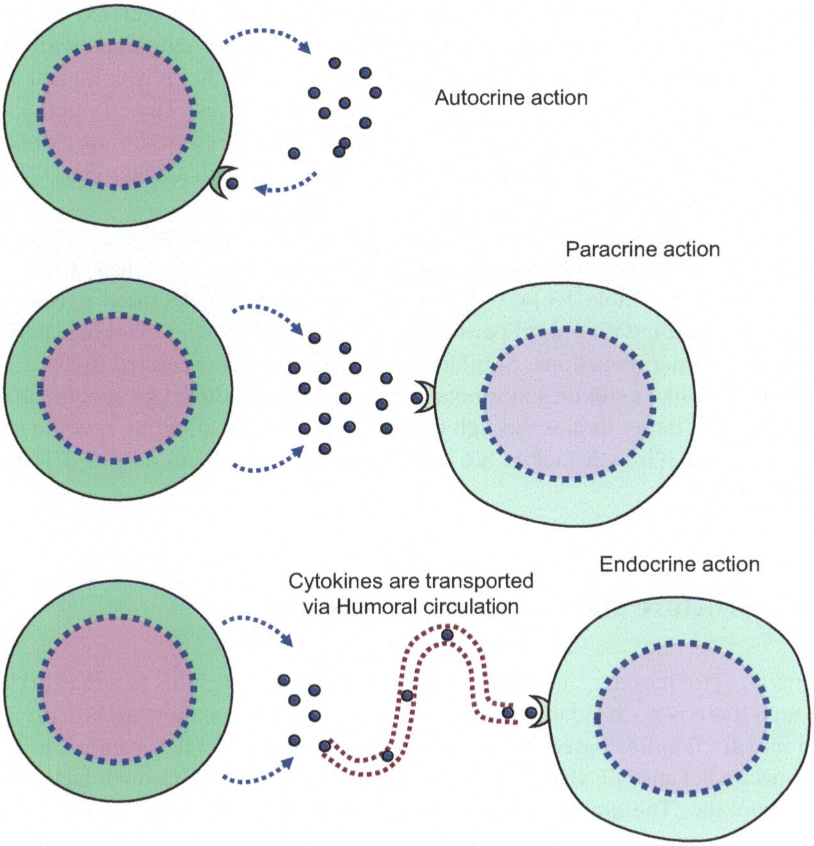

Fig. 10.2 Schematic diagram of the mode of cytokine action

10.2 History and Coverage of Cytokine Research

The role of cytokines was discovered in the 1960s when it was found that the supernatant of lymphocyte cultures contained components that had a role in regulating cell proliferation, differentiation and maturation. These substances were later found to be produced by lymphocyte after activation. However, due to the low content of these substances, it was difficult to obtain purified cytokines because there was no proper detection method, so it was not until the 70s to 80s that, with the development of gene cloning and expression technology, purified cytokines were obtained, thus confirming that cell growth mainly relied on the action of cytokines. In addition, due to the wide use of monoclonal technology, immunoassay techniques developed using monoclonal antibodies were able to detect cytokines very easily, which greatly promoted the process of cytokine research.

Cytokines cover lymphokines, monokines, interleukins, and chemokines. Lymphokines are cytokines secreted by lymphocytes, monokines are cytokines secreted by monocyte-macrophages, interleukins are cytokines secreted by leukocytes and act among leukocytes, chemokines are a class of cytokines with small molecular weights that have effects on cell chemotaxis or other physiological properties and play an important role in inflammatory responses. Thus, lymphokines, monokines, interleukins, and chemokines are collectively referred to as cytokines. Some cytokines have additional commonly-used names, such as interferons (IFNs) and tumor necrosis factors (TNFs).

Although cytokines, hormones, and growth factors are all substances secreted by cells and act on target cells by binding to receptors, there are clear differences between the three (Table 10.1). Cytokines are secreted by cells after activation in response to stimuli, with a short period of secretion, and most of cytokines perform autocrine or paracrine actions. Similarly, hormones are also secreted by cells after stimulation, Unlike cytokines, hormones are generally produced by specific glands and often reach target tissues through the body circulation, affecting specific target organ or tissues. Growth factors are continuously secreted and do not depend on external stimuli.

10.3 Structure and Function of Cytokines

In general, cytokines are small molecules with molecular weights less than 30 kDa. Although there is no standard for classification of cytokines, they are broadly classified into six families based on function, namely interferon (IL), interferon (IFN), tumor necrosis factor (TNF), Colony stimulating factor (CSF), Growth factor (GF), and chemokine. The amino acid sequence homology of the same cytokine varies among different species, only about 12–55% similarity between avian and mammalian cytokines. For example, the amino acid sequence homology between avian and mouse cytokines are: 15.33% in IL-2, 22.2% in IL-4, 38.33% in IL-10, 37.55% in IL-12p35, 66.8% in IL-12p40, 48.17% in IL-17A, 20.71% in IFN-α, 12.32% in IFN-β, 23.97% in IFN-γ, 54.34% in TGF-β1, 13.33% in TNF-α, and so on. (see Appendix 8 for details). The amino acid sequence homology between avian and human cytokines are: 17.11% in IL-2, 18–23% in IL-4, 43.89% in IL-10, 12.82% in

Table 10.1 Comparison of cytokines, hormones, and growth factors

Comparison	Cytokine	Hormones	Growth factor
Secretion needs to be stimulated	Need	Need	Unneeded
Source of secretion	Cell	Specific organs	Cell
Major mode of action	Endocrine/autocrine/paracrine effects	Endocrine effects	Endocrine effects
Secretion time	Transcient or short term	Transcient or short term	Long term
Target tissue	Multipe types of cells	Specific target cells or tissues	Multipe types of cells

IL-12p35, 43.39% 12p40, 43.75% in IL-17A, 19.7% in IFN-α, 19.9% in IFN-β, 29.45% in IFN-γ, 54.84% in TGF-β1, 15.42% in TNF-α, and so on. (see Appendix 9 for details). The amino acid sequence homology of swine cytokines with that of human and mouse cytokines is relatively close, at approximately 40–95% (see Appendices 10 and 11 for details). Structure determines function, and it is not yet clear how the large differences in amino acid sequences between avian and mammalian cytokines affect function, but a special detection kit for poultry cytokines must be used when assaying avian cytokines to obtain reliable data.

In practice, structurally similar cytokines are often classified into a subfamily. This chapter mainly introduce some representative mammalian cytokine that have been well-studied in recent years, such as IL-10, IL-12, IL-17, IFN, TNF and other subfamilies, which are mainly related to the regulation of immune response and diseases.

10.3.1 IL-10 and Related Cytokines

IL-10 is an important immune regulatory cytokine. If cytokines structurally similar to IL-10 are classified as the IL-10 cytokine family, the current members of this family include 9 cytokines, and they are IL-10, IL-19, IL-20, IL-22, IL-24, IL-26, IFN-λ1 (IL-29), IFN-λ2 (IL-28A) and IFN-λ3 (IL-28B). The IL-10 cytokine family requires binding to cell membrane receptors, which are heterodimers consisting of two transmembrane peptide chains, one chain longer and the other shorter (Fig. 10.3).

Upon binding of the IL-10 cytokine family to their corresponding receptor, two transmembrane protein peptide chains come in close proximity to form a heterodimer, and the intracellular functional region of cytokine receptor is activated to recruit an adaptor molecule that acts through the Jak/STAT signaling pathway (Fig. 9.9).

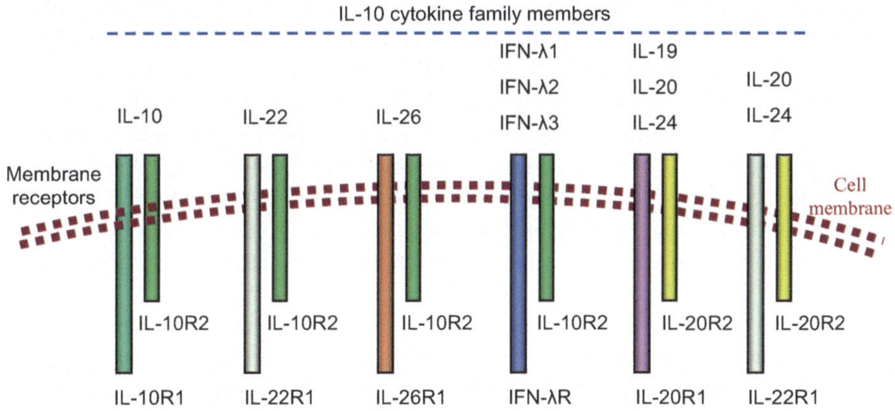

Fig. 10.3 Schematic diagram of IL-10 cytokine family and receptors

Based on function, IL-10 family cytokines can be broadly divided into three subfamilies.

1. IL-10 plays a major immune regulatory role.
2. The IL-20 subfamily includes IL-19, IL-20 and IL-24, which mainly play roles in antidefense (especially against bacterial and fungal infection).
3. Type III interferons, including IFN-λ1 (IL-29), IFN-λ2 (IL-28A), and IFN-λ3 (IL-28B), mainly act as antiviral factors.

IL-10 cytokines tend to play very different dual roles in different situations, sometimes playing immunosuppressive roles (suppressing immune damage as well as autoimmune diseases) and sometimes promoting inflammatory responses, but the main function is related to immune regulation.

10.3.2 IL-12 and Related Cytokines

The IL-12 cytokine family includes cytokines such as IL-12, IL-23, IL-27, and IL-35 IL-12 was first discovered in 1989 and the molecule has the ability to activate NK cells to produce IFN-γ. Further studies revealed that the molecule is a heterodimeric complex consisting of α (P35) and β (P40) subunits that induce an immune response in cells. There are currently four members of the IL-12 cytokine family, each of which is a heterodimeric structure consisting of α and β subunits but with differences in molecular size. These four members are IL-12 (p35/p40), IL-23 (p19/p40), IL-27 (p28/Ebi3), and IL-35 (p35/Ebi3) (Fig. 10.4).

The IL-12 cytokine family includes the cytokines IL-12, IL-23, IL-27, and IL-35, which are composed of alpha (p35, p28, p19) and beta (p40, Ebi3) subunits, respectively.

During pathogenic infection, PRRs recognize PAMPs, and dendritic cells (DCs) and macrophages (Mϕ) are activated and secrete large amounts of IL-12. With the progress in research on IL-12 in the mid-1990s, a line of experimental evidence showed that IL-12 has important immunomodulatory functions, mainly inducing a cell-mediated immune response. IL-23 and IL-27 were discovered in the early 21st century, and IL-23 shares a common beta (p40) subunit with IL-12 and therefore has similar biological functions. In addition, IL-23 induces the Th17 immune response. IL-27 inhibits the Th1 and Th17 immune responses and induces IL-10 production by Treg. IL-35 is a newly discovered member of the IL-12 cytokine family in recent year, and is now found to play an immunosuppressive role. All members of the IL-12 cytokine family exert their biological functions through binding to the corresponding membrane receptors. IL-12 cytokine family members (IL-12, IL-23, and IL-27) except for IL-35 have well-defined membrane receptors, and consist of two transmembrane protein peptide chains forming a heterodimer structure (Fig. 10.5).

The receptors for IL-12 cytokine family members are all heterodimeric structures consisting of subunits such as IL-12Rβ1, IL-12Rβ2, IL-23R, WSX-1, and gp130.

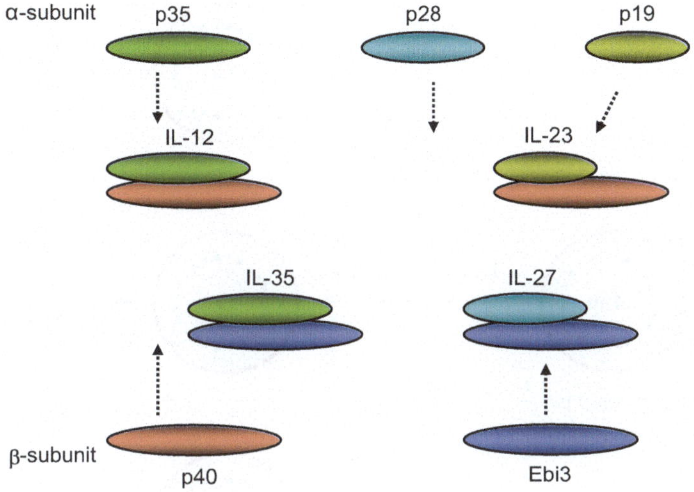

Fig. 10.4 Schematic diagram of IL-12 family cytokines

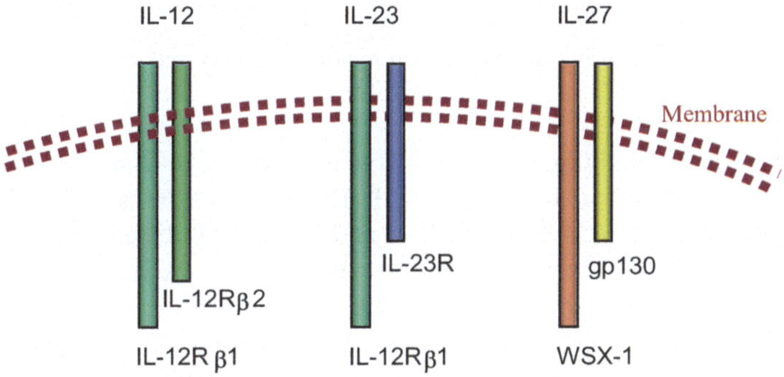

Fig. 10.5 Schematic diagram of the receptors for IL-12 cytokine family members

As mentioned above, members of the IL-12 cytokine family have common aspects in terms of function, but also have significant differences. All have different applications in clinical trials, such as the use of IL-12 as an adjuvant to enhance the immunization effect of vaccines in practice, and the use of IL-27 with IL-35 to suppress the autoimmune inflammatory response. In immune response studies, IL-12 is often used to induce Th0 to differentiate into Th1 and produce a Th1 immune response (Fig. 10.6).

IL-12 is an important pro-inflammatory response cytokine and a key regulatory molecule in both the innate and adaptive immune responses (Fig. 3.1), and IL-12 acts on target cells via binding to its receptor to exert effects through the Jak/STAT signal transduction pathway (Fig. 10.7).

After activation (e.g. stimulated with anti-CD3 and CD28 antibodies), the naive CD4T cells (Th0) differentiate into Th1 cells if IL-12 and anti-IL-4 antibodies are

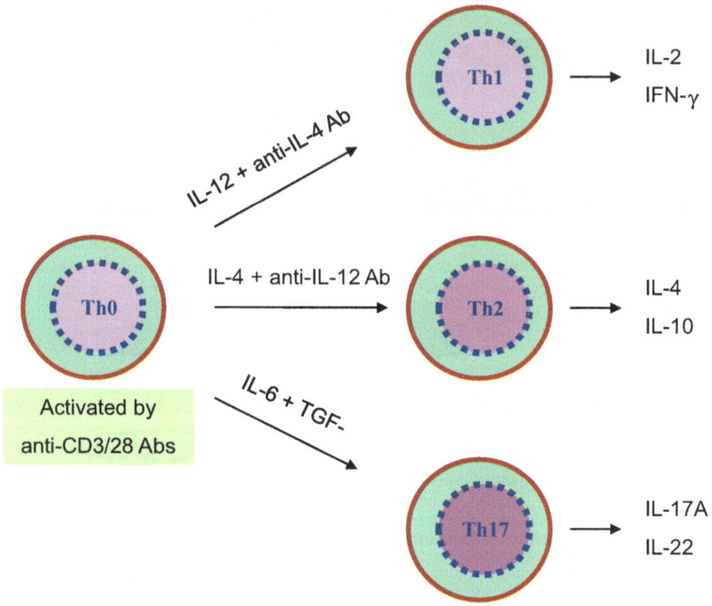

Fig. 10.6 Schematic diagram of the differentiation of naive CD4T cell into different subsets of Th cells

added to the culture medium, producing large amounts of IL-2 and IFN-γ. Th0 differentiates into Th2 cells if IL-4 and anti-IL-12 antibodies are added to the culture medium, producing large amounts of IL-4 and IL-10. Th0 differentiates into Th17 cells if IL-6 and TGF-β (transforming growth factor-β) are added to the culture medium, producing large amounts of IL-17A and IL-22.

After IL-12 binds to the receptor IL-12β1/2 heterodimer, β2 recruits Jak1, and β1 recruits Tyk2, which phosphorylates the tyrosine of Jak1/Tyk2 kinase, adaptor molecules. The activated Jak1/Tyk2 further recruits and phosphorylates STAT1, STAT3, STAT4 and STAT5, and the activated STAT1/3/4/5 molecules enter the nucleus through the nuclear membrane gap and bind to the promoters of specific genes to initiate the expression of the relevant genes.

10.3.3 IL-17 and Related Cytokines

The IL-17 cytokine family is a group of important pro-inflammatory cytokines that have been discovered in recent years. The family currently has six members, IL-17A, IL-17B, IL-17C, IL-17D, IL-17E, and IL-17F. Among these cytokines, IL-17A is commonly referred to as IL-17, and IL-17E also known as IL-25. IL-17A is the first member of the family to be discovered, along with the discovery of Th17 cells in 2005. Because this molecule induces the secretion of pro-inflammatory

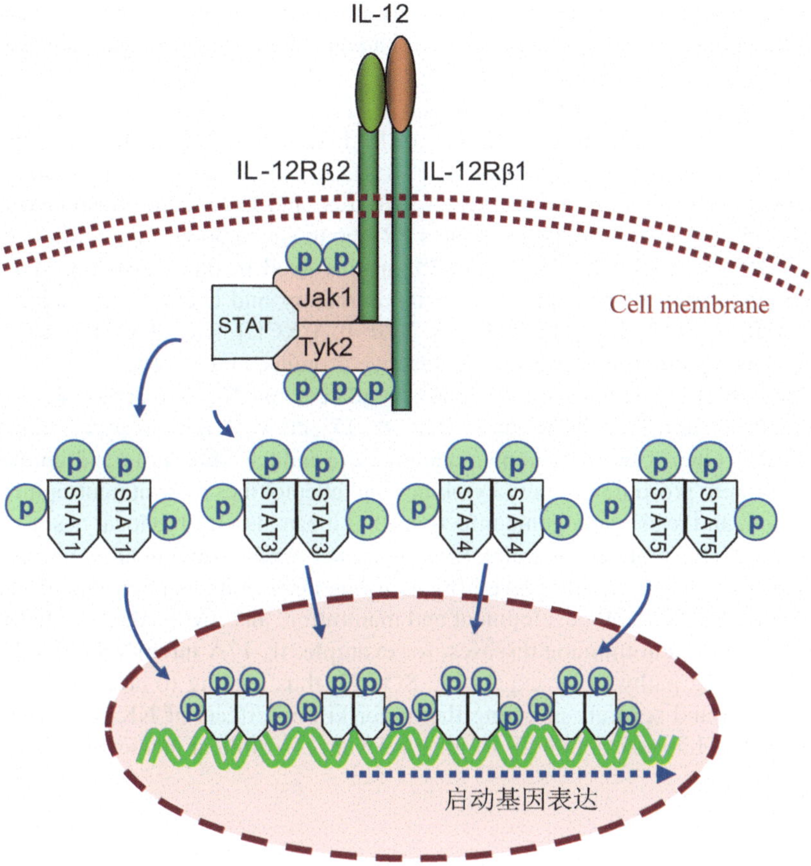

Fig. 10.7 Schematic diagram of IL-12-mediated signal transduction pathway

cytokines by a variety of cells, IL-17A expression is tightly regulated. Although multiple cells can express IL-17A, Th17 and γδT are the major cells that produce IL-17A. IL-17F has the highest homology with IL-17A, with 50% amino acid sequence homology, and is also functionally similar to IL-17A, but differs in that IL-17F is also associated with allergic inflammatory responses, and Th17 and γδT cells are also the major cells that produce IL-17F. IL-17E (IL-25) is more different from IL-17A, induces mainly Th2 immune responses, and is therefore associated with allergic reactions and anti-helminthic immunity. This cytokine is mainly secreted by epithelial cells, eosinophils, basophils and mast cells. Less is known about other IL-17 cytokines such as IL-17B, IL-17C, and IL-17D, which are the focus of future research.

IL-17 cytokines exert their biological effects by binding to target cell membrane receptors (IL-17R), and five members of the IL-17R family have been identified, namely IL-17RA, IL-17RB, IL-17RC, IL-17RD, and IL-17RE. IL-17RA is the first receptor identified in this family, and IL-17RA itself can form a homodimer or a heterodimer with IL-17RB or IL-17RC, and the dimer acts as a receptor that binds to the ligand and recruits Act1 and TRAF6 adaptor molecules to signal downward and initiate the immune response (Fig. 10.8).

IL-17RA recognizes both IL-17A and IL-17F, but IL-17A binds to IL-17RA with much greater affinity than IL-17F to IL-17RA. IL-17RC is the receptor for IL-17F. IL-17RB is the receptor for IL-17E (IL-25). IL-17 receptors recognize the corresponding of ligands and recruits Act1/CIKS, and have them phosphorylated. The phosphorylated Act1/CIKS activates TRAF6 to signal downstream. Molecules, such as NF-κB, ERK1/2, JNK, and p38, are involved in this signal transduction process and regulate the expression of inflammatory and chemotactic molecules, such as IL-4, IL-5, IL-6, IL-9, IL-13, TGF-β, CXCL1, CCL2 CCL7, CCL20, β-defensins, antimicrobial peptides, C-reactive proteins, and so on.

Although IL-17 expression is tightly regulated, the IL-17 receptor is widely expressed in multiple cells in vivo. Thus, IL-17 acts with pleiotropy on multiple target cells. One of the best known actions is to induce fibroblasts and epithelial cells to secrete pro-inflammatory cytokines and chemokines, recruit neutrophils for infiltration, and resist pathogenic infections, but it can also lead to local tissue damage because large numbers of granulocytes can cause severe inflammatory responses. Although IL-17 has anti-infective effects, it has been demonstrated that IL-17 is also involved in NK cell development and maturation, and involved in the development of various autoimmune diseases, for example, IL-17A inhibits STAT5 in the IL-15 signaling pathway by expressing SOCS3, thus leading to arrested NK cell development and severely affecting the tumor-killing effects of NK cells. Further elucidation of biologic activity is of great relevance to clinical medicine.

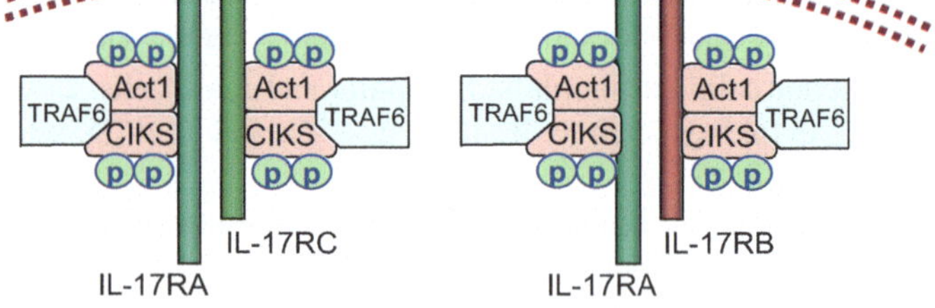

Fig. 10.8 Schematic diagram of IL-17-mediated signal transduction

10.3.4 Interferons and Related Cytokines

Interferon (IFN) plays an important role in the early immune response and is the first line of host defense against viral infection. Interferons exert their biological effects by binding to the membrane receptors of target cells, and depending on the binding membrane receptors, interferons are classified into three types, type I, type II and type III interferons. It is currently known that type I interferons include subtypes IFN-α, IFN-β, IFN-ε, IFN-κ, and IFN-ω. Among IFN-α subtypes, there are 12 highly structurally similar members, and no additional members among the other subtypes have been identified. The recognition receptor for type I interferons is the IFN-α/β receptor (IFNAR), which consists mainly of two parts, IFNAR1 and IFNAR2, is widely expressed in a variety of cells, and initiates the immune response signaling process after engagement with type I interferons (Fig. 10.9). Type II interferon has only one member, IFN-γ, which is mainly expressed by activated NK, NKT and T cells. The receptor for IFN-γ is IFNGR, which is a heterodimer and is widely expressed in a variety of cells. Type III interferons are IFN-λ family molecules discovered in recent years, and are also classified as members of the IL-10 family because of their structural similarity to IL-10. Type III interferons include IFN-λ1 (IL-29), IFN-λ2 (IL-28A), and IFN-λ3 (IL-28B), and the receptor for IFN-λ is also a heterodimer, consisting of IL-10R2 and IL-28Rα (Fig. 10.9), and because IL-28Rα is expressed only in epithelial cells, the action of IFN-λ is tissue-specific.

Interferons play an important role in anti-defense and immune regulation. Type I interferon is an important component of innate immunity, especially in combating virus and intracellular bacteria, and IFN-α has been shown to be effective in the clinical treatment of hepatitis C. Type II interferon (IFN-γ) is an important immune regulator that participates in the entire process of immune response, regulates the course of immune response (Fig. 3.1), and has an important role in promoting cell-mediated immunity in particular (Fig. 9.1), and IFN-γ can be detected clinically for disease diagnosis (e.g., tuberculosis). Because type III interferon receptors are restricted to expression in epithelial cells and certain subpopulations of dendritic cells (plasmacytoid DCs), the action of type III interferon is tissue-specific. The preliminary information available suggests that type III interferon may be associated with allergic reactions and autoimmune diseases.

10.3.5 Tumor Necrosis Factor Superfamily and Corresponding Cell Receptors

The tumor necrosis factor superfamily (TNFSF) and its corresponding receptor members are numerous, with more than 50 identified, and these molecules are evolutionarily conserved, expressed in all mammals, and play important roles in organ development, metabolism, and immune response. Many members of the TNF

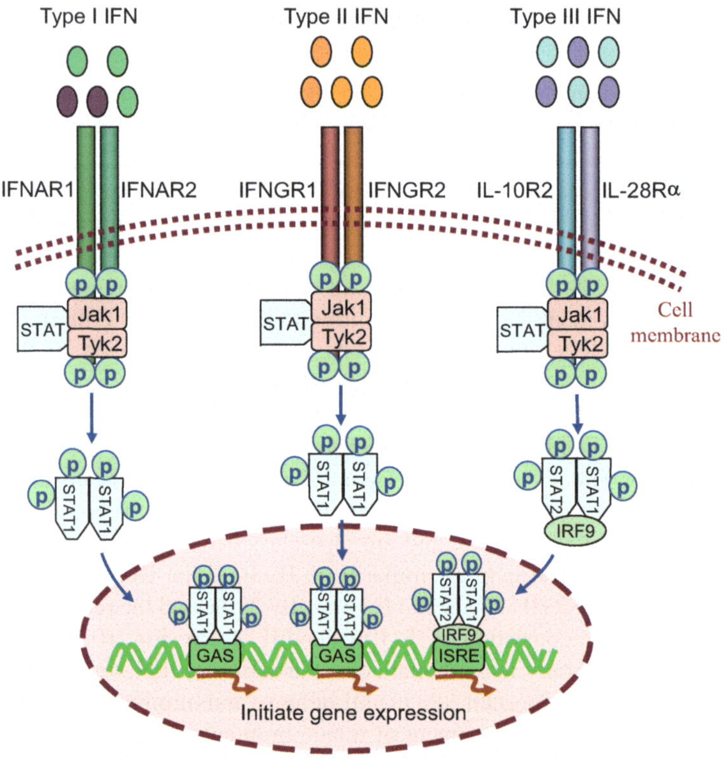

Fig. 10.9 Schematic diagram of interferon-mediated signal transduction. The type I interferon receptor is composed of two parts, IFNAR1 and IFNAR2, and is expressed widely in multiple types of cells. The type II interferon (IFN-γ) receptor is IFNGR, which belongs to a heterodimer and is widely expressed in a variety of cells. The type III interferon (IFN-λ) receptor is also a heterodimer, consisting of IL-10R2 and IL-28Rα, but IL-28Rα expression is restricted to epithelial cells, making the action of IFN-λ tissue-specific. All interferon receptors mediate immune response via the Jak-STAT signaling pathway. Type I and type II interferon receptors receive signals, recruiting and phosphorylating STAT1 via Jak/Tyk adaptor molecules. Phosphorylated STAT1 enters the nucleus as a dimer and binds to promoters containing specific GAS sequences to initiate expression of the corresponding genes. The type III interferon receptor receives the signal and recruits STAT1 and STAT2 via the Jak/Tyk adaptor molecules, the phosphorylated STAT1/STAT2 heterodimer recruits IRF9, and STAT1/STAT2/IRF9 co-enters the nucleus, binds to a promoter containing a specific ISRE sequence, and initiates the expression of the corresponding gene. *Jak1* Janus kinase 1, *Tyk2* tyrosine kinase 2, *STAT* signal transducer and activator of transcription, *GAS* gamma-activated sequence, *IRF* interferon-regulatory factor, *ISRE* interferon-stimulated response elements

family share similar structures and biological functions, and the basic structures are shown in Fig. 10.10.

Because TNF is a proinflammatory cytokine involved in a variety of inflammatory responses, special attention has been paid to the biological role of members of the TNF family by inhibiting the action of TNF in clinically chronic inflammation,

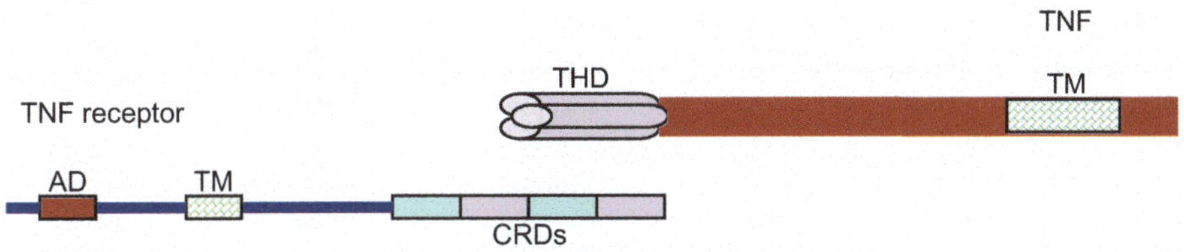

Fig. 10.10 Schematic diagram of TNF ligand and receptor. The TNF ligand family has 18 genes encoding 19 type II transmembrane proteins (N-terminal for intracellular and C-terminal for extracellular, with the C-terminal region conserved as THD), with 20–30% homology of THD among members. *THD* TNF homology domain, *TM* transmembrane domain, *CRDs* cysteine-rich domains, *AD* activation domain

especially in the treatment of arthritis. The TNF superfamily and associated diseases are shown in Table 10.2.

Among the TNF family, TRAIL (TNF-related apoptosis inducing ligand) has been investigated with a great enthusiasm since this molecule can induce apoptosis in tumor cells but not in normal somatic cells. Thus, people cherish a great hope for the role of TRAIL in tumor therapy, and for this reason the molecular mechanism of TRAIL action has been intensively studied. TRAIL is currently found to be expressed in a variety of cells, and the receptors on which TRAIL acts are TRAIL-R1 (DR4), TRIAL-R2 (DR5), TRIAL-R3 (DcR1), TRAIL-R4 (DcR2), OPG (Osteoprotegerin) (Fig. 10.11). Since TRIAL receptors (DR4, DR5) have an extracellular recognition region (mainly cysteine-rich region, referred to as CRD) intracellular activation region (referred to as AD), the CRD at the receptor is engaged by TRAIL, which generate a signal transduced into the intracellular region to activate AD, and the activated AD recruits adaptor molecules for signal transduction, resulting in biological effects. However, if TRAIL binds to the DcR, the DcR is unable to transduce signals as it has no intracellular AD and therefore cannot produce biological effects. As such, target cells receive or block TRAIL-mediated apoptotic signal by selectively regulating the expression of DRs or DcRs. Because of the predominant expression of DcRs in normal cells and the predominant expression of DRs in tumor cells, these differential expressions determine that TRAIL has a selective killing effect on tumor cells, giving a new hope for effective tumor treatment. Among the TNF family, FasL, LIGHT, LT-α, TNF-α, and RANKL all have multiple receptors, and differential expression of the corresponding receptors determines the biological effects of these ligands when they recognize target cells.

TRAIL has 5 receptors, TRAIL-R1 (DR4), TRAIL-R2 (DR5), TRAIL-R3 (DcR1), TRAIL-R4 (DcR2), Osteoprotegerin (OPG). DR4 (death receptor 4) is mainly found in humans, while DR5 (death receptor 5) is mainly found in mice. Target cells receive or block TRAIL-mediated apoptotic signals by selectively regulating the expression of DRs or DcRs.

Table 10.2 TNF superfamily and associated diseases (examples)

TNF family ligands (aka)	TNF family expression cell	TNF family receptors (aka)	TNF family receptor expression cell	Related diseases	Applications
TRAIL (APO2L, CD253, TNFSF10)	NK cells, T cells	TRAILR1 (DR4, CD26 TNFRSF10A), TRAILR2 (DR5, CD262, TNFRSF10B), TRAILR3 (DCR1, CD263, TNFRSF10C), TRAILR4 (DCR2, CD264, TNFRSF10D), OPG (TNFRSF11B)	Multiple cell expression, including tumor cells and normal cells	Neoplastic diseases	Eight products have entered clinical phase I–II
TWEAK (TNFSF12)	Monocytes, Dendritic Cells, NK Cells	FN14 (TWEAKR, CD266, TNFRSF12A)	Epithelial cells, endothelial cells, immune cells, multiple tumor cells	Neoplastic diseases, autoimmune diseases	Three products in clinical phase I–II
APRIL (CD256, TNFSF13)	Tumor cell	TACI (CD267, TNFRSF13B), BCMA (CD269, TNFRSF17)	Tumor cells, especially B-lymphocytoma	Neoplastic diseases, autoimmune diseases	One product in clinical phase I–II
RANKL (CD254, TNFSF11)	T-cell	RANK (CD265, TNFRSF11A), OPG (TNFRSF11B)	Osteoblasts, dendritic cells, Langerhans cells	Osteoporosis, inflammatory response	One product approved and 3 products in clinical phase I
OX40L (CD252, TNFSF4)	Dendritic cells, B cells, macrophages, T cells, NK cells, mast cells, endothelial cells, smooth muscle cells	OX40 (CD130, TNFRSF4)	T cells, NK cells, NKT cells, neutrophils	Inflammatory response, autoimmune disease, neoplastic disease	One product in clinical phase I

(continued)

Table 10.2 (continued)

TNF family ligands (aka)	TNF family expression cell	TNF family receptors (aka)	TNF family receptor expression cell	Related diseases	Applications
4-1BBL, (TNFSF9)	Dendritic cells, B cells, macrophages, T cells, NK cells, mast cells, smooth muscle cells, hematopoietic precursors, osteoblast precursors	4-1BB (CD137, TNFRSF9)	T-cell NK cells, NKT cells, neutrophils, mast cells, dendritic cells endothelial cells, eosinophils, osteoclast precursor cells	Inflammatory response, autoimmune diseases	2 more products in clinical phase I–II
CD27L (CD70, TNFSF7)	Dendritic cells, B cells, macrophages, T cells, NK cells, mast cells, smooth muscle cells, thymic epithelial cells	CD27 (TNFRSF7)	T cells, B cells, NK cells, NKT cells, blood-producing cell precursors	Tumors, autoimmune diseases	Five products have been approved and three more clinical phase I–II
TL1A (TNFSF15)	Dendritic cells, B cells, macrophages, T cells, endothelial cells	DR3 (TNFRSF25)	T cells, B cells, NK cells, NKT cells	Asthma, rheumatoid arthritis, colitis	
CD30L (CD153 and TNFSF8)	T cells, B cells, mast cells, monocytes, neutrophils, eosinophils	CD30 (TNFRSF8)	T cells, B cells, NK cells, macrophages, eosinophils	Diabetes, asthma, transplant rejection, colitis	One product has been approved and three more clinical phase I–II
CD40L (CD154 and TNFSF5)	T cells, B cells, eosinophils mast cells, monocytes, macrophages, NK cells, endothelial cells, smooth muscle cells, epithelial cells	CD40 (TNFRSF5)	Dendritic cells, B cells, macrophages, basophils, epithelial cells, endothelial cells, smooth muscle cells, fibroblasts	Inflammatory response, autoimmune disease	Seven products are in clinical phase I–II

(continued)

Table 10.2 (continued)

TNF family ligands (aka)	TNF family expression cell	TNF family receptors (aka)	TNF family receptor expression cell	Related diseases	Applications
LIGHT (CD258 and TNFSF14)	T cells, B cells, dendritic cells, NK cells, granulocytes, monocytes	HVEM (CD270 and TNFRSF14), LTβR, DcR3	T cells, dendritic cells, B cells, macrophages, NK cells, neutrophils, mucosal epithelial cells	Autoimmune disease	Two products in clinical phase I–II
LTα 1β 2	T cells, B cells, NK cells	LTβ R (TNFRSF3)	Stromal cells, epithelial cells, dendritic cells, macrophages, fibroblasts	Autoimmune diseases	One product in clinical phase I–II
GITRL (TNFSF18)	Dendritic cells, B cells, macrophages, T cells, endothelial cells	GITR (CD357. TNFRSF18)	T cells, B cells, NK cells, NKT cells, macrophages, dendritic cells	Autoimmune disease	One product in clinical phase I
BAFF (CD257, TNFSF13B) APRIL (CD256, TNFSF13)	Neutrophils, eosinophils, dendritic cells, B cells, macrophages, T cells	TACI (CD267, TNFRSF13B), BAFF-R (CD268, TNFRSF13C), BCMA (CD269, TNFRSF17)	B cells, T cells.	Autoimmune diseases, neoplastic diseases	Five products are in clinical phases I-III and one product has been approved

aka also known as, *APRIL* a proliferation-inducing ligand, *BAFF* B cell activating factor, *BAFFR* BAFF receptor, *BCMA* B cell maturation antigen, *CD27L* CD27 ligand, *CD30L* CD30 ligand, *DCR3* decoy receptor 3, *FASL* FAS ligand, *GITR* glucocorticoid-induced TNFR-related protein, *GITRL GITR ligand* HVEM, herpes virus entry mediator, *LTα* lymphotoxin-α, *LTβ* lymphotoxin-β, *LTβR* LTβ receptor, *OPG* osteoprotegerin, *OX40L* OX40 ligand, *RANK* receptor activator of NF-κB, *RANKL* RANK ligand, *SLE* systemic lupus erythematosus, *TACI* transmembrane activator and CAML interactor, *TNF* tumour necrosis factor, *TNFR1* TNF receptor 1, *TRAIL* TNF-related apoptosis-inducing ligand, *TRAILR1* TRAIL receptor 1, *TWEAK* TNF-related weak inducer of apoptosis

10.3.6 Tumor Growth Factors and Corresponding Cell Receptors

Tumor growth factor (TGF)-β, one of the most important immunoregulatory molecules, can be expressed by a variety of cells, with the highest expression in intestinal mucosal tissues, and is associated with mucosal immune tolerance. Mature TGF-β exists as a homodimer with a molecular mass of 25 kDa (12.5 kDa for monomers). TGF-β is cleaved from a 55 kDa pre-pro-TGF-β protein (Fig. 10.12)

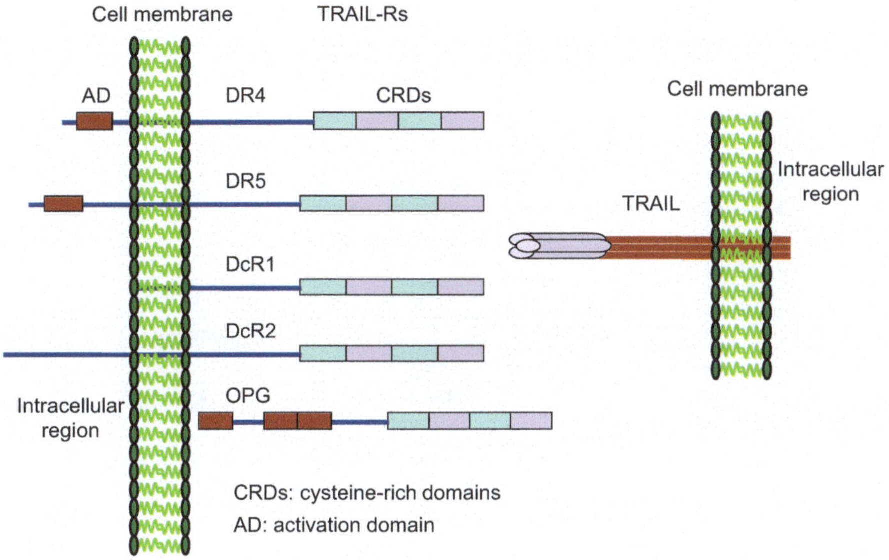

Fig. 10.11 Schematic diagram of TRAIL and its receptors

TGF-β is separated from LAP and exists as a homodimer. This TGF-β is active and binds to the corresponding membrane receptor for signaling and exerts a biological role (Fig. 10.13). TGF-β plays a negative regulatory role in the immune response, inhibits the immune response from overreaction so that the immune response is controlled within appropriate levels, thus playing a key role in immune tolerance. The most important experimental evidence is that TGF-β knockout mice develop autoimmune disease, indicating that TGF-β is important in immune regulation.

TGF-β is functionally pleiotropic, inducing the differentiation of naive T cells into Treg or Th17 subsets depending on the local microenvironment. Treg suppresses the immune response and elicits immune tolerance, whereas Th17 promotes the immune response, in particular, an inflammatory response (see the immune tolerance in Chap. 13 for details).

Pre-pro-TGF-β is a macromolecule consisting of signal peptide, LAP (latency-associated protein), and TGF-β with a molecular mass of 55 kDa. Under the action of protease, the signal peptide is cleaved off to form pro-TGF-β, subsequently, under the action of Furin convertase, the LAP-TGF-β heterodimer complex is formed. Through matrix protein modification, LAP is disassociated with TGF-β to form their respective homodimers, which are secreted outside the cell to function

The homodimeric TGF-β binds to the corresponding membrane receptor TGF-βRII, which further binds to and phosphorylates the receptor TGF-βRI, the phosphorylated TGF-βRI recruits and phosphorylates Smad2/3, and phosphorylated Smad2/3 form a heterodimer with Smad4, which enters the nucleus, binds to the

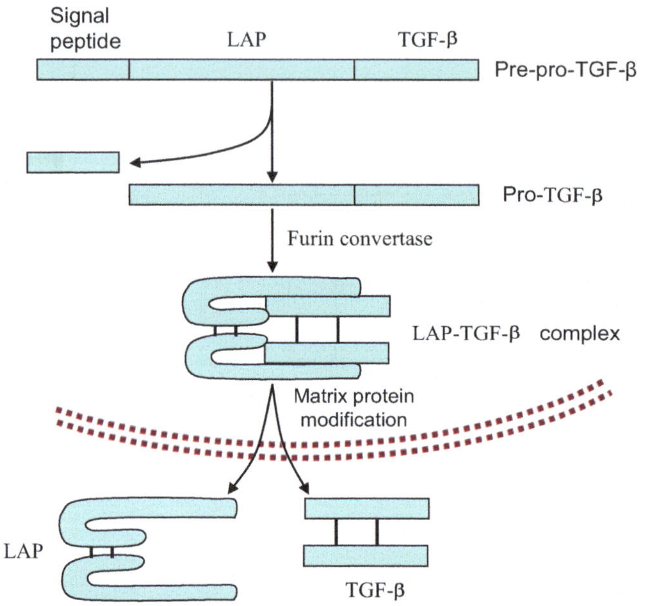

Fig. 10.12 Schematic diagram of the process of TGF-β formation

genomic promoter, initiates the expression of the corresponding genes, primarily TGF-β and IL-10, for immunoregulation.

10.4 Application of Cytokines in Practice

Because cytokines play an important role in immune regulation, cytokines or their antagonists can be used to enhance immunity or suppress immune response. The use of cytokines in the treatment of diseases has aroused much interest, such as the clinical use of cytokines for anti-infection, anti-tumor, and treatment of autoimmune diseases, or as adjuvants or immune enhancer to promote immune response.

In addition to injectable applications, cytokines can also be sprayed for treatment of respiratory diseases, such as spray of IFN-α, IFN-β, IFN-γ, GM-CSF and other cytokines for treatment of respiratory diseases, and the spray of cytokines have been approved for clinical use, mainly for the treatment of lung diseases (infections or tumors) or systemic diseases. Clinical use of IFN-α spray-type reagents for the treatment of viral infections of the respiratory tract or lungs (influenza, syncytial virus-induced bronchitis, tuberculosis, etc.) and lung tumors has yielded good results. Injection of IFN-α via subcutaneous route has also been used clinically to treat hepatitis B and C, and anti-cytokine monoclonal antibodies such as anti-IL-2, anti-IL-4, and anti-IL-5 are used to treat asthma. Appropriate and rational use of

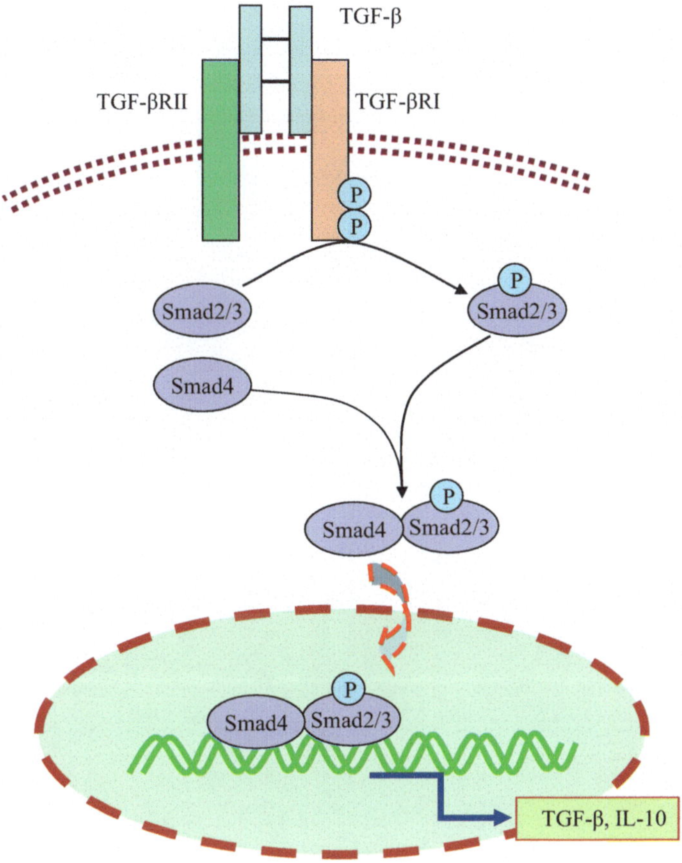

Fig. 10.13 Schematic illustration of TGF-β regulation of the immune response via binding to its receptor

cytokines can achieve a better adjuvant therapeutic effect and facilitate disease recovery.

Study Questions

1. What is the concept of cytokines?
2. What are the modes of action of cytokines?
3. What are the biologically active roles of cytokines?
4. How can naive CD4T cells be induced to differentiate into Th1, Th2, and Th17 cells?
5. What are the functions of cytokines such as IL-10, IL-12, IL-17, and TGF-β?
6. How many types of interferon are there? What are the main functions of each type of interferon?

Chapter 11
The Complement System

Overview:
The complement system contains a variety of soluble and membrane-bound natural immune protein molecules in the form of zymogen that, upon activation, are involved in innate and adaptive immune responses. The complement response is characterized by the activation of complement components that produce a series of cascade reactions that exert biological effects, including lysis of target cells and pathogens (bacteria and viruses), opsonin-mediated phagocytosis, immune complex clearance, and immune regulation. This chapter focuses on the pathways of action of the complement system and the relationship to disease.

11.1 History of the Discovery of Complements

Research on complement dates back to the year of 1890, when Jules Bordet of the Pasteur Institute in France discovered that antiserum to *Vibrio cholerae* from sheep lysed the bacteria, but lost its lytic activity when heated. However, if the heat-inactivated serum was added to a newly prepared serum that did not contain specific antibodies, it was able to restore the lytic activity of the bacteria. Bordet therefore hypothesized that there might be two components in the sheep's Vibrio cholera antiserum, a specific antibody against the bacteria that would be heat resistant and a substance that would lyse the bacteria that would not be heat resistant. To test this hypothesis, Bordet devised the erythrocyte hemolysis test, in which erythrocytes are immunized against another species of animals to obtain anti-erythrocyte antibodies, and erythrocytes binding to the antibodies subjected to hemolysis in response to complement. At the same time, Paul Ehrlich discovered the same phenomenon and referred to the serum active substance that assists the antibody in completing the hemolysis as complement, meaning that it helps the antibody to take action. It was later discovered that complement action is actually the result of the combined action of several protein molecules.

S. J. Zheng, *Veterinary Molecular Immunology*,
https://doi.org/10.1007/978-981-99-8929-4_11

11.2 Biological Properties of Complement

11.2.1 *Overview of the Biological Activities of Complement*

The complement system has more than 30 proteins, including both soluble and membrane-bound protein molecules, with biological activities spanning both innate and adaptive immunity. When activated, the complement system generates a cascade of reactions to exert biological effects, including those as follows.

1. Lysis of target cells and pathogens (including bacteria and viruses).
2. Opsonization, promoting phagocytosis of pathogens by phagocytes.
3. Clearance of immune complexes, promoting macrophage clearance of antigen-antibody complexes deposited in the circulatory system and tissues and organs.
4. Immunoregulation, certain complement binds to the membrane receptors of immune cells and influences cytokine secretion, thus exerting an immune regulatory effect (Fig. 11.1).

11.2.2 *Generation and Nomenclature of Complement Components*

The complement system is composed of a variety of components, whose main chemical constituents are proteins or glycoproteins. Complement components are mainly synthesized and secreted by hepatocytes, but monocytes-macrophages and mucosal epithelial cells also produce large amounts of complement. Complement components are found mainly in the body fluid circulatory system, in the form of zymogens, which are active only after cleavage and participate in cascade reactions.

The naming of complement components has not been unified, and currently includes C1 through C9, factor D, and other names. When a complement component is cleaved, the small fragment is usually named the a-fragment of the component and the larger one the b-fragment, e.g., C3 cleaves to become C3a and C3b, with the exception of C2a and C2b, where the C2a molecule is larger than C2b. Typically, after complement activates by cleavage, large fragments bind to target molecules and small ones bind to cell receptors to cause local inflammatory responses. Complement fragments or complexes superscripted with horizontal lines are used to indicate enzymatic activity, and complement-activated fragments bind together to form functional complexes, such as $C\overline{4b2a3b}$ and $C\overline{3bBb}$.

11.2.3 *Complement Activation Pathway*

Currently there are three complement activation pathways that are known: the classical pathway, the alternative pathway, and the lectin pathway.

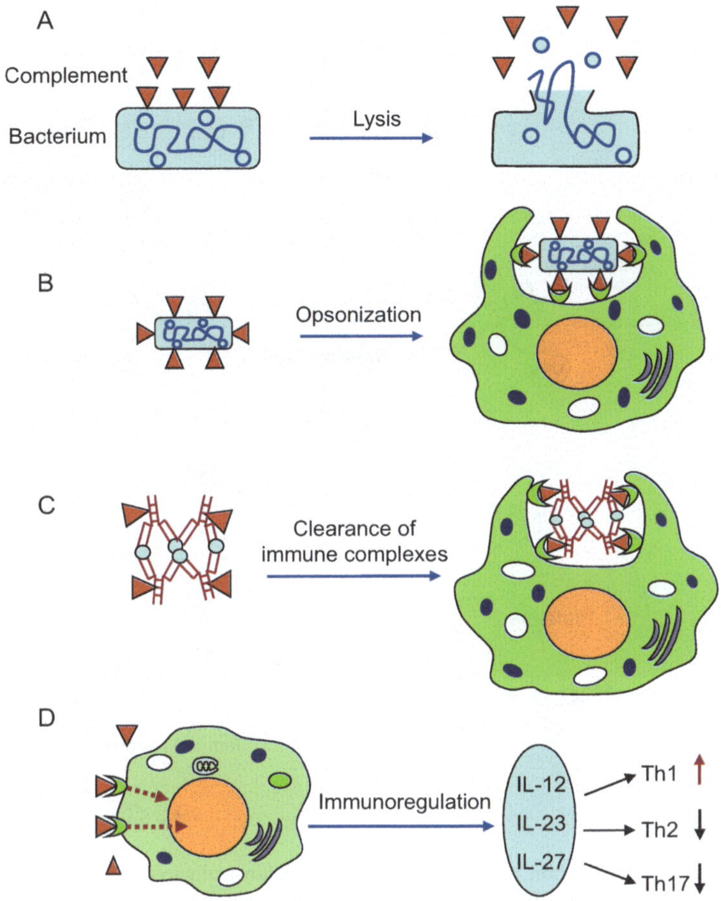

Fig. 11.1 Schematic illustration of the biological effects of complement. (**a**) Complement molecules act directly on pathogens or target cells to lyse them. (**b**) Complement has an opsonization effect, because phagocytes have complement receptors, after complement binds to pathogens, the complement promotes phagocytosis of pathogen by phagocytes through membrane receptor binding of complement-pathogen complexes. (**c**) Complement promotes phagocytosis of antigen-antibody complexes by phagocytes through binding to the complement-binding site in the Fc fragment of Ab-Ag complex and binding to membrane receptor of phagocyte. (**d**) Certain complement- cleaved fragments (e.g., C3a, C5a) bind to immune cell membrane receptors to activate cells to secrete immunoregulatory molecules such as IL-12, IL-23, and IL-27, promoting Th1 immune responses while inhibiting Th2 and Th17 immune responses

11.2.3.1 The Classical Pathway of Complement Activation

The classical pathway of complement activation is activated primarily by antigen-antibody binding to form complexes. Upon antibody-antigen binding, the antibody conformation is altered, exposing the complement binding site in the CH2 region of the Fc fragment, which binds to C1q, a subunit of complement C1, initiating the complement activation process (Fig. 11.2).

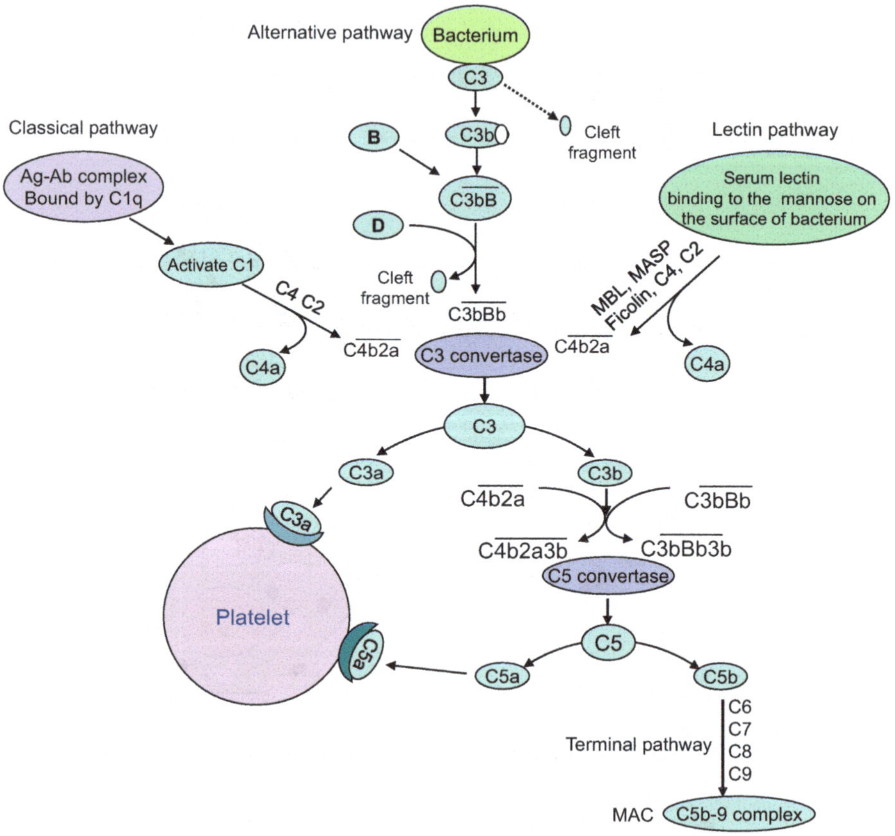

Fig. 11.2 Schematic diagram of the complement activation pathways. Complement activation pathways include the classical pathway, alternative pathway and lectin pathway, each with different activation initiation signals, but all eventually form C5b-9 membrane-attack complexes (MAC) through the terminal pathway, causing membrane damage. In the classical pathway, the primitive activation signal originates from C1q binding to the complement binding site in the CH2 region of antibody Fc fragment in the antigen-antibody complex, or to C-reactive protein (CRP) or to dead cells, which subsequently recruits and cleaves C4 and C2 complements to form $\overline{\text{C4b2a}}$ (C3 convertase). In the alternative pathway, the original signal can be a C3b fragment formed by cleavage of C3 by lipopolysaccharide, glycoprotein or cell membrane hydrolase of the bacteriophage, forming $\overline{\text{C3bBb}}$ (C3 convertase) in the presence of B- and D-factors. The initial signal for activating lectin pathway (MBL/MASP pathway) is the binding of mannose-binding lectin (MBL) as well as fibrin (ficolin) to mannose residues on the surface of the pathogen, which activates the MBL-associated protease (MASP1-3) that cleaves C4 and C2 complements to form $\overline{\text{C4b2a}}$ (C3 convertase).C3 convertase cleaves C3 to form C3a and C3b two fragments, C3b binds to C3 convertase to form C5 convertase ($\overline{\text{C3bBb3b}}$, $\overline{\text{C4b2a3b}}$), C5 convertase cleaves C5 to form C5a and C5b two fragments, C5b recruits C6, C7, C8, C9 to form C5b-9 membrane-attack complex to lyse target cells. The C4a fragment is involved in allergic response and inflammatory response, while C3a and C5a bind to specific receptors on platelet membranes and are involved in local inflammatory response. *MAC* membrane-attack complex, *MBL* mannose-binding lectin, *MASPs* MBL-associated serine proteases

Complement C1 is a macromolecular complex present in serum, consisting of 1 C1q, 2 C1r, 2 C1s and other subunits to form polymerized molecule (C1qr2s2), the short peptide of each subunit contains two functional domains, a binding domain and a catalytic domain. C1q binding to the complement binding site in the CH2 region of antibody Fc fragment causes a conformational change in C1r to form a catalytically active serine protease C1r, C1r cleaves C1s to become the active catalase C1s, the catalytic substrate of this enzyme is complements C4 and C2, and after catalysis, the product C4b2a is formed, serving as C3 convertase. Complement C3 is a heterodimer consisting of α and β chains. C3 convertase cleaves C3 to form C3a and C3b, C3b binds to C3 convertase to form a trimeric enzymatically active macromolecule C4b2a3b (C5 convertase), C5 convertase cleaves C5 to form C5a and C5b, initiating the terminal pathway where C5b binds to C6, 7, 8, and 9 to form a membrane-attack complex that directly destroys pathogens or target cells.

In addition to antigen-antibody complexes, anionic phospholipid analogs, liposome A, and E. coli can activate the classical pathway of complement.

11.2.3.2 Alternative Pathways of Complement Activation

The alternative pathway of complement activation (alternative pathway) is not antibody-dependent. The activation process of this pathway mainly involves C3, B-factor and D-factor. As mentioned earlier, C3 is a heterodimer composed of α and β chains connected by disulfide bonds. C3 is structurally unstable and spontaneously undergoes hydrolysis to form C3a and C3b, which binds to cell membrane components (including bacteria or the host's own cells), but because the host cell surface has a large amount of sialic acid to inactivate C3b while the bacterial surface has only a small amount of sialic acid that cannot inactivate C3b, thereby recruiting and binding of B-factor to form the C3bB complex, which is a substrate for serum D-factor. In response to D-factor, B in C3bB is cleaved to form C3bBb (C3 convertase), which cleaves C3 to form C3a and C3b, C3b binds C4b2a and C3bBb to form C5 convertase (C4b2a3b, C3bBb3b), which cleaves C5 to form C5a and C5b, initiating the terminal pathway.

11.2.3.3 Lectin Pathway

Lectin pathway or mannose-binding lectin (MBL) pathway is also antibody-independent. The initial activation of this pathway is due to lectins in the host serum binding to mannose on the surface of the organism. For example, the glycoprotein molecules on the surfaces of Salmonella, Listeria, Candida albicans, and Cryptococcus contain mannose residues that bind to the acute response protein lectin to form MBL, which recruits and activates serine proteases MASP-1 and MASP-2, which are structurally and functionally similar to C1r and C1s. The activated MASP-1 and MASP-2 cleave and activate C4 and C2, thereby forming C4b2a (C3 convertase), and the subsequent process is identical to the classical pathway.

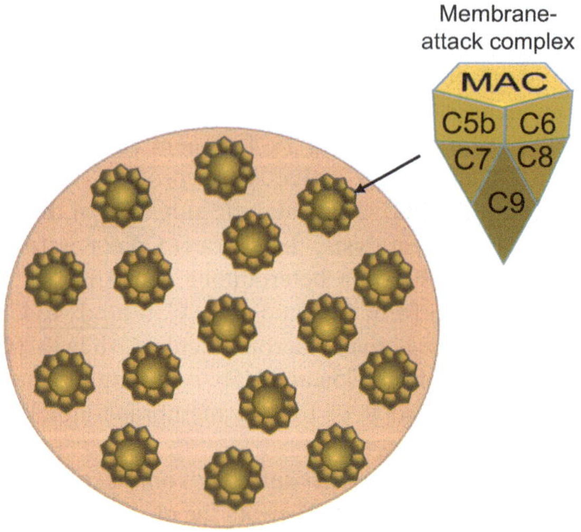

Fig. 11.3 Schematic illustration of cell membrane disruption by complements. After complement activation, the membrane-attacking complexes (MAC) are formed at the terminal phase, forming circular pores in the target cell membrane, disrupting the cell membrane, causing cells to leak a large amount of intracellular materials, and finally resulting in lysis of the cell

All three activation pathways of the complement system end up in the terminal pathway, where C5b recruits and bind to C6, C7, C8, and C9 to form the membrane-attack complex (MAC), causing membrane damage (Fig. 11.3).

11.2.4 Biological Effects of the Complement System After Activation

Activation of the complement system produces a variety of biological effects, such as lysis of target cells, antigen-opsonized phagocytosis, clearance of immune complexes, and immunoregulation of inflammatory responses. The exertion of these effects is dependent on the active fragments or membrane attack complexes (MACs) that are produced by the complement system during the activation process. A variety of cells in animals express complement receptors that bind to the active fragment after complement cleavage, and the binding of complement to specific receptors produces biological effects (Table 11.1).

The membrane-attack complex formed in the terminal pathway after activation of the complement system can lyse a wide variety of cells, such as Gram-negative bacteria, parasites, viruses, erythrocytes, and nucleated animal cells. Since the alternative and lectin pathways are not antibody-dependent during activation, and the classical pathway relies on antibody-antigen binding for activation as only one of many situations in which this pathway is activated, the complement system is part of innate immunity. The complement system plays an important role in host anti-defense response. As viral envelope is derived from the membrane of infected cells, it is susceptible to complement cleavage. The viruses that are known to be susceptible to complement cleavage include herpesviruses, orthomyxoviruses, paramyxoviruses and retroviruses.

Table 11.1 Complement receptors and the corresponding biological effects

Complement receptor	Complement components (ligands)	Cells expressing receptors	Biological effect
C3a/C4a receptor	C3a, C4a	Mast cells, basophils, granulocytes	Induces cell degranulation; promotes local inflammatory response
CR1 (CD35)	C3b, C4b	Blood erythrocytes, basophils, eosinophils, monocytes, macrophages, dendritic cells, certain T cells	Inhibits formation of C3 convertase; binds to immune complexes to enhance phagocytosis
CR2 (CD21)	C3d, C3dg	B cells, dendritic cells, certain T cells	Component of the BCR; combined with EBV
CR3 (CD11b/18) CR4 (CD11c/18)	iC3b	Monocytes, macrophages, neutrophils, NK cells, certain T cells	Binds to adhesion molecules on neutrophils and promotes cell migration to sites of inflammation; binds to immune complexes and enhances phagocytosis
C5a receptor	C5a	Mast cells, basophils, granulocytes, monocytes, macrophages, platelets, endothelial cells	Induces cell degranulation; promotes local inflammatory response

Note: iC3b is the remaining portion of C3b after it has been cleaved out of a small fragment

Most Gram-negative bacteria are susceptible to complement lysis, but some bacteria develop resistance to complement lysis by certain mechanisms, such as Escherichia coli and Salmonella, which develop resistance to complement lysis after growth change from a rough to a smooth colony type, mainly because the cell wall lipopolysaccharide (LPS) on the surface of the smooth type has many long polysaccharide side chains that prevent the membrane-attack complex of the complement from contacting the cell membrane. Gram-positive bacteria cell walls are less susceptible to lysis by membrane-attack complex due to the large number of peptidoglycan components that prevent the contact of the membrane-attack complex with the cell membrane. In addition, some pathogenic microorganisms have evolved multiple mechanisms to resist complement attack, such as pneumococci that prevent C3b from contacting the bacterium through the capsule. Some bacteria secrete some lipase to inactivate C3a and C5a, preventing them from causing an inflammatory response. Some pathogens secrete some complement inhibitory factors that inhibit the activation of the complement system.

11.2.5 Regulation of the Complement System

Although activation of the complement system is an important part of the innate immune response and plays an important role in pathogenic clearance, excessive complement activation can also have damaging effects on self tissues, so the complement system has evolved over time to develop fine regulatory mechanisms that play an important role in maintaining homeostasis. Factor H (involved in the activation of the complement alternative pathway) competes with C1q in the classical pathway to bind to anionic phospholipids, liposome A, and *E. coli* (all three of which activate the classical pathway of complement) and prevents classical pathway activation. Factor H binds to factor I as a complex to degrade C3b to iC3b and promote degradation of C3 convertase (C3bBb), thereby inhibiting the complement alternative pathway. Since all three complement activation pathways produce C3b, insufficient C3b production increases animal susceptibility, while excessive C3b production causes immune damage. Thus, the regulation of C3b by H- and I-factors is important, and these two factors serve as important regulators for the complement system. In addition, certain proteins outside the complement system also regulate the activation of complement system, such as vitronectin, which binds to the C5bC6C7 complex to inhibit membrane-attack complex formation, and molecules such as CD35, CD46, CD55, and CD59, which inhibit C3b deposition in the tissue. A variety of complement regulators have been identified, and their roles and the mode of action are shown in Table 11.2.

The lysis of cells is stronger if the complement is derived from a different species, a phenomenon associated with two proteins that inhibit the formation of the membrane-attack complex. These two proteins are HRF (homologous restriction factor) and MIRL (membrane inhibitor of reactive lysis, CD59) and are present in the membranes of a variety of cell types. HRF and MIRL bind to C8 to prevent C9 from forming a membrane-attack complex and prevent nonspecific cell lysis. However, complement inhibition by HRF and MIRL occurs only when the target cells are derived from the animals of the same species, and is abolished if the complement and target cells are derived from different species of animals. This is important in the clinical context of organ transplantation studies.

11.2.6 Immunoregulatory Effects of Complement

Complement fragments generated during complement activation play different roles in immune regulation, e.g. C3a fragment inhibits humoral immune response, while C5a promotes humoral immune response. The fragment C3d cleaved from C3b promotes B cell activation upon binding to B cell membrane protein CD21 (belonging to complement receptor 2, CR2), thus C3d can be used as an adjuvant to enhance the efficacy of vaccines. CD46 is a membrane receptor that cleaves C4b, C3b, or C3bBb and inhibits complement activation, and T cells also express CD46, which mediates immunoregulatory signaling and inhibits IFN-γ production by Th1 cells in favor of IL-10 production, suppressing the immune response and avoiding immune damage.

C3a, C4a and C5a are also known as allergic toxins, mainly because of the strong inflammatory response that these small molecules can induce. However, recent

Table 11.2 Regulators of the complement system and their functions

Complement regulator protein	Protein features	Protein function	Regulated complement pathways
C1 inhibitor (C1inh)	Soluble	Is a serine protease inhibitor that dissociates $C1r_2$ s_2 from C1q	Classic pathway
C4b binding protein (C4bBP)	Soluble	Binds to C4b to block formation of C3 convertase; works in association with factor I to cleave C3b	Classical pathway, lectin pathway
H-factor	Soluble	Block C3 convertase formation; cleave C3b; promote C3b binding to membranes	Alternative pathway
CR1 (complement receptor type I); MCP (membrane-cofactor protein, CD46)	Membrane-bound	Prevents formation of C3 convertase by binding to C3b and C4b; promotes cleavage of C4b, C3b or $\overline{C3bBb}$	Classical pathway, alternative pathway, lectin pathway
DAF (decay-accelerating factor, CD55)	Membrane-bound	Promote dissociation of C3 convertase	Classical pathway, alternative pathway, lectin pathway
I-factor	Soluble	Is a serine protease that acts with $\overline{C4bBP}$, CR1, H-factor, DAF, or MCP to cleave C3b and C4b	Classical pathway, alternative pathway, lectin pathway
S protein	Soluble	Binds to C5b67 and prevents it from inserting into the cell membrane	Terminal pathway
HRF (homologous restriction factor); MIRL (membrane inhibitor of reactive lysis, CD59)	Membrane-bound	Binds to C5b678 and prevents C9 binding	Terminal pathway
Allergic toxin inactivator	soluble	Inactivation of allergic toxins C3a, C4a and C5a	Phase of effect

studies have revealed that these allergic toxins are involved in the activation process of dendritic cells and macrophages, and have an important regulatory role in the T-cell immune response.

11.2.7 Complement and Disease

If an animal is genetically defective in a complement component, it will be more susceptible to the pathogenic infection and will be more susceptible to systemic lupus erythematosus (SLE) due to compromised clearance of immune complexes. Any defect in complement regulatory proteins will cause the corresponding clinical diseases, e.g., a defect in DAF or MIRL will cause paroxysmal nocturnal hemoglobinuria (PNH), leading to chronic hemolytic anemia in the long term. DAF and MIRL are both protein molecules on the cell membrane that have the ability to inhibit cell lysis by complements, but the underlying mechanism of their actions differs. DAF inhibits cell lysis by dissociating or inactivating C3 convertase in the classical, alternative and lectin pathways, while MIRL inhibits cell lysis by binding to C8 and preventing C9 from forming a membrane-attack complex. Thus, if DAF or MIRL is defective, complement-mediated lysis of cell are deregulated, allowing massive erythrocyte lysis and causing PNH, a commonly-seen related symptom being venous atherosclerosis, which severely affects liver function and causes progressive bone marrow failure, with patients surviving for an average of 10–15 years. In addition, defects in any component of the classical pathway of complement (C1q, C1r, C1s, C4, C2) cause similar clinical symptoms, such as the diseases associated with immune complexes SLE, glomerulonephritis, arteritis, etc. Patients also have recurrent septic bacterial (staphylococcal, streptococcal) infections, which are also resistant to complement-mediated cell lysis.

Patients with C3 deficiency develop recurrent bacterial infections and immune complex-related disorders, and C1 inhibitor (C1Inh) deficiency causes edema and inflammatory exudation in the respiratory or digestive tract. Strong light can activate the complement system at the base of the eye, and if the activation of the complement is not regulated in time, it can cause age-related macular degeneration (AMD), which is now found to be associated with C1Inh. CD46 deficiency causes SLE, asthma, and rheumatoid arthritis. Patients with hemolytic uraemic syndrome (HUS) are often associated with mutations in factor H, CD46, factor I, C3, and factor B. CR1 is a receptor for complement C3b and C4b, and present on a variety of cell membranes, and the receptor is polymorphic (polymorphism) in the population and associated with the development of Alzheimer's disease. Factor H is polymorphic and associated with the geographical distribution of human races.

Study Questions

1. What is the definition of complement?
2. What are the biological roles of complement?
3. What are the characteristics of the classical, alternative, and lectin pathways of complement activation?
4. What immunoregulatory effects do complements have?

Chapter 12
Hypersensitive Reactions and Autoimmune Diseases

Overview:
In the process of antigenic removal, the body sometimes develops abnormally excessive immune response that causes damage to self tissues and organs, which is called a hypersensitive reaction, or allergy. Autoimmune disease is a disease in which the immune system is unable to tolerate self tissues and organs and produces an immune response that causes damage to self tissues and organs. The immune damage caused by hypersensitivity reaction is due to overreaction, i.e. the damage caused by immune response is "unintentional", while the immune damage caused by autoimmune disease is due to intolerance to own tissues, i.e. the immune system actively recognizes and attacks self tissue antigens, and the damage caused by this immune response is "intentional". The main cause of hypersensitive reactions is a problem in the regulation of the immune response, whereas in autoimmune diseases, immune tolerance is broken. Both hypersensitive reactions and autoimmune diseases are pathological damage processes caused by abnormal immune responses.

12.1 Hypersensitive Reactions

In the early twentieth century, jellyfish stings often occurred on European beaches, and Prince Albert of Monaco met French professor Charles Richet on a cruise ship and suggested that he study jellyfish toxins. Richet, together with Georges Richard and Paul Portier, extracted the jellyfish synapses with glycerol, and injected the glycerol-soluble toxin into animals to successfully reproduce the symptoms of jellyfish stings. After returning to France, Richet used actinia as experimental materials to extract similar toxins due to the unavailability of Jellyfish. Together with Portier, Richet succeeded in obtaining the toxin from actinia. They calculated the lethal dose by administering the toxin to the dogs, while some of toxin-treated dogs did not die in

S. J. Zheng, *Veterinary Molecular Immunology*,
https://doi.org/10.1007/978-981-99-8929-4_12

the test, and those that survived generally lost their symptoms after 3–4 days, recovered fully after another 3–4 weeks, and could be used again in lethality tests. However, when Richet administered the toxin again to dogs that survived the first trial, it was found that very low doses of the toxin caused animals to develop unusually severe symptoms, including vomiting, bloody diarrhea, fainting, respiratory distress, and even death. Richet and Portier coined the word "anaphylaxis" to describe this overreaction. Anaphylaxis is derived from the Greek word phylaxis, which means protection. Richet was awarded the 1913 Nobel Prize in Medicine for his discovery of allergic reactions.

12.1.1 Types of Hypersensitive Reactions and the Underlying Mechanisms

As research on hypersensitive reactions has progressed, several different mechanisms have been found to cause hypersensitive reactions. P. G. H. Gell and R. R. A. Coombs proposed to classify hypersensitive reactions into four types. The first three of the hypersensitive reactions are caused by humoral immunity, i.e., they are related to antigen-antibody reaction, whereas the fourth type is related to cellular immunity and is also known as delayed hypersensitive reaction. This classification facilitates research, but there are no strict boundaries among the four types of allergic reactions in terms of immune damage. The Gell and Coombs classification of hypersensitive reactions is currently still followed. The four types of hypersensitive reactions include: type I, an IgE-mediated allergic reaction; type II, an antibody-mediated allergic reaction; type III, an immune complex-mediated allergic reaction; and type IV, a cell-mediated allergic reaction, also known as delayed hypersensitive reaction (DTH).

12.1.1.1 Type I Hypersensitive Reaction

Type I hypersensitive reaction, also known as IgE-mediated allergic reaction, is an immune response induced by an allergen. Unlike the normal immune response, allergens induce plasma cells to produce antibodies, mainly IgE, and the Fc fragment of IgE binds to Fc receptors on the surface of mast cells and basophils, and the process of IgE binding to cells is called sensitization. However, when sensitized mast cells and basophils encounter the same allergen again, IgE on the cell surface binds to the allergen, leading to cell degranulation, and the released granules contain large amounts of active substances that have the ability to cause vasodilation and smooth muscle contraction (Fig. 12.1).

Allergens activate B cells, the activated B cells differentiate into plasma cells that secrete IgE, IgE binds to the Fc receptors of mast cells or basophils, which is the sensitization process. When the sensitized cell encounters with allergens again, it releases active mediators such as histamine, heparin, 5 hydroxytryptamine, eosinophil chemotactic factor, neutrophil chemotactic factor, and proteases, causing local

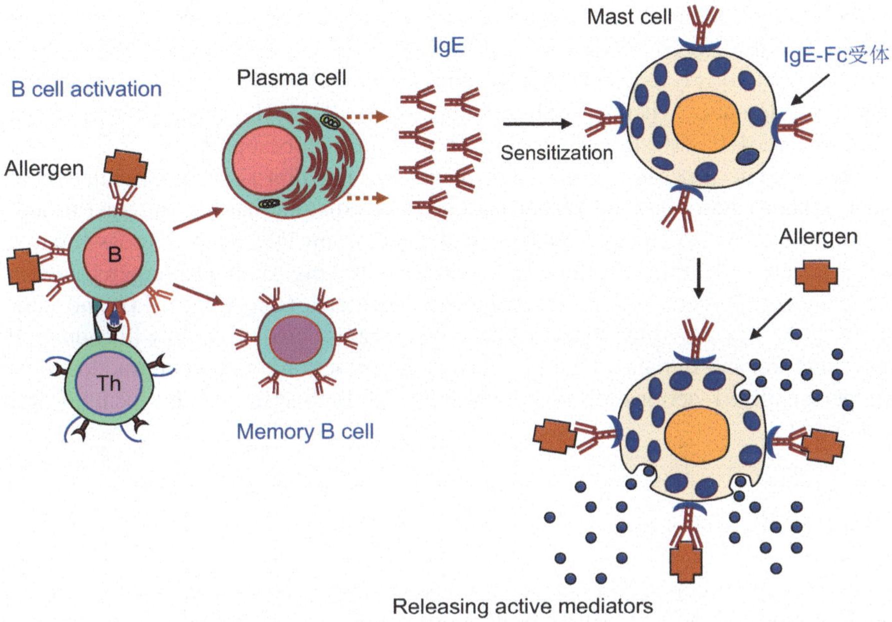

Fig. 12.1 Schematic illustration of the mechanism by which type I hypersensitive reactions occur

Table 12.1 Types of commonly-seen allergens

Allergen type	Commonly-seen allergens
Protein	Heterologous animal serum, vaccines.
Botanical pollen	Rye, ragweed, Phleum pratense (Timothy), birch
Medicines	Penicillin, sulfonamide, local anesthetics, salicylates
Food	Peanuts, nuts, seafood, eggs, milk, wheat, peas, soybeans
Insect products	Bee venom, ant venom, dust mites
Other	Animal hides and dander, mold spores

or systemic vasodilation with increased permeability, smooth muscle contraction, recruitment of inflammatory cells, damage to mucosal integrity, etc.

The allergens that cause type I hypersensitive reactions are proteins, plant pollen, drugs, foods, insect products, and mold spores (Table 12.1). The molecular structure and characteristics of allergens have long been studied, but still largely unknown. Most allergens are small-molecule proteins or small-molecule compounds that can bind to protein carriers, but there is no clear answer as to whether immunogens can be defined by structure. There is growing evidence that allergic reactions may be more related to individual constitution as well as genetics.

Mast cells and basophils express two IgE receptors, one is a high-affinity receptor, FcεRI, which binds to IgE with an affinity of Kd = 1×10^{-9}, and the other is a low-affinity receptor, FcεRII (also known as CD23), which binds to IgE with an affinity of Kd = 1×10^{-6}. There is a 1000-fold difference in affinity for IgE binding between the two IgE receptors. FcεRI binds to the CH3/CH3 as well as CH4/CH4 functional domains, whereas FcεRII binds only to CH3/CH3. It has been analyzed that there are 4–9 × 10^4 IgE receptor molecules on each basophil.

The ultimate process of type I hypersensitive reaction is the degranulation of mast cells and basophils, which releases active mediators that act on target tissues to produce effects. Cell degranulation is dependent on IgE receptor cross-linking, and a variety of factors can cause receptor cross-linking, such as allergens, certain compounds, allergenic toxins (complement fragments C3a, C4a, C5a), and anti-FcεR antibodies. Anything that can cause IgE receptor cross-linking will cause cell degranulation and initiate a type I hypersensitive reaction. In clinical practice, type I hypersensitive reactions are very common, with both systemic and local allergic reactions.

Systemic Allergic Reactions

A systemic allergic reaction occurs with a very rapid progression, usually causing death within minutes. Among animals, guinea pig is the most sensitive to allergic reactions, followed by the dog. If a guinea pig is inoculated with the antigen OVA, and if it is re-inoculated with OVA by the intravenous injection 2 weeks later, the animal will develop respiratory distress, decreased blood pressure, and incontinence within 1 min, and die of asphyxia in 2–4 min. Anatomic examination reveal pathological changes such as generalized massive edema, tracheal and bronchial tightness, and shock in the animal.

Localized Allergic Reactions

The Localized allergic reactions are mainly confined to specific tissues and organs, such as the skin or mucosal surfaces that mostly contact with allergens. A high percentage of the population has suffered from localized allergies, taking more than 20%, and there is a genetic linkage to the allergy. Commonly-seen localized allergies include allergic rhinitis, asthma, eczema, and food allergies.

1. Allergic rhinitis, also known as Hay Fever, is the most common localized allergic reaction, belonging to a type I hypersensitive reaction caused by allergens floating in the air. The sensitized mast cell degranulation causes the conjunctiva, nasal mucosa vascular expansion and increased capillary permeability, and the symptoms include conjunctivitis, runny nose, sneezing and coughing. According to the American Epidemiological Survey, allergic rhinitis ranks fifth among chronic diseases and is very common in the population, with annual allergic rhinitis medical costs amounting to tens of billions of dollars.

2. Asthma, is also a common localized allergic reaction. Allergens that cause asthma can be airborne or bloodborne, such as pollen, dust, smoke, insect secretions, and viral antigens. In some cases, asthma can also be triggered by exercise or a cold. The main cause of asthma is the contraction of tracheal or bronchial smooth muscle caused by active mediators (released by mast cell degranulation), which causes narrowing of the airways. In addition, respiratory edema and increased secretions exacerbate dyspnea. Asthma is often treated clinically as an inflammatory condition and is divided into two response phases, early and late phases. The early response refers to the release of mediators such as histamine, leukotrienes, and prostaglandins within minutes of the body's exposure to allergens, causing vasodilation, airway constriction, and increased mucus, and the late response occurs several hours later with the release of more inflammatory mediators, including IL-4, IL-5, IL-16, TNF-α, eosinophil granulocyte chemotactic factor (ECF), platelet-activating factor (PAF), and others. These inflammatory mediators increase endothelial cell adhesion and recruit inflammatory cells such as eosinophils and neutrophils to accumulate in the tracheal tissue. Eosinophils and neutrophils cause tissue damage by releasing toxic enzymes, oxygen free radicals, cytokines, etc. Tissue injury causes more severe inflammatory exudation, including increased mucus, epithelial cell shedding, and cellular debris, which predisposes to the formation of tracheal emboli, and the thickening of submucosal basal layer due to the inflammatory cell infiltration of the tracheal wall, which aggravates narrowing the airway. According to statistics, 5% of the US population has asthma, and this figure has been rising in recent years. The incidence is the highest in children, and is especially high in people of African descendants living in the downtown. Although some genetic studies have demonstrated a genetic link to asthma, genetic factors account for 40–60% of the occurrence of the disease. Genome sequencing has shown a linkage to asthma in the 5q31-33 region of chromosome 5, a region with genes encoding cytokines such as IL-3, IL-4, IL-5, IL-9, IL-13, and GMCSF, while IL-4, a key cytokine that promotes IgE expression, and a polymorphism in the promoter that regulates IL-4 expression, is associated with asthma development. Clinical investigations in the United States have shown that much of the asthma in urban children is associated with cockroach allergens, which indicates the importance of public health.
3. allergic eczema is an inflammatory reaction that occurs in the skin and is associated with family history, commonly in infants and young children. The diseased children display elevated IgE concentration in the serum, and reddening skin that might show swollen patches or even ruptures and becomes pus-filled. Th2 lymphocytes and eosinophils infiltrate the affected areas of the skin with dermatitis.
4. Food allergies are caused by a variety of foods and are specific immune responses that are harmful to the body, and occur repeatedly as a result of exposure to a specific food. Numerous data indicate that food allergies occur in up to 10% of the population. Common allergy-causing foods include food allergens that bind to IgE-sensitized mast cells in the digestive tract, causing cell degranulation, and the active substances in the granule cause local vasodilation, increased exudation of body fluids, and smooth muscle contraction, resulting in vomiting or diarrhea.

Due to increased permeability of the intestinal mucosa, allergens may enter the bloodstream and enter other tissues, causing local allergic symptoms, such as asthma in some individuals due to food allergy, cellulitis in others, and skin edema or eczema in others.

The Method for Testing Type I Hypersensitivity

A commonly used test for type I hypersensitivity is the local skin test, in which a small amount of antigen is injected intradermally and the local skin inflammatory response is examined within 30 min to determine whether the body is allergic to the antigen. If local papules and redness occur, the body is allergic to the antigen. Alternatively, it is possible to determine whether the organism will develop a type I hypersensitivity reaction by testing for specific IgE in the serum. a commonly used serological technique for detecting IgE is the radioimmunoassay.

Prevention and Control of Type I Hypersensitive Reactions

Prevention and control of type I hypersensitive reactions should first avoid contact with allergens, such as removing dust mites from the room, not touching small animals, not touching allergic food, etc. However, it is difficult to completely avoid contact with pollen in the air for pollen allergy. Immunotherapy can take the method of injection with allergens multiple times, and the injection dose should be gradually increased to achieve a state of desensitization, for example, treatment of allergic rhinitis by this method is very effective. The main mechanism is that multiple subcutaneous injections of allergen shift the immune response to produce IgG or Th1 immune response, with IgG neutralizing the allergen and eventually being cleared by macrophages and Th1 immune response inhibiting IgE production. Another immunotherapeutic approach is to neutralize IgE by injecting humanized antibodies against IgE. In addition to immunotherapy, chemical drugs are commonly used clinically to block histamine receptors, which inhibit histamine activity. In anaphylaxis (e.g., asthma), emergency treatment is provided with epinephrine injections, which bind to β-adrenergic receptors on tracheal smooth muscle to promote tracheal smooth muscle relief and relaxation, and also bind to β-adrenergic receptors on mast cells to inhibit cell degranulation. Glucocorticoids suppress the immune response and can also be used to treat allergic reactions.

12.1.1.2 Type II Hypersensitive Reaction

Type II hypersensitive reaction is antibody-mediated damage to target cells, including antibody binding to target cells that causes antibody-dependent cell-mediated cytotoxicity (ADCC, see Chap. 6), antibody-dependent opsonin-mediated phagocytosis, and cell injury following complement activation upon antibody binding to target cells.

Hemolytic Transfusion Reactions (HTRs)

The cell membranes of red blood cells contain a variety of protein and glycoprotein molecules, and blood is clinically classified into four types (A, B, AB, and O) depending on the type of membrane protein antigens expressed in different populations. Antibodies against A, B, and O antigens are called isohemagglutinins. As a result of exposure to environmental microorganisms inside and outside the body during growth, some microorganisms have antigenic epitopes similar to those of the A and B antigens, so isohemagglutinins against different blood types are present in the body. For example, human red blood cells of blood type A contain A antigen, but the serum contains isohemagglutinin against B antigen, and when blood of type B blood is transfused, hemolysis occurs, but blood of type O blood is acceptable. Similarly, human red blood cells of blood type B contain B antigen, but the serum contains isohemagglutinin against A antigen, and when blood of type A blood is transfused, hemolysis occurs, but blood of type O blood is acceptable. Human red blood cells of blood type AB contain A and B antigens, the serum does not contain isohemagglutinin and can accept blood of other blood types, whereas human red blood cells of blood type O do not contain A or B antigen, but its serum contains isohemagglutinin against A or B antigen (see Table 6.2 in Chap. 6), can only accept blood of the same blood type. When the recipient's blood contains isohemagglutinins that recognize the antigens of the transfused red blood cells, hemolysis occurs in response to the action of complement. In addition to A, B, and O antigens, red blood cells have other antigens, among which Rh, Kidd, Kell, and Duffy antigens are commonly-seen, and the patient would develop antibodies to these antigens post repeated multiple transfusions and subsequently develops delayed hemolytic transfusion reactions.

Hemolytic Disease of the Newborn

Neonatal hemolytic disease is a disease caused by maternal antibodies (mainly IgG) entering the infant through the placenta or colostrum, binding to the infant's red blood cell antigens, and causing hemolysis by the action of complement. The disease usually occurs in infants whose mothers are Rh− and whose offspring are Rh+, usually in second or subsequent children. During the first delivery of a child by Rh-mother, some of the infant's umbilical cord blood enters the mother's body through the uterus, Rh+ red blood cells cause an immune response in the mother, which produces anti-Rh antibodies and memory cells. When the mother is re-exposed to Rh+ red blood cells in the following delivery, a rapid recall response is generated and there are large amounts of Rh antibodies in the blood and breast milk especially in colostrum, and IgG can enter the fetus directly through the placenta, causing haemolysis and anemia in the fetus. In addition, the underdeveloped intestinal wall of the newborn infant can absorb maternal antibodies directly into the circulatory system, where Rh antibodies bind to Rh+ red blood cells, leading to hemolysis under the action of complement. Hemolysis in infants can lead to serious consequences, with the conversion of hemoglobin to bilirubin in the body, and the accumulation of too much bilirubin causes brain damage. Infants with hemolysis develop varying degrees of jaundice, which can cause death in severe cases. Low doses of ultraviolet radiation are commonly used clinically to accelerate the breakdown of bilirubin and avoid brain damage. Neonatal hemolysis is commonly seen clinically in infants born to mothers with blood type O who have infants with blood type A or B. If the mother is Rh− and the fetus is Rh+ by testing in advance, the risk of neonatal hemolysis in subsequent fetuses can be reduced by administering Rh antibodies to the mother within 1–2 days of birth to clear Rh+ red blood cells that enter the mother upon delivering to avoid immune response.

Drug-Induced Hemolytic Anemia (DIA)

Some antibiotics readily bind to protein molecules on the red blood cell membrane in the body, resembling a hapten-carrier complex, and, in some individuals, cause an immune response, producing antibodies against the hapten. The resulting antibodies bind to the hapten on the red blood cell surface, causing hemolysis under the action of complement. Drugs that commonly cause hemolytic anemia include penicillin, cephalosporins, and streptomycin. Drug-induced hemolysis is a gradual process, and stops when the drug is discontinued. Some drugs cause not only type II hypersensitive reactions, but also other types of hypersensitive reactions, such as penicillin (Table 12.2).

12.1.1.3 Type III Hypersensitive Reaction

Antibody-antigen binding forms a complex, which is normally cleared by phagocytes. However, in some cases, large amounts of antigen-antibody complexes cannot be cleared rapidly and are deposited in certain tissues causing local inflammatory reactions that result in tissue damage. The main mechanism of type III

Table 12.2 Penicillin-induced hypersensitive reactions of various types

Type of hypersensitive reaction	Antibodies or cells involved in hypersensitive reactions	Clinical symptom
Type I	IgE	Urticaria, systemic allergy
Type II	IgE, IgG	Hemolytic anemia
Type III	IgG	Serum sickness, glomerulonephritis
Type IV	T_{DTH} cells	Contact dermatitis

hypersensitive reaction is the deposition of antigen-antibody complexes in tissues, activation of the complement system, production of C3a, C4a, C5a and other allergic toxins, causing local mast cell degranulation, resulting in increased local vascular permeability and recruitment of neutrophils causing local inflammatory reactions. The direct cause of local tissue damage is the release of lytic enzymes by neutrophils during the clearance of immune complexes. C3b plays a regulatory role in binding to antigen-antibody complexes, and neutrophil membrane receptors bind to C3b, enhancing its phagocytic effect. However, if antigen-antibody complexes are deposited in the basement membrane, phagocytosis is impeded and neutrophils release large amounts of lytic enzymes without phagocytosis, damaging local tissues. In addition, the activated complement system also forms membrane-attack complex (MAC) causing local injury, and activated complement fragments recruit large numbers of platelets to aggregate, forming local microenvironmental vascular emboli and exacerbating local injury. Deposition of antigen-antibody complexes in the kidney is a common cause of glomerulonephritis.

Type III hypersensitive reactions can occur not only locally but also systemically.

Local Type III Hypersensitive Reaction

If the animal has high levels of anti-allergen antibodies, then intradermal or subcutaneous injection of the allergen will cause local deposition of immune complexes, infiltration of neutrophils, local inflammatory reactions such as edema and hemorrhagic spots in the local tissue, which shows redness, swelling or even necrosis in the local tissue, 4–8 h after injection. This local allergic reaction caused by the injection of immunogens is called Arthus reaction. As in the case of mosquito bites, allergic individuals rapidly develop local type I allergic reactions, and local Arthus reactions appear after 4–8 h, with obvious local symptoms of hemorrhagic spots and edema. Certain bacterial spores, fungi, or dry protein-containing dust, inhaled into the lungs, will cause Arthus reactions in the lungs, for example, when farmers turned moldy straws, mold spores were inhaled and caused Arthus reactions in the lungs, commonly known as “farmers’ lung”. A variety of allergens can be inhaled into the respiratory tract to cause Arthus reactions in the lungs.

Generalized Type III Hypersensitive Reaction

When a large amount of antigen enters the body, it is difficult to be cleared by phagocytes because the antigen is in excess and the antigen-antibody cannot form large complexes. These small complexes are deposited in the tissues causing tissue damage. Most commonly, after injection with antitoxin serum from another species of animals, some individuals develop a variety of symptoms a few days later, including fever, generalized rash, edema, swollen lymph nodes, arthritis, glomerulonephritis, etc. This disease caused by serum injection is called "serum sickness". The severity of serum sickness is determined by the size and number of antigen-antibody complexes and the site of deposition. In addition, antigen-antibody complexes that are not cleared in time can cause a variety of diseases, including certain autoimmune diseases such as systemic lupus erythematosus, rheumatoid arthritis, and pulmonary hemorrhagic-renal syndrome.

12.1.1.4 Type IV Hypersensitive Reaction

Type IV hypersensitive reaction is characterized by the release of cytokines from specific Th cells exposed to the antigen, causing a massive infiltration of inflammatory cells (mainly macrophages) and resulting in a local inflammatory response. In contrast to the first three types of hypersensitive reactions, the pathogen triggers a relatively delayed type IV hypersensitive reaction and tissue damage, hence called delayed-type hypersensitivity (DTH). The area of inflammation in the DTH reaction recruits predominantly macrophages, which is distinct from the type III hypersensitive reaction (predominantly neutrophil infiltration). Although the DTH response causes tissue damage, it is beneficial in limiting, killing, and clearing pathogens. The pathogens that cause the DTH response are mainly intracellular pathogens such as *Mycobacterium tuberculosis*, *Listeria monocytogenes*, *Brucella abortus*, *Candida albicans*, *Histoplasma capsulatum*, *Cryptococcus novelis*, *Leishmania protozoan*, *herpes simplex* virus, *poxvirus*, and *measles* virus. In addition, certain contact antigens can also cause DTH reactions, such as trinitrochlorobenzene, nickel salts, and poison ivy.

The sensitization phase of the DTH reaction is the process in which $CD4^{+}T$ cells are activated to recognize antigens presented by MHC-II molecules of APCs (mainly macrophages, Langerhans cell, and vascular endothelial cells) when the body is first exposed to the antigen. In some cases, $CD8^{+}T$ cells recognizing antigens presented by MHC class I molecules are activated and are also involved in the DTH response. When the activated T cells encounter the antigen again, they rapidly proliferate into effector T cells, usually within 1 to 3 days, and this phase is the effector phase of the DTH reaction. During this phase, effector CD4T cells secrete Th1 cytokines (IL-3 and GM-CSF) and chemokines (monocyte chemotaxis and activation factor (MCAF) and migration inhibitory factor (MIF)) to recruit mononuclear macrophages, while IFN-γ and TNF-α induce macrophage activation. Activated macrophages produce pro-inflammatory cytokines that act on vascular endothelial cells, causing a large number of monocytes and inflammatory granulocytes in the circulatory system to migrate to local tissues, and these inflammatory cells are activated and then secrete inflammatory factors, attracting more inflammatory cells to participate in the inflammatory response, whereby the inflammatory response continues to amplify and the inflammatory cells accumulate, forming granulomas or causing tissue damage (Fig. 12.2).

Invasion by pathogenic bacteria (e.g., Mycobacterium tuberculosis) causes a local DTH response, and the local response has both advantages and disadvantages. The advantage is that it limits the spread of the pathogen, and the disadvantage is that it causes local tissue damage.

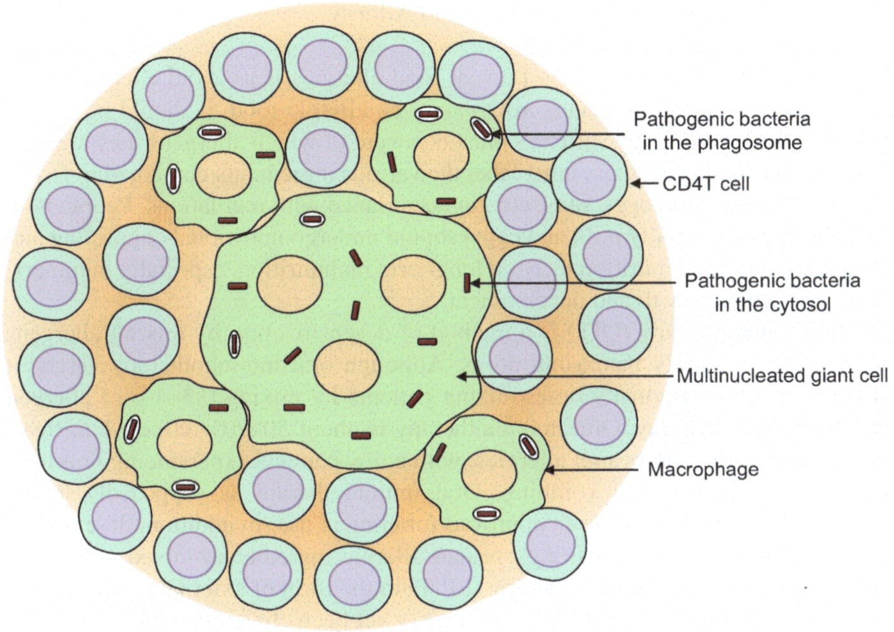

Fig. 12.2 Schematic diagram of nodular granuloma caused by type IV allergic reaction

Contact dermatitis is a DTH reaction and is caused by substances such as formalin, trinitrophenol, nickel salts, turpentine, active substances in certain cosmetics, and poison ivy. These small molecular compounds readily bind to skin protein molecules to form complex antigens, which are phagocytosed and presented by Langerhans cells, consequently activating Th cells. When the body's skin is re-exposed to the same compound, it causes a Th1 immune response and a DTH reaction. Afflicted local skin appears red, swollen or abscessed.

Depending on the characteristics of the DTH reaction, the skin test can be performed, e.g. in the case of tuberculosis testing, by intradermal injection of tuberculin and observation of the local reaction for 48–72 h. If the skin is red and swollen, this indicates a previous infection with Mycobacterium tuberculosis. Alternatively, the method can be used for brucellosis infection surveillance.

12.1.2 Treatment of Hypersensitive Reactions

In daily life, food allergies and allergic rhinitis are the most common-seen type of hypersensitive reactions. In the population, allergies are more likely to occur in younger ages than in adults. Food allergies and eczema in children have been on the rise over the past few decades, and the common routes that cause allergies include ingestion (allergens in food), inhalation (exposure to airborne allergens), and bites (venomous allergens), and treatment of allergies has always been a high concern for the medical profession.

12.1.2.1 Prevention and Control of Food Allergy

The most straightforward way to prevent food allergies is to avoid allergic foods, but if an allergic reaction occurs after eating an allergic food, it should be treated immediately. Food allergy sufferers must be aware of which allergens they are sensitive to, and food processors must label the various ingredients contained in foods, such as peanuts, milk, eggs, nuts, etc., in accordance with regulations. People with food allergies (especially milk and eggs) should undergo nutritional counseling and pay attention to food nutrition to avoid long-term malnutrition, especially in minors, which can affect growth and development.

Oral immunotherapy (OIT) is a method of desensitization by mixing allergens into food in gradually increasing doses. Although oral immunotherapy varies, it basically involves gradually desensitizing patients by mixing 0.3–4 g of allergen into their food each day. Oral immunotherapy is about 50–90% effective in treatment of egg, milk, and peanut allergies, with a few patients experiencing oral itching, abdominal pain, and vomiting. Oral immunotherapy is supposedly in the research phase and has not been approved for clinical use. In addition, desensitization by injection of allergens (e.g., peanut allergic ingredients) mixed with adjuvants or epicutaneous immunotherapy (EPIT) in the form of external ointment is in the research phase of clinical trials and has not yet been clinically applied.

Food allergy should be treated immediately with epinephrine injection. However, in real life, when food allergy occurs, it is often not possible to inject epinephrine immediately, mainly because the allergic person does not carry the medication with them, especially most of the underage allergic persons are not used to carrying epinephrine with them, so they often miss the best time to rescue and save the life.

12.1.2.2 Immunotherapy for Allergy Caused by Exposure to Aeroallergens

Airborne allergens (e.g., flower pollen, grass pollen, dust, dust mites, etc.) often cause hypersensitive reactions in the respiratory system, the most common type of hypersensitive reaction being allergic rhinitis. Allergic rhinitis is very common in the population and is a common chronic condition and a potential risk factor for asthma. Allergen immunotherapy (AIT) can modulate the Th2 immune response and reduce clinical symptoms. There are two types of allergen immunotherapy, namely: subcutaneous allergen immunotherapy and sublingual allergen immunotherapy.

Subcutaneous Allergen Immunotherapy (SCIT)

In the early twentieth century, Noon and Freeman carried out the treatment of chytrid fever with subcutaneous injections of hay powder extract. Constantly increasing doses of allergen injected are the process of desensitization and are now a common clinical treatment for allergic rhinitis, asthma, and mosquito bite allergy, and are effective. The effect usually lasts for a period of time (about 2 years) after the allergen injections are discontinued. It should be noted that subcutaneous injection of allergens tends to cause local scratching and redness, and occasionally very few patients will have systemic reactions that should be treated promptly with epinephrine injections.

Sublingual Allergen Immunotherapy (SLIT)

Sublingual allergen immunotherapy is the main treatment for upper respiratory allergy in Europe, and in recent years the placement of allergy-treating tablets under the tongue has significantly reduced allergic rhinitis caused by hay dust and dust mites. This approach has caused mild itching and edema in the oral mucosa, epiglottis, and larynx in a small number of patients, and also some degree of gastrointestinal reaction. This treatment is not yet commonly used in other areas.

12.1.2.3 Treatment of Allergies Caused by Bites (Toxins)

Allergic reactions to the bites of hymenopterous insects (such as wasp and honey bees) can be serious. Most people who have been bitten by honey bees or wasps will tolerate re-exposure to bee venom (bite or subcutaneous desensitization with bee venom). However, a few will develop an allergic reaction with systemic reactions and should be treated promptly.

12.2 Autoimmune Diseases

Autoimmune disease is a disease caused by self-reactive lymphocytes that attack self tissues. Normally, the immune system is tolerant to self-tissue antigens, and if autoimmune tolerance is broken, auto-reactive lymphocytes produce an immune response against self antigens, leading to autoimmune disease. There are three main ways in which the body avoids autoimmune responses. (1) central tolerance: auto-reactive lymphocytes are induced to apoptosis in negative selection and subsequently cleared by macrophage during development and maturation; (2) peripheral tolerance: a small number of auto-reactive lymphocytes escaping from the central immune organs are exposed to auto-antigens in the periphery, but are unresponsive (in anergy) due to the lack of a second costimulatory signal or are suppressed; (3) clonal ignorance: self-reactive lymphocytes are not exposed to their own antigens due to the closure of self antigens (blood-brain barrier, blood-testis barrier, eyeball barrier, etc.) and are in a state of "ignored". Some self-reactive lymphocytes exist in the peripheral circulatory system of normal individuals, and these cells are in a state of suppression, anergy or inaccessibility to self-antigens under immune regulation, which all constitute self-tolerance. However, if the self-tolerance is broken, it will result in activation of self-reactive lymphocytes, cause damage to the tissue and organs, and lead to autoimmune diseases.

12.2.1 Main Types of Autoimmune Diseases

Either auto-reactive antibodies or auto-reactive T cells or both can cause autoimmune diseases. Autoimmune diseases can occur either locally in specific tissues or in multiple tissues leading to systemic damage (Table 12.3).

12.2.2 Causes and Mechanisms of Autoimmune Diseases

The mechanisms of different types of autoimmune diseases vary, but there are common patterns of occurrence. Clinical studies have shown that there is a gender-tendency in the occurrence of autoimmune diseases, such as Hashimoto's thyroiditis, systemic lupus erythematosus, multiple sclerosis, rheumatoid arthritis and scleroderma, which occur more frequently in women. This may be related to physiology and hormonal secretion. The causes and mechanisms of autoimmune diseases are broadly summerized as follows: (1) autoimmune diseases caused by release of segregated antigens; (2) autoimmune diseases caused by molecular mimicry; (3) autoimmune diseases caused by sensitization of self-reactive T cells by expression of MHC-II molecules in specific tissues; and (4) autoimmune diseases caused by polyclonal activation of lymphocytes.

12.2.2.1 Autoimmune Diseases Caused by Release of Segregated Antigens

During lymphocyte development and maturation, those self-reactive lymphocytes that bind to autoantigens are removed during negative selection. However, certain autoantigens are in a segregated state that do not enter the circulatory system or are not expressed in thymic epithelial cells and cannot come into contact with developing lymphocytes, so self-reactive lymphocytes that target these segregated antigens are able to mature and enter the circulatory system. Once these segregated antigens are released for some reason, they activate self-reactive lymphocytes in the peripheral and cause autoimmune disease. For example, sperm, ophthalmic proteins, and cardiac antigens released into the bloodstream as a result of bacterial or viral infection or mechanical injury can activate self-reactive B cells to produce self-reactive antibodies, causing infertility, eye disease, and myocardial infarction. Another example is myelin basic protein (MBP), a protein in the brain and spinal cord, which does not come into contact with the immune system because of the blood-brain barrier. However, if MBP is mixed with adjuvant and injected directly into mice, it can cause experimental autoimmune encephalomyelitis (EAE). If MBP is injected into the thymus of mice so that it comes into contact with T cells that develop in the thymus and the T cells that react with MBP are cleared, the mice are not susceptible to EAE even if they are injected with MBP with adjuvant.

Table 12.3 Major types of autoimmune diseases

Name of disease	Self-antigen	Immune response	Tissue damage	Disease characteristics
Pulmonary hemorrhage-nephritis syndrome (Goodpasture's syndrome)	Basement membrane of the kidney and lungs	Autoantibody	Specific organizations (lungs, kidneys)	Simultaneous lung and kidney injury, hemoptysis, inflammatory infiltration in the lungs, glomerulonephritis
Addison's disease	Adrenal cells	Autoantibody	Specific organizations (adrenal)	Low blood pressure, weight loss, lack of appetite, sometimes bronze-like melanosis of the skin
Autoimmune hemolytic anemia	Erythrocyte membrane protein	Autoantibody	Specific organizations (red blood cells)	Hemolytic anemia
Grave's disease	Thyrotropic receptor	Autoantibody	Specific organizations (thyroid)	Goiter, hyperthyroidism
Hashimoto's thyroiditis	Thyroid cells and proteins	Autoantibody	Specific organizations (thyroid)	Enlarged thyroid gland, hypothyroidism
Type I diabetes (insulin-dependent diabetes mellitus)	β cells in the pancreas	Autoantibodies, DTH-reactive T cells	Specific organizations (pancreas)	High blood sugar, diabetic symptoms
Myasthenia gravis	Acetylcholine receptor	Autoantibody	Specific organizations (muscle)	An autoimmune disease in which transmission at the nerve-muscle junction is impaired by antibodies against acetylcholine receptors. It causes the muscles to lose their contractile function
Idiopathic thrombocytopenic purpura	Platelet membrane protein	Autoantibody	Specific organizations (platelets)	Thrombocytopenia, bleeding spots throughout the body
Post streptococcal glomerulonephritis	Kidney	Antigen-antibody complexes	Specific organizations (kidney)	Nephritis (urine protein, hematuria, even uremia)
Congenital infertility	Spermatozoa	Autoantibody	Specific organizations (sperm)	Infertility

(continued)

Table 12.3 (continued)

Name of disease	Self-antigen	Immune response	Tissue damage	Disease characteristics
Sjogren's syndrome	Salivary gland, liver, kidney, thyroid	Autoantibody	Systemic injury	It is a systemic autoimmune disease that damages the exocrine glands, mainly invades the lacrimal and salivary glands, manifests as dry keratoconjunctivitis, oral dryness signs and other rheumatic diseases such as rheumatoid arthritis, and causes damage to the respiratory, digestive, urinary tract, blood, nerves, muscles and joints
Ankylosing spondylitis	Spine	Immune complexes	Systemic injury	Inflammatory injury in the joints and periarticular tissues of the spine
Scleroderma	Nucleus, heart, lungs, gastrointestinal tract, kidneys	Autoantibody	Systemic injury	Connective tissue, skin fibrosis
Multiple sclerosis	Brain and white pulp	Autoantibodies, self-reactive T cells	systemic injury	It is a disease of the brain and spinal cord that clinically presents with multiple sclerosis, muscular dystrophy or polio-like paralysis leading to disability
Rheumatoid arthritis	Connective tissue, IgG	Autoantibodies, immune complexes	Systemic injury	Characterized by acute or chronic connective tissue inflammation that can recur and involve the heart. Clinically, it presents as wandering joint and muscle aches and pains and is a hypersensitive disease
Systemic lupus erythematosus	DNA, nuclear proteins, red blood cells, platelet membranes	Autoantibodies, immune complexes	Systemic injury	Lupus erythematosus, multiple tissues such as skin, blood vessels, heart, liver, kidney, brain and other organs and tissue damage

12.2.2.2 Molecular Mimicry Causes Autoimmune Diseases

In the early 1980s, it was discovered that some antiviral monoclonal antibodies could bind to certain protein molecules in host cells, cross-reacting with self-antigen. Michael Oldstone proposed that if certain components or spatial structures in the antigen of a pathogen are similar to certain components in the host cell, then the immune response induced by the pathogen will also react against the host cell, causing an autoimmune disease. This similarity of pathogenic antigens to host tissue proteins is called molecular mimicry. For example, there is molecular mimicry between streptococcal antigens and myocardial antigens, renal glomerular antigens, and skin antigens, so streptococcal infections predispose to autoimmune glomerulonephritis, myocarditis, and psoriasis. Then again, herpes virus stromal keratitis (HSK), an autoimmune reaction caused by reencountering virus-specific proteins in mice infected with herpes simplex virus type I, causes corneal damage and leads to blindness, mainly because herpes simplex virus type I proteins have molecular mimicry with corneal proteins.

12.2.2.3 Tissue-Specific Expression of MHC-II Molecules to Sensitize Self-Reactive T cells

In general, healthy tissue cells express only MHC class I molecules and not MHC class II molecules. However, in some cases of autoimmune diseases, damaged tissues highly express MHC-II molecules, e.g., beta cells of type I diabetic patients highly express MHC-II molecules, whereas beta cells of healthy individuals do not express MHC-II molecules. As another example, thyroid follicles in patients with Grave's disease highly express MHC-II-like molecules. These specific tissues highly express MHC-II molecules, present autoantigens, activate self-reactive Th cells, and cause autoimmune disease.

Experimental evidence suggests that large amounts of IFN-γ promote the expression of MHC-II molecules in tissue cells, which predispose self-reactive Th cells to sensitization and induce autoimmune disease. In addition, IFN-γ regulates the immune response toward a Th1 immune response, which promotes self-reactive CTL killing target cells and damaging self tissues.

12.2.2.4 Autoimmune Diseases Caused by Polyclonal Activation of Lymphocytes

Certain pathogens encode thymus-independent antigens that induce the simultaneous activation of multiple B-cell clones, and if self-reactive B cells get activated due to polyclonal activation of B cells, autoantibodies are produced. If the pathogen contains superantigens, which activate polyclonal Th cells, it will also activate self-reactive Th cells, causing autoimmune disease. For example, Gram-negative bacteria, cytomegalovirus, and EBV can cause activation of polyclonal lymphocytes, and the infected patients produce autoantibodies against red blood cells and platelets.

12.2.3 Treatment of Autoimmune Diseases

Autoimmune diseases are attributed to the immune response of the immune system to self antigens. If the autoimmune response is suppressed, the anti-defense and immune surveillance functions of the immune system are also suppressed, making it susceptible to infection and neoplastic disease. Therefore, the ideal treatment would be to suppress only the autoimmunity without affecting the normal immune response. However, no such ideal treatment has been found so far.

All current treatments for autoimmune diseases are symptomatic but not curative, and only provide some relief from certain symptoms, such as the use of immunosuppressive glucocorticoids, azathioprine, and cyclophosphamide, which suppress the immune response and reduce the clinical symptoms of autoimmune disease, but these approaches make the patient more susceptible to infection and tumor development. Cyclosporin A and Pulcolax (FK506) inhibit TCR-mediated signaling, suppress activated T cells, and alleviate the clinical symptoms of autoimmune disease. In addition, in patients with myasthenia gravis, symptoms may be relieved by removal of the thymus gland. In some cases of autoimmune diseases, symptoms can also be temporarily relieved by hemodialysis and removal of antigen-antibody complexes, such as Grave's disease, myasthenia gravis, rheumatoid arthritis, and systemic lupus erythematosus. CTLA-4Ig fusion protein was prepared by expression of CTLA-4 with the Fc fragment of Ig using molecular biology techniques, and since the affinity of CTLA-4 binding to B7 is higher than that of CD28 to B7, CTLA-4Ig can bind to the co-stimulatory molecule B7 of APC to block T cell activation, and this method has been used for clinical trial in treatment of psoriasis with promising results, cherishing some hopes for clinical application.

Study Question

1. What is the definition for hypersensitive reaction?
2. What is the definition for autoimmune disease?
3. What is the difference between hypersensitive reaction and autoimmune disease?
4. Gell and Coombs classified hypersensitive reactions into four types, and what are the mechanisms of each type of hypersensitive reaction?
5. In what ways does the body avoid autoimmune disease?
6. What are the causes and mechanisms of autoimmune diseases?
7. What should be considered in terms of treating autoimmune diseases?

Chapter 13
Mucosal Immunity and Immune Tolerance

Overview:
The mucosa is in direct contact with the outside and is the first line of defense of the immune system, undertaking the complex immune task of both tolerance to beneficial commensal microorganisms and food and immunity to pathogenic microbial infection. The mucosal immune system recognizes pathogen-associated molecular patterns (PAMPs) through pattern recognition receptors (PRRs) to generate an immune response, clear pathogens, and maintain homeostasis in the organism. Immune tolerance is the unresponsiveness of the adaptive immune system to an antigen, owing to the clearance or suppression of antigen-specific lymphocytes or their conversion to regulatory T cells (Treg). Immune tolerance determines that the immune system does not produce an immune response to self-antigens, however if immune tolerance is broken, the immune system will attack self tissues, resulting in autoimmune disease. The digestive, respiratory, and genitourinary tracts all have mucosal tissues, and this chapter focuses on the basic principles of mucosal immunity and immune tolerance, using the intestinal mucosa as an example.

13.1 Intestinal Mucosal Tissue and Mucosal Immunity

The intestine, the pathway through which the intake food runs, is divided into the small and large intestine portions. The small intestine is divided into the duodenum, jejunum and ileum portions. The most important function of the intestine is to digest food and absorb nutrients (including water). The gut has a large number of microorganisms that live in symbiosis with the host. These symbiotic bacteria have the function of breaking down and transforming the intake substances and are an important source of nutrients such as proteins and vitamins for the host. However, the intestinal mucosa is also an important gateway for pathogenic invasion. Mucosal immunity is a protective mechanism of host that has formed during evolution as host continues to fight pathogens.

S. J. Zheng, *Veterinary Molecular Immunology*,
https://doi.org/10.1007/978-981-99-8929-4_13

13.1.1 Symbiotic Bacteria of Intestines Have an Important Role in Mucosal Immunity

Intestinal commensal bacteria produce biofilms that are effective in preventing pathogenic infection. Germ-free (GF) animals have significant defects in the structure, function and renewal of intestinal epithelial cells, and also have fewer cells in the intestinal lymphoid tissue, and reduced antibody production.

13.1.2 Mucosal Epithelial Cells of Intestines

The mucosal epithelial cells of intestines are interconnected by a membrane protein called connexin (CX), which binds to each other to form channels through which cells transmit signals. The intestinal epithelial cells form an important natural immune barrier. The barrier consists of villi and crypts that, under the microscope, resemble range upon range of heavy mountains and, if spread out, cover an area of up to 300 m^2. The intestinal epithelium is a collective term for a variety of cells (Fig. 13.1), which are divided into three main categories according to their functions: (1) enterocytes, which are specialized for nutrient absorption; (2) secretory cells, including goblet cell, Paneth cell, enteroendocrine cell, and tuft cell; (3) Microfold cell, which transport antigens (The intestinal mucosal epithelium has a

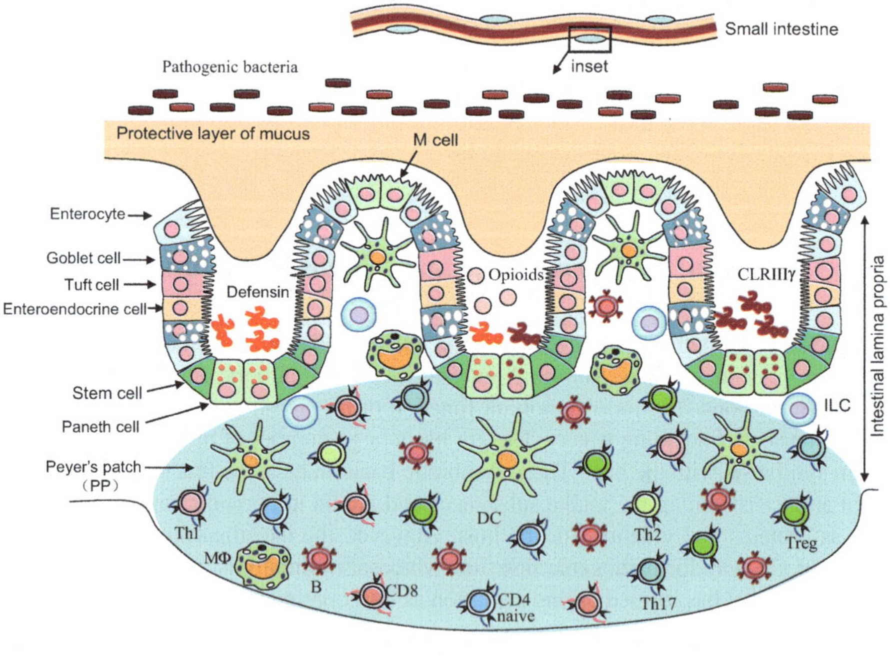

Fig. 13.1 Schematic illustration of the basic structure of the intestinal mucosa

type of cell with little cytoplasm and a microfold (M) surface, called an M cell, which has the function of efficiently obtaining antigens from the intestine for transportation to dendritic cells). The goblet cells secrete large amounts of glycosylated mucin, which forms a protective layer of mucus. Paneth cells are present only at the base of the small intestinal villi crypt and secrete a large number of antimicrobial peptides, including defensins and C-Type Lectin Reg IIIγ (CLR IIIγ), which is highly potent against Gram-positive bacteria, and thus Paneth cells mainly exert anti-infective effects. Tuft cells recognize compounds and secrete opioids.

The intestinal mucosal epithelium is covered with a protective layer of mucus, and the intestinal mucosal epithelium consists of a variety of cells, mainly including enterocyte, goblet cell, tuft cell, enteroendocrine cell, stem cell, Paneth cell, and M cell. There are antimicrobial peptides on the surface of the intestinal mucosa, such as defensin, C-Type Lectin RegIIIγ (CLRIIIγ), and so on, which recognize compounds and secrete opioids. The mucin is secreted mainly by goblet cells, the antimicrobial substances (antimicrobial peptides such as defensins, CLRIIIγ) are secreted mainly by Paneth cells, and the M cells transport antigens from the intestine to the dendritic cells. The lamina propria of the intestinal mucosa is scattered with immune cells, including $CD4^{+}$T cells (including naive $CD4^{+}$T cells, Th1, Th2, Th17, and so on), Treg (nTreg, iTreg, Tr1Treg, Th3, $CD8^{+}LAP^{+}$Treg), $CD8^{+}$T cells, B cells, ILCs, DCs and macrophages that are aggregated in the Peyer's patch (PP) that are the main site of mucosal immune response.

In addition to the cells mentioned above, the intestinal mucosa is scattered with immune cells including innate lymphocytes (ILCs) in the lamina propria, and immune cells accumulate in large numbers to form Peyer's patch. The important immune cells and functions of the intestinal immune system are shown in Table 13.1.

13.1.3 Intestinal Mucosal Immunity

Intestinal mucosal immunity is also divided into innate and adaptive immunity. Innate immunity is primarily carried out by innate immune cells, and adaptive immunity by the tertiary lymphoid tissue at Peyer's patch (PP), the intestinal immune cell aggregation zone. (Fig. 13.1).

13.1.3.1 Innate Immunity of the Intestinal Mucosa

Innate immunity of the intestinal mucosa is primarily performed by innate immune cells, which include intestinal mucosal epithelial cells, mononuclear phagocytes, dendritic cells, and innate lymphocytes.

1. Mucosal epithelial cells of intestine. The mucosal epithelium of intestine forms a natural mechanical barrier to resist entry. The transport of antibodies is carried out via two important receptors, the poly-immunoglobulin (Poly-Ig) receptor,

Table 13.1 Important mucosal immune cells

Cell type	Marker	Basic characteristics	Role of TGF- β
Naive T cells	TCRα⁺/β⁺, CD62L^hi	T cells that have not encountered with Specific antigens, have not been activated.	With action of TGF-β and other cytokines, naïve T cells can differentiate into different types of T cells
Treg	TCRα⁺/β⁺, CD4⁺, FoxP3⁺	Primarily inhibits the immune response and protects the tissues from damage	TGF-β induces FoxP3 expression, causing naive $CD4^+$ T cells to become Treg. Treg exerts immunosuppressive effects by secreting TGF-β
Th1	TCRα⁺/β⁺ CD4⁺, T-bet⁺	A subpopulation of helper T cells that secrete IL-2, IFN-γ, whose primary role is to activate macrophages, promote cellular immune responses, and clear intracellular pathogens	TGF-β suppresses naive $CD4^+$ T cells becoming Th1 cells
Th2	TCRα⁺/β⁺ CD4⁺, GATA3⁺	A subpopulation of helper T cells that secretes IL-4, IL-5, IL-10, and IL-13 and whose primary role is to promote humoral immune response and clearance of extracellular pathogens	TGF-β suppresses naive $CD4^+$ T cells becoming Th2 cells
Th17	TCRα⁺/β⁺ CD4⁺, RORγ t⁺	A subpopulation of helper T cells that secrete IL-17 and IL-22 and whose primary role is to promote inflammatory responses and clear fungal infections	TGF-β together with IL-6 induces RORγt expression, driving the naive $CD4^+$ T into Th17 cells
Th3	TCRα⁺/β⁺ CD4⁺	A subpopulation of helper T cells that secretes mainly TGF-β	Th3 cell produces TGF-β, which plays an immunosuppressive role in immune tolerance to oral antigens
$CD103^+$ DC	CD11c⁺ CD103⁺	Present in the intestine as well as in mucosa-associated lymphoid tissue, induces Treg formation	$CD103^+$ DC secretes large amounts of TGF-β, driving naive $CD4^+$ T into Treg
Macrophage	CXCR1⁺	Promotes Treg proliferation	Secrete large amounts of TGF-β, which plays a role in immune tolerance

which is responsible for the transport of IgA or IgM from the body to the intestinal mucosal surface, and the neonatal Fc receptor (FcRn), through which cells capture antigen-antibody complexes for transport to dendritic cells or macrophages. This transport of macromolecules through the receptor-mediated endocytosis is called transcytosis. Mucosal epithelial cells of intestine express pattern recognition receptors (PRRs) that recognize pathogens and their secreted toxins to initiate an innate immune response and enhance killing and clearance of pathogenic microorganisms. TLRs and NLRs play an important role in the immune response of mucosal epithelial cells.

The role of TLRs in mucosal immunity: TLRs are transmembrane proteins with recognition domain on the cell surface or inside endosomes and signal transduction domain in the cytoplasm. TLRs recognize pathogenic molecules and recruit adaptor molecules through intracellular domain to initiate the immune response, including the secretion of large amounts of antimicrobial peptides and the production of inflammatory cytokines. Due to the large number of microorganisms and the toxins they produce in the gut, and to avoid unnecessary overreaction, TLRs are unevenly distributed on the gut surface, e.g. TLR2, TLR4, and TLR5 are predominantly distributed in the intestinal villous crypts with minimal or no expression by cells in the apical villi (see Chap. 3 for details).

Role of NLRs in mucosal immunity: NLRs, mainly NOD1 and NOD2, are present in the cytoplasm and specifically recognize intracellular microbes and elicit inflammatory responses. However, the role of NLRs in mucosal immunity and tolerance is not well understood, and experiments have confirmed that NOD2 is involved in IL-10 production and plays an important role in defensin production by Paneth cells.

2. Mononuclear phagocytes and dendritic cells. Mononuclear phagocytes and dendritic cells migrate through the blood circulation to the intestinal mucosa and play an important role in mucosal immune tolerance. Macrophages and dendritic cells in the intestinal mucosa secrete large amounts of the anti-inflammatory cytokine IL-10, which suppresses the immune response and plays a role in homeostasis. Dendritic cells in the intestinal mucosa are specifically characterized by the expression of the integrin CD103. $CD103^+$DCs in the mucosa continuously capture antigens transported by goblet cells, secrete IL-10 and TGF-β, and migrate to the mesenteric lymph nodes (MLN), activating naive T cells of the MLN to become Treg. Thus $CD103^+$DCs play a crucial role in immune tolerance. There is also a group of $CD11b^+CD103^-$DCs in the mucosa, and these DCs elicit an inflammatory response.
3. Innate lymphoid cells (ILCs). ILCs are a new group of lymphocytes discovered in recent years, also derived from lymphoid precursor cells, which, unlike conventional lymphocytes, do not contain antigen receptors generated by genetic recombination. ILCs are broadly divided into five subpopulations. Each subpopulation expresses specific transcriptional regulators and cytokines or specific molecules. Except for NK and LTi cells that have specific roles, other ILCs mainly play immunomodulatory roles, similar to the corresponding three subpopulations of CD4T cells (Th1, Th2, Th17). For example, ILC-I expresses T-bet, mainly secreting IFN-γ and TNF-α; ILC-II expresses GATA3, mainly secreting IL-4, IL-5, IL-9 and IL-13; ILC-III expresses RORγt, mainly secreting IL-17 and IL-22. (See Chap. 2 for details).
4. Mucosal associated innate T cells (MAIT): these cells have a constant α/β-TCR, with the α-chain containing a combined Vα7.2-Jα33 fragment, which in combination with the Vβ2 or Vβ13 of the β-chain constitutes the TCR, and thus detection of $V\alpha7.2^+$ $CD161^+$ serves as a marker for identifying MAIT cells. The TCRs of MAIT cell recognize pathogenic antigens similar in structure to riboflavin (vitamin B2), folic acid (vitamin B9) metabolites presented by MR1 (MHC-I

like related molecule), and these antigens are normally present in intestinal bacteria. The folic acid metabolite 6-formylpterin (6-FP) is an antigen recognized by the TCR of MAIT cells. IL-12 and IL-18 can act directly on MAIT cells to promote their activation. Activated MAIT cells produce cytokines such as IFN-γ, TNF-α, IL-10, and IL-17.

13.1.3.2 Adaptive Immunity of the Intestinal Mucosa

Mucosal adaptive immunity occurs primarily in mucosal-associated lymphoid tissue (MALT). MALT is a collective term, which includes the Harderian gland, tonsils, and Peyer's patch (PP). PP belongs to the tertiary lymphoid tissue of the intestinal mucosa and aggregate $CD4^{+}$T cells, $CD8^{+}$T cells, B cells, DCs, and macrophages that perform adaptive immunity (Fig. 13.1). Thus, the PP is a complete functional structure capable of mounting an immune response. Antigens in the gut are transported to MALT primarily via M-cell, and activate B cells to elicit humoral immune response, but can also be captured by DCs to activate the corresponding T cells. Activated T cells can also rapidly migrate to the mesenteric lymph nodes (MLN) to elicit a more robust immune response.

B cells in MALT differentiate into plasma cells and secrete antibodies in response to antigen stimulation. Since dendritic cells and $CD4^{+}$T cells in MALT secrete large amounts of TGF-β (Th3 immune response), the activated B cells in the process of differentiation into plasma cells secrete antibody with heavy chain class-switching to Cα, and thus the isotype of secreted antibody is IgA. IgA is linked by the J-chain as a dimer, which is transported to the surface of mucosa through poly-Ig receptor of epithelial cells. IgA is an important immunoglobulin for mucosal protection, and there are two isotypes of human IgA, IgA1 and IgA2. IgA1 is readily degraded by bacterial proteases and is mainly found in body fluids, IgA2 is resistant to bacterial proteases and is the main antibody on the mucosal surface for mucosal protection. High-affinity IgA can directly neutralize toxins and pathogens, and low-affinity IgA can inhibit pathogens from adhering to mucosa, making IgA an important component of mucosal immunity against infection (Fig. 13.2).

Secretory IgA is a dimeric structure formed by two monomeric IgA via covalent linkage by J-chain. The secretory piece is a poly-immunoglobulin structure (Poly-Ig) containing five Ig repeat regions, and the fifth Ig region of the secretory piece is bound to the heavy chain of dimeric IgA by disulfide bonds to form a complex. The mucosal epithelial cell membrane contains the Poly-Ig receptor through which the IgA complex is transported to the mucosal surface.

13.2 Mechanisms of Immune Tolerance

Immune tolerance refers to the unresponsiveness of the adaptive immune system to an antigen. The primary cause of immune tolerance is the clearance or suppression of antigen-specific lymphocytes or their conversion to regulatory T cells (Treg).

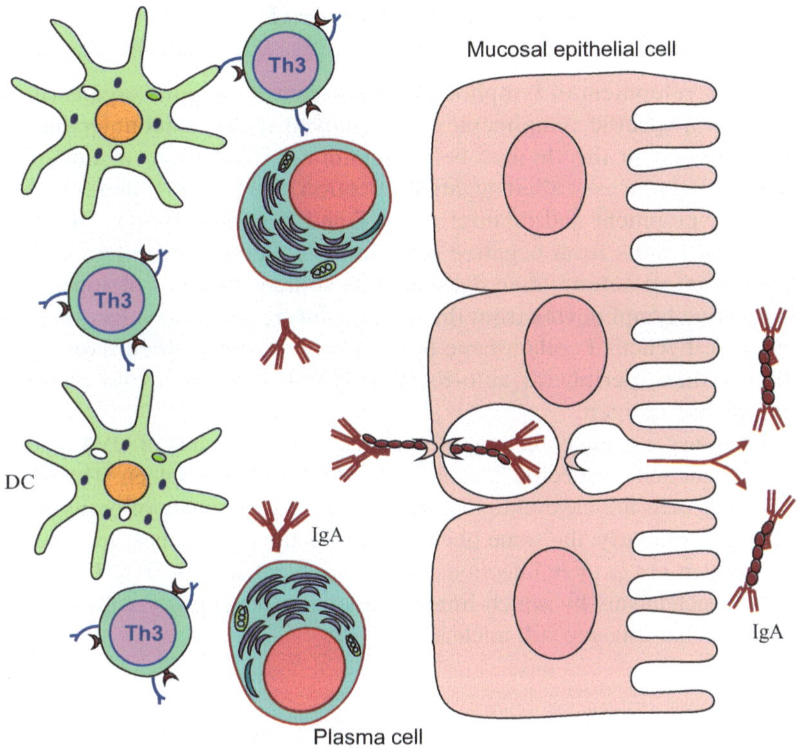

Fig. 13.2 Schematic illustration of the production and transport of secretory IgA

Immune tolerance ensures that the immune system does not produce an immune response to self antigens, however, if immune tolerance is broken, the immune system attacks self tissues, leading to autoimmune diseases (see Chap. 12 for more information on autoimmune disease).

Oral tolerance was first discovered in 1911 when Wells and Osborne fed guinea pigs diets containing ovalbumin, purified ovalbumin allergen, or oats, and their results indicated that the guinea pigs had a reduced immune response to these proteins or allergens and did not produce an allergic response. It was not until the mid-1990s that a preliminary understanding of the mechanism of oral tolerance was gained, and it was found that oral intake of antigens induces the production of a specific population of $CD4^{+}T$ cells, which produce mainly TGF-β, IL-4 and IL-10, which have immunosuppressive effects. The population of $CD4^{+}T$ cells characterized by the production of TGF-β is called Th3 cells. In recent years, it has been found that Treg plays a major role in oral tolerance. Oral tolerance is not dependent on natural Treg (nTreg), but is closely related to the production of induced Treg (iTreg, $CD4^{+}CD25^{+}FoxP3^{+}T$ cells).

13.2.1 Specific Lymphocytes Are Removed

During the development of lymphocytes in the central immune organs (thymus, bone marrow), specific lymphocytes are removed if they encounter the corresponding antigen in the thymus before maturation (deletion of auto-reactive lymphocytes), a process called negative selection (see Chaps. 4 and 5 for details about the development and maturation of T and B lymphocytes). The immune tolerance that results from negative selection is called central tolerance, which safeguards against autoimmune disease at its source. Because of some leakage of self-reactive lymphocytes from the thymus during negative selection, a small number of self-reactive cells escape to peripheral tissues, which requires iTreg to suppress these peripheral auto-reactive lymphocytes to avoid an immune response to self-antigen.

In OVA transgenic mice, feeding of mice with large amounts of OVA resulted in a massive reduction of specific T cells in the PP, MLN, and spleen. This suggests that specific T cells are cleared in the presence of large amounts of orally administered antigen. Similarly, the same phenomenon occurs in mice post treatment with massive antigen via *ip* or *iv* injection. Thus, peripheral specific T-cell clearance is one of the mechanisms by which immune tolerance develops, yet the underlying molecular mechanisms are still unclear.

13.2.2 Anergy

Specific lymphocytes are in anergy in the absence of co-stimulatory signals, which is one of the main reasons for the development of immune tolerance. T cells in anergy could be restored to the immune response state under the action of IL-2, which is one of the hallmarks that distinguish anergy from immunosuppression.

13.2.3 Immunosuppression by Regulatory T cell

The intestinal mucosal microenvironment contains high levels of TGF-β and IL-10, which are secreted by a variety of cells. $CD103^{+}$DCs in the lamina propria of the intestinal mucosa migrate to the mesenteric lymph nodes (MLN) after capturing antigens transported by goblet and M cells, secreting TGF-β and retinoic acid, which induce the naive T cells in the MLN to become iTreg. Subsequently the iTreg migrates to the intestinal mucosa or other immune organs to play immunosuppressive roles. Thus, $CD103^{+}$DCs play a crucial role in immune tolerance, and the MLN is an important site for the formation of immune tolerance. If the MLN is surgically removed, mucosal immune tolerance will not exist, which fully illustrates the importance of MLN in the formation of mucosal immune tolerance. The fact that removal of PP does not affect mucosal immune tolerance suggests that PP are not required in mucosal tolerance and that CD103$^+$DCs in the lamina propria outside the PP can likewise migrate to the MLN to function. However, this mechanism may only be present in mammals, as avian species do not have lymph nodes (no MLN either) and do not have immune tolerance problems, suggesting that there is big difference between species.

In addition to the secretion of TGF-β by mucosal cells to induce naive CD4$^+$T cells to become iTreg, FoxP3 expression in naive CD4$^+$T is significantly elevated if there is helminth or clostridial infection in the gut, suggesting that some pathogenic infections can also induce the formation of iTreg and suppress the immune response to establish persistent infections (see Chap. 14 for more information on immunosuppression and persistent infection).

It should be noted that the above mechanisms of immune tolerance are not independent and often work together. For example, in transgenic mice, a single immunization of mice with a high dose of antigen (>200 mg) causes a decrease in lymphocytes, mainly due to the apoptosis caused by activation of the Fas-FasL apoptotic signaling pathway by the high dose antigen, or lymphocyte activation lacking a second signal, or immune response negative feedback causes activation of immune checkpoints, and these factors result in reduced numbers of lymphocytes or in anergy. However, if mice are repeatedly inoculated with low doses of antigen (0.2–1 mg), iTreg is induced to cause immunosuppression (see Chap. 9 for iTreg function).

13.3 Balance Between Mucosal Immunity and Immune Tolerance

The first and second sections of this chapter address mucosal immunity and immune tolerance, respectively, and the cytokine TGF-β plays an important role in the regulation of the mucosal immune response. TGF-β is pleiotropic, i.e., acts on different target cells to produce different effects. TGF-β not only induces naive CD4$^+$T cells to form iTreg suppressing the immune response, but also induces naive CD4$^+$T cells to differentiate into Th17 cells promoting an inflammatory response. Thus, TGF-β plays an important regulatory role in mucosal immunity and immune tolerance.

13.3.1 Mucosal Immunity Against Infection

During the infection of the intestinal mucosa by pathogens (either virulent or mild strain), intestinal mucosal epithelial cells recognize pathogens and their secreted substances through pattern recognition receptors (PRRs), which activate the innate immune signaling transduction to initiate the expression of pro-inflammatory cytokines including TNF-α, IL-1, IL-6, IL-12, etc. through the NF-κB signaling pathway, and since a variety of cells in the intestinal microenvironment secrete TGF-β, TGF-β acts together with these pro-inflammatory response cytokines and participates in the regulation of the adaptive immune response (Fig. 13.3).

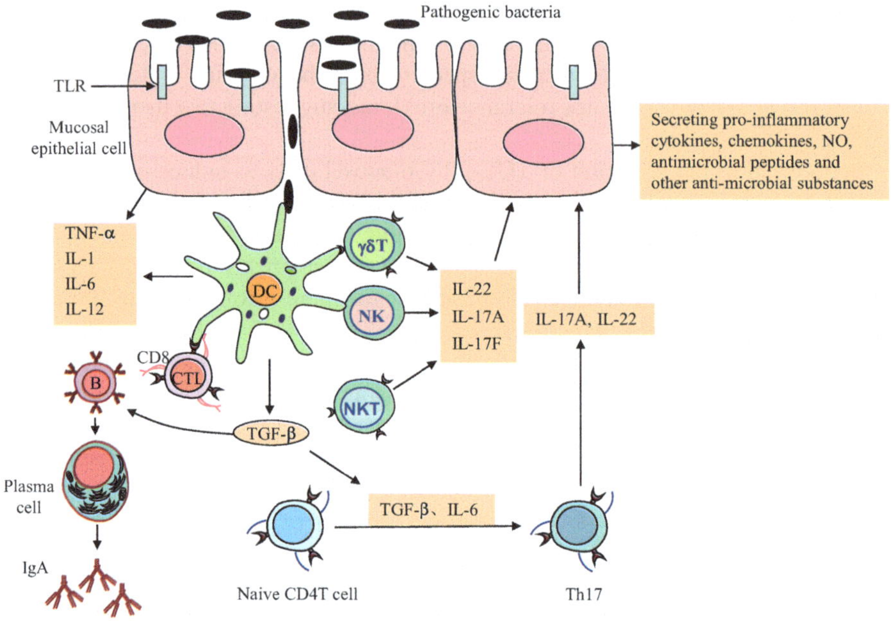

Fig. 13.3 Schematic illustration of the mucosal immunity of the intestinal mucosa against pathogenic infection

TGF-β acts together with IL-6 to induce the differentiation of naive CD4⁺T cells into Th17 cells, while inhibiting the conversion of naive CD4⁺T cells to Th1 or Th2 cells. Th17 cells secrete the pro-inflammatory cytokine IL-17, but other cells such as NK, NKT and γδ T cells, also secrete IL-17 family molecules like IL-17A, IL-17F, IL-22, etc. Mucosal epithelial cells produce pro-inflammatory response cytokines, chemokines, NO, antimicrobial peptides and other anti-infective substances in response to IL-17 and IL-22 to protect local tissues of the intestinal mucosa. In addition, TGF-β is an important cytokine that induces IgA production by B cells. Experimental results have confirmed that without TGF-β, B cells do not produce IgA. IgA is an important immunoglobulin that protects the mucosa, so TGF-β is very important for mucosa immunity against infection (Fig. 13.3).

Pathogenic bacteria adhere to epithelial cells or enter the body through the intercellular space, pattern recognition receptors recognize the pathogen and elicit an innate immune response, and innate immune cells secrete pro-inflammatory cytokines (TNF-α, IL-1, IL-6, and IL-12). Dendritic cells (DCs) secrete TGF-β after capturing pathogens and activate NK, NKT, and γδ T cells by contact, causing them to secrete cytokines such as IL-17A, IL-17F, IL-22, and so on. Naive CD4⁺T cells differentiate into Th17 and secrete IL-17A and IL-22 under the joint action of TGF-β and IL-6. Epithelial cells secrete pro-inflammatory cytokines, chemokines, NO, antimicrobial peptides and other anti-microbial substances under the joint action of IL-17 and IL-22. B cells generate an immune response after antigenic recognition and secrete IgA under the action of TGF-β to protect mucosal tissues. (* Of note, this paragraph should be the legend for Fig. 13.3)

13.3.2 Immune Tolerance to Oral Antigens

Oral tolerance, is the inhibition of systemic cellular and humoral immune responses to an antigen due to the oral administration of that antigen. Immune tolerance of the digestive track mucosa is a complex biological process. As the gastrointestinal tract passes 70–100 g of protein per day, three factors are required for the immune system to develop immune tolerance against these proteins, including the physiological barrier of the gastrointestinal tract, the digestive degradation of the antigen, and the immunosuppressive environment.

13.3.2.1 Physiological Barrier of the Gastrointestinal Tract

The physiological barrier of the gastrointestinal tract consists mainly of mucosal epithelial cells and their secretions, which contain mucin oligosaccharides and IgA forming a hydrophobic layer, while IgA binds antigens to prevent them from crossing the epithelium and protects the intestinal mucosa.

13.3.2.2 Digestive Degradation of Food Antigens

Food is degraded by various digestive enzymes as it passes through the gastrointestinal tract and is broken down into small molecules of 2-peptide or 3-peptide or less, which lose their antigenic structure for immune recognition, neither bind PRRs, nor elicit an immune response, and therefore have no inflammatory response and are actively transported into the body as nutrients through the intestinal mucosa cells. However, if the food antigen is not completely degraded into small short peptide molecules and retain antigenic epitopes, these undigested and degraded antigens will be taken up by M (microfold) cells and intestinal epithelial cells (IECs), and transported to dendritic cells, or activate lamina propria lymphocytes to produce an immune response. The lamina propria dendritic cells can also protrude directly into the intestine to take up undegraded antigens. Dendritic cells take up food antigens to induce differentiation of naive $CD4^{+}T$ cells into iTreg ($CD4^{+}Foxp3^{+}$), which produces immunosuppression (see Chap. 9 for iTreg function). If the iTreg fails to suppress the immune response induced by food antigen, an allergic reaction will occur.

13.3.2.3 Immunosuppressive Arena in Gut

It is generally believed that mucosal tolerance is associated with early microbial mucosal colonization and that there are approximately 500 types of commensal bacteria in the intestine, with total number over 10^{14}. Commensal bacteria inhibit NF-κB signaling in intestinal mucosal epithelial cells and induce intestinal mucosal cells to express TGF-β, retinoic acid (RA), thymic stromal lymphopoietin (TSLP) and other molecules to promote differentiation of tolerogenic dendritic cell. Dysbacteriosis promotes Th2 immune response, which predisposes to allergic disease. In animal models of respiratory and food allergy, germ free (GR) animals or antibiotic-treated animals are more prone to allergic reactions, and gut microflora play an important role in restoring mucosal tolerance.

The immunosuppressive arena is an important factor in safeguarding food immune tolerance, and food allergic immune responses (immune tolerance broken) are generally unrelated to central tolerance and are primarily related to immune regulation by the iTreg (see Chap. 9 for iTreg function). If the iTreg fails to suppress the immune response induced by the food antigen, an allergic response will occur. Thus, iTreg plays an important role in suppressing the mucosal immune response to guarantee immune tolerance to oral antigens. Dendritic cells are key cells to induce formation of iTreg cells in immune tolerance to oral antigens. $CD103^{+}DCs$ in the mucosal lamina propria play an important role in the induction of iTreg formation. $CD103^{+}DCs$ secrete large amounts of TGF-β, which together with retinoic acid induce the differentiation of naive $CD4^{+}T$ cells into iTreg ($CD4^{+}Foxp3^{+}$), and iTreg suppress an immune response of specific lymphocytes to antigenic epitopes in food to generate an immune tolerance to food (Fig. 13.4). Thus, $CD103^{+}DCs$ are the dominant players for immune tolerance.

Food proteins are actively transported into the body by intestinal mucosal epithelial cells after degraded into short peptides, and $CD103^{+}DCs$ in the lamina propria of the mucosa capture antigenic peptides and secrete large amounts of TGF-β and retinoic acid (RA), which work together to induce differentiation of naive $CD4^{+}T$ cells into iTreg that secrete large amounts of immunosuppressive cytokines inhibiting the immune response of specific lymphocytes to antigenic epitopes in food.

It has recently been found that the proportion of Treg in the intestinal mucosa is about 5% of lymphocytes, and that in addition to $CD4^{+}FoxP3^{+}T$ cells (nTreg, iTreg), there are other types of regulatory cells that exert immunosuppressive effects, such as IL-10-producing Tr1Treg, TGF-β-producing $LAP^{+}Th3$, and $CD8^{+}LAP^{+}Treg$. All of these cells are associated with mucosal tolerance.

Mammalian immune tolerance to food antigens involves an important role of the mesenteric lymph nodes (MLN) in addition to the intestinal mucosa, including the Peyer's patch, where $CD103^{+}DCs$ capture antigenic peptides and migrate to the MLN to induce differentiation of naive $CD4^{+}T$ cells into iTreg by secreting TGF-β and retinoic acid (RA), which secretes a large number of immunosuppressive cytokines that inhibit the immune response of specific lymphocytes to antigens in food (Fig. 13.5).

Food proteins are actively transported into the body by intestinal mucosal epithelial cells after enzymatic degradation into short peptides, and $CD103^{+}DCs$ in the lamina propria of the mucosa capture antigenic peptides and secrete immunosuppressive cytokines (TGF-β and IL-10) in the mucosa, further migrate to the mesenteric collecting lymph nodes (MLN) and secrete large amounts of TGF-β and retinoic acid (RA). TGF-β and RA work together inducing naive $CD4^{+}T$ cells to differentiate into iTreg, which secretes large amounts of immunosuppressive cytokines (TGF-β and IL-10) and inhibits an immune response of specific lymphocytes to antigens in food.

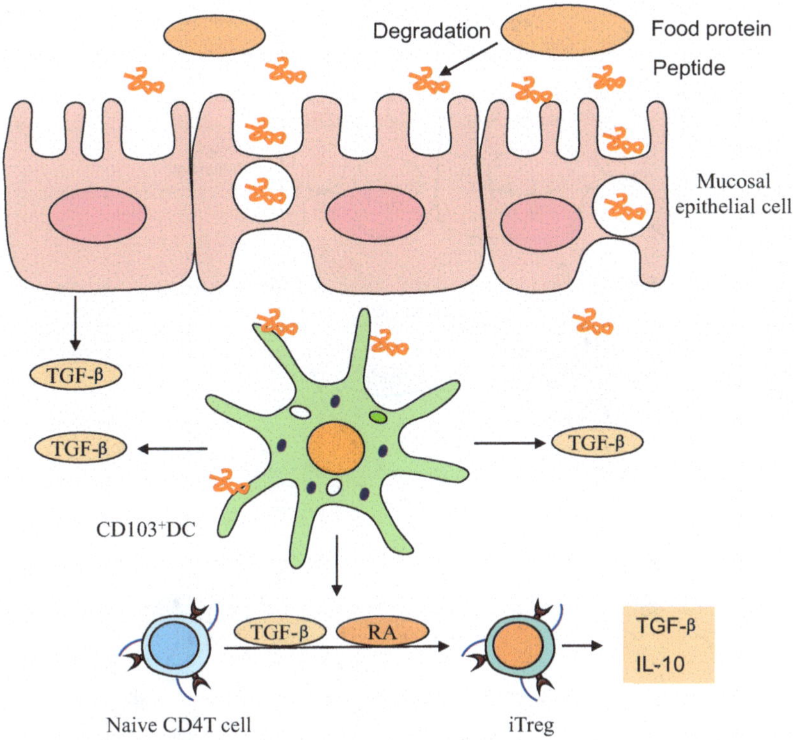

Fig. 13.4 Schematic diagram of oral tolerance-induced iTreg

13.3.3 Immune Tolerance Break by Oral Antigen

The development of immune tolerance to oral antigens is a normal immune phenomenon, and experimental evidence suggests that the addition of adjuvants to oral antigens, such as cholera toxin (CT) or staphylococcus enterotoxin B (SEB), can break immune tolerance and drive the immune response in a Th2 direction, causing intestinal mucosal inflammation or allergic reactions. If the body has been exposed to the antigen by non-oral routes, oral administration of the antigen is likely to break immune tolerance and cause an immune response.

In prevention and control of infectious diseases, establishing mucosal immunity can effectively prevent and control pathogenic infections. Thus, it is critical to obtain a good mucosal immune response to vaccination. Vaccination of animal with attenuated live vaccines by oral route or spray can elicit better mucosal immunity, while inactivated vaccines or polypeptides need special adjuvants or antigenic delivery-system to achieve good immune response. The antigenic delivery-system of mucosal vaccine is crucial for the vaccine to take effect. The ideal mucosal vaccine should have characteristics that include (1) antigenic stability without destruction under enzymatic digestion and low pH environment, (2) effective entry

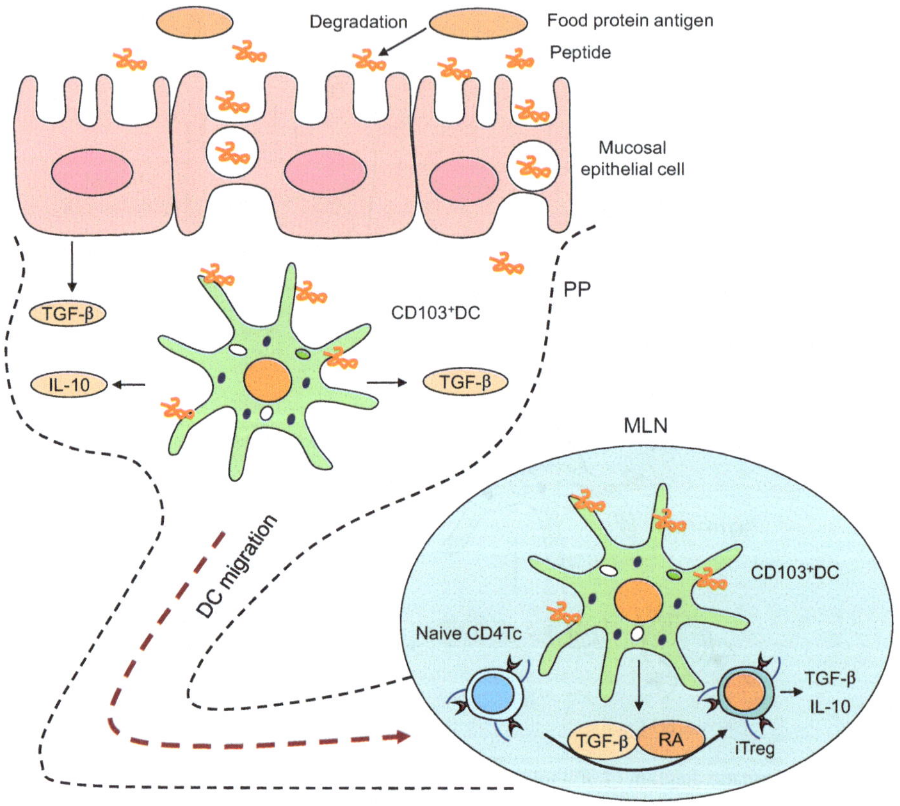

Fig. 13.5 Schematic diagram of the mechanism of oral food-induced immune tolerance

into the body through the mucosal barrier, and (3) breaking of mucosal tolerance to induce an immune response. It has been established that in the development of mucosal vaccine, vaccine designs using intestinal mucosal M cells and intestinal mucosal epithelial cells as target cells as well as combining immunogens with the ligands of the membrane receptors of target cells or with monoclonal antibodies, could achieve high levels of IgA antibodies by oral immunization. The development of novel antigenic delivery systems using receptor-mediated trans-endocytosis of antigen by M cells or intestinal mucosal epithelial cells is an important trend of research to improve the effectiveness of mucosal immunization.

Study Questions

1. What is immune tolerance? What are its causes?
2. How is secretory IgA produced and transported?
3. What is the mechanism of immune tolerance?
4. What is the mechanism by which immune tolerance to oral antigens develops?
5. What are the bottlenecks in the development of oral vaccines?

Chapter 14
Immunosuppression and Immune Escape

Overview:
The immune system is an important defense system of the organism that has evolved over time. Innate immunity is an immune response in animals through recognition of pathogens by pattern recognition receptors (PRRs), which rapidly suppress and clear pathogens early during pathogenic infection. In addition to innate immunity, the immune system of higher animals has evolved adaptive immunity, which is characterized by the production of humoral and/or cellular immune responses to pathogens, more efficient clearance of pathogenic infections, and the generation of immune memory, as well as the elimination of mutated cells in the body to maintain homeostasis. However, certain pathogens, after infecting the host, suppress the immune response and reduce host immunity, resulting in persistent, secondary and mixed infections. If the animal is immunocompromised, tumor cells will evade immune surveillance, resulting in neoplastic disease. This chapter does not involve immunodeficiency due to genetic causes, mainly focusing on the mechanisms underlying pathogen-induced immunosuppression and immune escape of tumor cell from host immunity.

14.1 Major Ways in Which Pathogens Suppress the Immune Response

Pathogenic infections induce an immune response, which once initiated is like an accelerating train that runs until the end, and in the end, the pathogen is removed and homeostasis is restored. Some pathogens have acquired the ability to suppress host immunity during their long struggle with the host, establishing persistent infection. Such pathogens have an advantage of survival in terms of evolution. *Human immunodeficiency virus* (*HIV*) is a typical immunosuppressive pathogen that suppresses the immune response by infecting and damaging $CD4^{+}$T cells, establishing persistent infection that leads to the death of the patient from secondary and/or

S. J. Zheng, *Veterinary Molecular Immunology*,
https://doi.org/10.1007/978-981-99-8929-4_14

mixed infections due to immunocompromise. Among animal diseases, there are many pathogens with immunosuppressive effects, such as *foot-and-mouth disease virus (FMDV), Porcine reproductive and respiratory syndrome virus (PRRSV), Porcine circovirus type 2 (PCV2), Hog cholera virus (HCV), Pseudorabies virus (PRV), Infectious bursal disease virus (IBDV), Marek's disease virus (MDV), Avian leukemia virus (ALV), Chicken infectious anemia virus (CIAV), Avian reovirus (ARV), Reticuloendotheliosis virus (REV), Brucella, Salmonella, Mycobacterium tuberculosis, Staphylococcus, Feline immunodeficiency virus (FIV), Feline leukemia virus (FLV), Canine distemper virus (CDV), canine parvovirus (CPV), Rickettsia, Leishmania*, and fungi. All these pathogens suppress the normal immune response in the host by different ways.

Pathogens can suppress the immune response in a number of ways, and although the ways in which different pathogens suppress the immune response may vary, the main ones are summarized as follows.

14.1.1 Destruction of Immune Cells

Pathogenic infection of certain important immune cells induces apoptosis, such as HIV infection of $CD4^+$ T cells, measles virus infection of DC, PRRSV targeting mononuclear macrophages (especially lung alveolar microphages), IBDV targeting bursal B cells, and PCV2 infection of lymphocytes. These immune cells are infected and undergo apoptosis, leaving the host immune response severely suppressed. Pathogenic toxins, such as Bacillus anthracis edema toxin (ET), lethal toxin (LT) and staphylococcal exfoliative toxin, severely destruct phagocytes and DCs, causing cell death.

Viruses are strictly intracellular pathogens that have acquired the ingenuity to use host cell apoptosis as a way of survival to proliferate in the long term evolution. IBDV induced apoptosis not only by using viral proteins, but also by inducing avian cells to express microRNA (miR)-16-5p to inhibit BCL-2 expression, i.e. IBDV induced apoptosis at both protein and RNA levels to destroy target cells (Fig. 14.1)

14.1.2 Inhibition of Antigen Presentation

A variety of cells can participate in the immune response by presenting antigens via cytosolic and/or endosomal pathways (see Chap. 7 for details). Some viral genes encode proteins that directly affect cellular protein function in the antigen presenting pathway, affecting antigen presentation and inhibiting the immune response, e.g., some viruses in the herpesviridae family inhibit MHC-I/II expression, and measles virus infection of DCs inhibits antigen presentation by HLA-I molecules, so does human papillomavirus infection, thereby circumventing CTL killing of infected cells and evading NK cell immune surveillance. The EBNA1 protein encoded by EBV affects antigen degradation by the proteasome, while BNLF2a inhibits antigenic epitope transport by TAP, blocking antigen presentation by MHC-I molecules. The

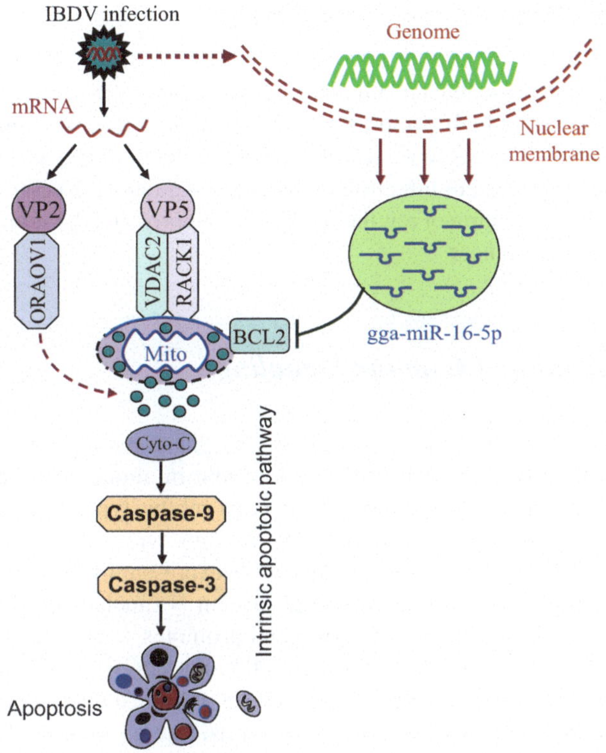

Fig. 14.1 Schematic illustration of the mechanism underlying infectious bursal disease virus (IBDV)-induced apoptosis in host cell. IBDV is a double-stranded RNA virus that expresses viral proteins upon infection of cells. At the protein level, IBDV induced apoptosis through the intrinsic apoptotic pathway by interaction of VP2 with ORAOV1 and by interaction of VP5 with VDAC2 and RACK1, which form a complex. At the RNA level, IBDV infection induces the expression of chicken microRNA-16-5p (gga-miR-16-5p) in host cell, which targets BCL-2 and induces apoptosis. Thus, IBDV infection induces apoptosis in host cells at both protein and RNA levels. *Mito* mitochondrion, *VP* viral protein, VP2 is a structural protein of IBDV, while VP5 is a non-structural protein, *ORAOV1* oral cancer overexpressed protein-1, which has an inhibitory effect on apoptosis, *VDAC2* voltage-dependent anion channel protein 2, which belongs to mitochondrial membrane protein channel protein and is associated with apoptosis, *RACK1* receptor of activated protein kinase C1, which interacts with VP5 and VDAC2 to form a complex that regulates apoptosis in IBDV-infected cells, *Cyto-C* cytochrome-C, *Caspase* cysteinyl aspartate-specific proteinase that normally exists as a procasapse and is involved in apoptosis upon cleavage. (See Chap. 2 for details of the apoptotic pathway)

mK3 protein encoded by the murine gamma herpesvirus 68 gene is an E3 ubiquitin ligase that catalyzes the degradation of MHC-I and TAP, affecting the antigen presentation of MHC-I molecules. Some DNA viruses, such as herpesviruses, encode miRNAs (miR-US4-1) that inhibit the expression of aminopeptidase (ERAP1), a key enzyme responsible for the assembly of antigenic epitopes into MHC-I molecules for antigen presentation, and inhibition of ERAP1 is tantamount to blocking endogenous antigen presentation. Cytomegalovirus inhibits MHC expression and suppresses the immune response in several ways. Salmonella inhibits the generation of specific CTL by inhibiting DC and preventing antigen presentation.

14.1.3 Induction of Regulatory T cells

Early after infection, Mycobacterium tuberculosis induces the generation of regulatory T cells (Treg), which suppresses the immune response and allows the Mycobacterium tuberculosis to establish persistent infection. Hepatitis C virus also establishes chronic persistent infection by inducing regulatory T cells. Helicobacter pylori induces the generation of regulatory T cells, suppressing the immune response and leading to gastric cancer.

14.1.4 Inhibition of Immune Signaling

Innate immunity is the first line of host defense against pathogenic infection, and type I interferon (IFN-alpha/beta) plays a key role in antidefense of the host. In order to establish persistent infection, virus must inhibit type I interferon expression. Foot-and-mouth disease virus (FMDV) inhibits type I interferon expression through VP1, PRRSV inhibits type I interferon expression through nonstructural proteins Nsp1, Nsp2, Nsp11, and structural protein N (nucleocapsid protein), and IBDV inhibits type I interferon expression and promotes viral replication by interacting with cellular target proteins through VP3 and VP4 (Fig. 14.2). In addition, both cytomegalovirus and human papillomavirus inhibit the expression of type I interferon. *Bacillus anthracis* secretes two toxins, edema toxin and lethal toxin, both of which enter cells and affect the activity of various kinases and inhibit immune signaling.

14.1.5 Encoding Immunosuppressive Cytokines

Many viruses have acquired the ability to regulate the host immune response during evolution, such as IL-10, which functions in a pleiotropic manner and is an important immunosuppressive cytokine. *Mycobacterium tuberculosis* inhibits the function of host macrophages and dendritic cells by increasing host IL-10 expression, thereby suppressing the immune response and favoring its infection. Some viruses suppress the host immune response by expressing viral proteins (vIL-10) that are highly similar to host IL-10 (cIL-10), e.g., the BCRF1 gene of EBV encodes vIL-10, which is highly homologous to host IL-10 and binds IL-10 receptors to act as an immunosuppressive agent. hCMV also encodes an IL-10 analogue (cmvIL-10), which is also immunosuppressive. In addition, most herpesviruses share this characteristic with some members of the poxvirus family.

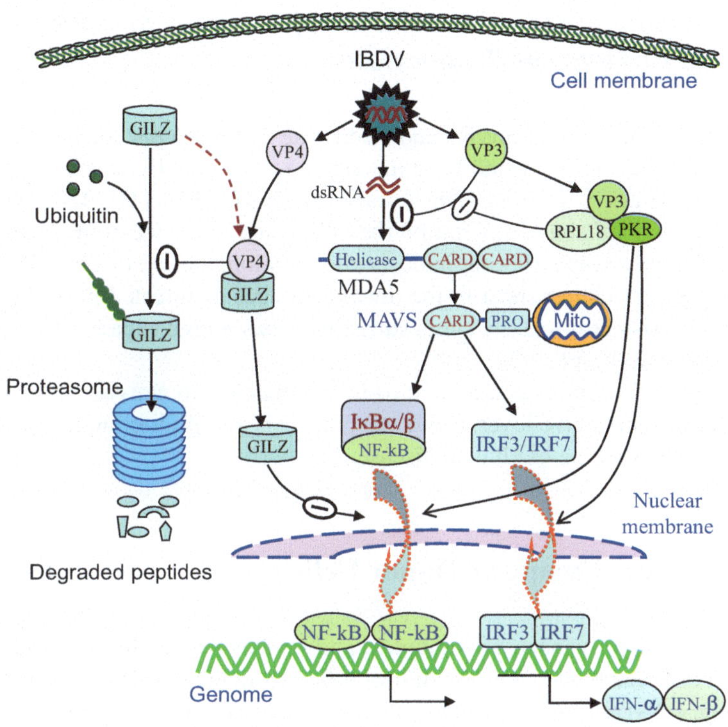

Fig. 14.2 Schematic illustration of the mechanism underlying IBDV-mediated inhibition of type I interferon expression. After IBDV infects host cells, viral genome double-stranded RNA (dsRNA) is recognized by host cellular MDA5 (chicken is genetically deficient of RIG-I), a member of RLRs, initiating the type I interferon expression signaling pathway. However, the virus inhibits type I interferon expression by expressing VP3 and VP4, preventing host antiviral immune response and resulting in immunosuppression. VP3 competitively binds to MDA5, blocking MDA5 recognition of dsRNA initiating immune response, thereby evading immune response. However, host cellular protein PKR and RPL18 interact with VP3 to form complexes to block immune evasion, ensuring the recognition of dsRNA by MDA5 to initiate NF-kB and IRF signaling transduction pathway. VP4 interacts with GILZ to inhibit ubiquitin-mediated proteasomal degradation of GILZ, and GILZ prevents NF-κB nuclear translocation to initiate the expression of type-I interferon. *Mito* mitochondrion, *VP* viral protein, VP3 is a structural protein of IBDV, while VP4, a non-structural protein, is an auto-cleaved protease of IBDV, *GILZ* glucocorticoid-induced leucine zipper; *IRF* interferon regulatory factor, IRF3/IRF7 forms a heterodimer after activation and translocate into the nucleus to initiate type I interferon expression, *MAVS* mitochondrial antiviral signalling protein, *MDA5* melanoma differentiation-associated gene-5, an important member of the family of pattern recognition molecules RLRs that recognize dsRNA to initiate immune response, *NF-κB* nuclear factor κ enhancer binding protein, *PKR* double-stranded RNA-activated protein kinase, *RPL18* ribosomal protein L18

14.1.6 Hiding in Host Cells or Foci of Inflammatory Response to Circumvent Phagocytosis

Staphylococcus aureus can survive and proliferate in the phagosome for a long time after phagocytosed by neutrophils or macrophages, eventually causing cell death. Intracellular parasites such as *Listeria monocytogenes*, *Brucella*, *Mycobacterium tuberculosis*, and *Salmonella* all inhibit the killing and clearance of invaders by host cells in different ways, harboring in host cells for a long period of time. The ability of Hog cholera virus to persist in the tonsillar tissues of infected pigs allows it to establish a persistent infection, suggesting that it has a high capacity for hiding in cells for immune escape.

Staphylococcus aureus can adhere to initial platelets to activate and aggregate them to form a protective layer, evade phagocytosis and proliferate to cause endocarditis, a common type of clinical disease.

14.2 Immune Escape of Tumor Cells

The causes of tumor genesis are complex, and it takes a long chronic process for a tumor cell to develop into a visible clinical tumor (roughly 0.5 CM or so, with about 100 million cells) as observed by imaging facility, usually taking about 7 years. The causes of tumor cell development may vary, but there are three main causes: (1) tumors caused by persistent viral infections; (2) tumors caused by carcinogenic chemicals; and (3) tumors formed by randomly mutated cells escaping immune surveillance. However, whatever the cause of tumor formation is, they all escape the surveillance of the immune system. There are several main ways of immune escape by tumor cells as summarized as follows.

14.2.1 Persistent Viral Infections Cause Tumors

Persistent viral infections cause tumors and may be associated with low immune response or anergy of T-cell as well as Treg-mediated suppression of immune responses. EBVs can cause B-cell lymphoma, with infection rates of 70–90% in the population. Human papillomaviruses can cause cervical cancer and are very common in clinical practice. These viruses suppress the cellular immune response by direct interaction of viral proteins with host cellular proteins, consequently resulting in immune escape. The mechanism by which these two viruses escape immune response is not well understood, but it may be related to Treg.

14.2.2 Suppression of the Cell-Mediated Immune Response Through Co-stimulatory Molecules

The co-stimulatory molecule CD28 of T cell binds to ligands (B7-1/CD80, B7-2/CD86) to promote cell activation, whereas CTLA-4 and PD-1 receptors are expressed only after T cell activation. CTLA-4 and PD-1 act as inhibitors of T cell immune response after binding to the corresponding ligand. CTLA-4 has the same ligand as CD28, but CTLA-4 has a stronger affinity for the ligand, competitively binding to ligands and acting as an immunosuppressive factor. The membrane receptor with immunosuppressive effects is called an immune checkpoint (see Chap. 4), and CTLA-4 is an important immune checkpoint. PD-1 has two ligands (PDL1/B7H1, PDL2/B7DC) that inhibit T-cell immune responses upon binding. Tumor cells express immunosuppressive molecules or inhibit immune signaling for the purpose of immune escape, for example, many tumor cells express PDL1 molecules in large quantities, which bind to the PD-1 receptor of T cells to inhibit T cell immune response. Tumor cells in a variety of tissues express PDL2 molecules, generating immunosuppressive signals and suppressing cell immune responses.

14.2.3 Suppression of Interferon

Type I interferon (IFN-α/β) and type II interferon (IFN-γ) bind to receptors and activate Jak/STAT signaling to induce the expression of a variety of proteins related to cell proliferation, antigen presentation, and apoptosis, and IFN-γ is an important regulatory molecule that promotes cell-mediated immune response. Suppressor of cytokines signaling (SOCS) play an inhibitory role in the interferon signaling pathway and suppress interferon-mediated cell responses. Tumor cells inhibit interferon-mediated immune signaling by suppressing interferon receptor expression and overexpressing SOCS.

14.2.4 Production of Immunosuppressive Cytokines

The function of IL-10 is pleiotropic, promoting humoral immune response and suppressive cell-mediated immune response. In addition, it has an inhibitory effect on macrophages and dendritic cells. vIL-10 produced by EBV suppresses T-cell immunity and creates the conditions for tumor formation. The inhibitory effect of IL-10 on dendritic cells may be related to T-cell anergy.

TGF-β is also a cytokine with pleiotropic functions. Most cells, including epithelial cells, endothelial cells and immune cells, produce TGF-β. TGF-β inhibits T cell proliferation and suppresses T cell differentiation to Th1 and Th2 cells, but promotes naive $CD4^{+}$T cell differentiation to Treg and Th17. TGF-β is secreted in large quantities by tumor cells, but the tumor cells themselves bypass its inhibitory effects on cell proliferation by mutation of proteins in the TGF-β signaling pathway, thus creating a microenvironment that inhibits T cells but not tumor cells. In addition, TGF-β has the effect of stimulating local tumor angiogenesis and tumor cell infiltration.

14.2.5 Chemokines and Immune Escape

Chemokines are a large class of cytokines with small molecular weight that have chemotactic effects. Chemokines bind to G protein-coupled receptors and attract immune cells and inflammatory cells to migrate to the site of infection or inflammatory foci. LMP-1 protein of EBV induces the expression of CCL17 (TARC) and CCL22 (MDC), two chemokines that attract $CCR4^{+}$Treg to assist the immune escape by tumor cells. Human cytomegalovirus (HCMV) expresses US28 protein, which serves as a receptor for several chemokines including CCL2 (MCP-1), CCL3 (MIP-1α), CCL4 (MIP-1β), CCL5 (RANTES), CX3CL1, and others. Transgenic mice with high expression of US28 promoted tumor formation. HCMV expresses IE86 protein, which not only inhibits the expression of type I interferon, but also inhibits cells from expressing CCL3, CCL5, CCL8 (MCP-2), CXCL9 (MIG), CXCL8 (IL-8), and so on. In summary, tumor cells attract immunosuppressive cells through chemokines and chemokine receptors, promoting local tumor angiogenesis and tumor cell infiltration.

14.3 Comparison of Persistent Viral Infection and Tumor Immune Escape

To establish persistent infection, virus must have the feature of immunosuppression or immune escape. The immune escape of tumor cell has similarities to that of virus, but are also different from viral immune escape (Table 14.1).

For a particular pathogen or tumor, immunosuppression or immune escape may occur in one or several ways (Fig. 14.1). Only by understanding the mechanisms of immune escape can intervention or treatment strategies be stipulated. Immunosuppressive pathogens in animals are numerous and harmful, and require intensive research to explore solutions. Therefore, studies on the mechanisms of immunosuppression and immune escape are a focus and central issue in the research areas of anti-defense and immunity against tumor.

Table 14.1 Comparison of persistent viral infection and tumor immune escape

Means of immune escape	Viral immune escape	Tumor immune escape
Antigen presentation	Inhibition of MHC expression by immunoregulatory proteins or miRNAs	Suppression of MHC expression by gene mutation or epigenetic modification
Co-stimulatory molecule	Promote expression of inhibitory co-stimulatory molecules (e.g. PDL1), and/or inhibit the expression of costimulatory molecules for cell activation (e.g. B7.1)	Promote expression of inhibitory co-stimulatory molecules (e.g. PDL1) Suppress the expression of costimulatory molecules for cell activation by epigenetic modifications (e.g. CD70) Cause abnormal expression of costimulatory molecules
Interferon	Inhibit interferon expression, receptor binding, signal transduction, etc.	Inhibit interferon receptors, inhibit STAT signaling, and enhance SOCS expression
Inhibitory cytokine	express vIL-10, induce production of TGF-β	induce expression of IL-10 and TGF-β, meanwhile tumor cells show mutations in TGF-β signaling protein
Chemotactic factor	Influence chemokine function through viral proteins	Induce migration of immunosuppressive cells to tumor sites by chemokines

Study Question

1. What is the primary way in which pathogens suppress the immune response?
2. What is the main mode of immune escape of tumor cells?
3. How do you understand the effects of hypersensitivity, immunosuppression, immune tolerance and immune escape on animal health?

Appendix A: Abbreviation of the Terms for Protein Molecules Involved in Immune Signal Transduction and Their Corresponding Functions

Abbreviation	Full name	Functions
AP-1	Activating protein-1	A transcriptional factor transducing signals and initiating downstream gene expression
APRIL	A proliferation-inducing ligand	A TNF family member promoting cell proliferation
BAFF	B-cell activating factor	ATNF family member, promotes B-cell activation
BCMA	B-cell maturation antigen	A BAFF receptor on the membrane of B and T cells, involved in cell proliferation
BLNK	B-cell linker	A cellular protein involved in signaling transduction
Btk	Bruton's tyrosine kinase	A kinase that phosphorylates adaptor proteins in signaling transduction
CARD	Caspase activation and recruitment domain	Upon activation, the CARD domain of RLRs and NLRs recruits adaptor proteins to transduce signals
CLRs	C-type lectin receptors	A class of pattern recognition receptors on host cell membrane, mainly sensing fungi
CTLD	C-type lectin-like domain	Sensing pathogen-associated molecular patterns (PAMPs)
CTLA4	Cytotoxic T lymphocyte-associated antigen-4	An immune checkpoint, inhibiting T-cell activation
DAG	Diacylglycerol	A second messenger in signaling transduction
DCR	Decoy receptor	A cell membrane receptor for TRAIL, fails to transduce TRAIL-mediated death signal due to the lack of intracellular portion for initiating apoptosis
DR	Death receptor	A receptor for TRAIL, belonging to TNFR family, engagement of DR results in apoptosis
Foxp3	Forkhead box P3	A transcriptional regulator regulating gene expression and acting as a marker for Treg, also serving as a crucial factor maintaining function of Treg

S. J. Zheng, *Veterinary Molecular Immunology*,
https://doi.org/10.1007/978-981-99-8929-4

Abbreviation	Full name	Functions
GADS	Grb2-like adaptor downstream of Shc	An adaptor protein involved in T-cell signaling transduction
GAS	Gamma-activated sequence	A promoter regulating IFN-γ expression
GEF	Guanine nucleotide exchange factor	Involved in signaling transduction in T-cell activation
GITR	Glucocorticoid-induced tumor necrosis factor receptor family related gene	A member of TNFR family, upregulated expression of GITR in Treg membrane, serving as a specific marker of Treg, promoting T-cell activation, proliferation, and producing cytokines after engaged by its ligand GITRL
HVEM	Herpes virus entry mediator	A member of TNFR family, expressed in multiple cell types, its ligand is LIGHT, involved in cell activation and proliferation
ICOS	Inducible costimulator	A suppressive receptor for T-cell activation, belonging to CD28 family, inhibiting cell activation after binding to ICOSL
IkB	Inhibitor of NF-kB	A cytosolic protein, inhibiting NF-kB from translocation into the nucleus
IKK	Inhibitor of NF-kB kinase	Phosphorylates IKB, and releases NF-kB for nuclear translocation
IP_3	Inositol 1,4,5-triphosphate	A second messenger in signaling transduction
IRAK	IL-1 receptor kinase	Phosphorylates adaptor protein for signaling transduction
IRF	Interferon regulatory factor	A transcriptional factor transducing signals and initiating interferon gene expression
ISRE	Interferon-stimulated response elements	A promoter regulating interferon expression
ITAM	Immunoreceptor tyrosine-based activation motif	An intracellular domain of immune receptor, recruiting adaptor protein to transduce signal
ITIM	Immunoreceptor tyrosine-based inhibitory motif	Inhibition of immune signaling
Jak1	Janus kinase 1	A tyrosine kinase, phosphorylating adaptor protein for signaling transduction
JNK	c-Jun-N-terminal kinase	Phosphorylating adaptor protein for signaling transduction
LAT	Linker of activated T cells	Involved in signaling transduction for T-cell activation
LAG-3	Lymphocyte activation gene-3	Involved in lymphocyte activation, acting as a specific marker for Treg
Lck	L-tyrosine kinase	A tyrosine kinase, phosphorylating ITAM for activating intracellular portion of CD3
LRR	Leucine-rich repeats	A recognition domain of PRR for sensing PAMP
LTα, β	Lymphotoxin-α; LT, lymphotoxin-β	A TNF family member, promoting cell proliferation
MAVS	Mitochondrial antiviral signaling protein	A mitochondrial membrane-bound protein involved in signaling transduction after recognition of PAMP by PRR

Abbreviation	Full name	Functions
MyD88	Myeloid differentiation factor 88	An adaptor protein, sensing signals generated by recognition of PAMP by PRR
NBS	Nucleotide-binding site	A binding domain of NLR for the formation of NLR trimer
NFAT	Nuclear factor of activated T cells	A transcriptional factor, transducing signals and initiating downstream gene expression
NF-kB	Nuclear factor kappa enhancer binding protein	A transcriptional factor, transducing signals and initiating downstream gene expression
NEMO	NF-kB essential modulator	Also called IKK-γ, involved in NF-kB signaling transduction by phosphorylate IKB
NGF	Nerve growth factor	A TNF family member, involved in inflammatory response
NIK	NF-kB-inducing kinase	A kinase, phosphorylating adaptor protein, involved in signaling transduction
NLRs	Nucleotide-binding oligomerization domain (NOD)-like receptors	Intracellular PRR, sensing intracellular PAMP (mainly bacteria)
OPG	Osteoprotegerin	A receptor for TRAIL, belonging to TNFR family, involved in apoptosis and bone metabolism
PD-1	Programmed death-1	Negative regulator on the surface of T cell, inhibiting T-cell activation after binding to PD-L
PGN	Peptidoglycan	A component of bacterial cell wall, specifically recognized by TLR2 and NLR
PKC	Protein kinase C	A kinase, phosphorylating adaptor protein, involved in signaling transduction
PIP2	Phosphoinositol bisphosphate	Involved in signaling transduction in cells
PI3K	Phosphoinositol-3 kinase	A kinase, phosphorylating adaptor protein, involved in signaling transduction
PLCγ	Phospholipase Cγ	Involved in signaling transduction in cells
RANK	Receptor activator of NF-κB	A receptor for RANKL, belonging to TNFR family, involved in inflammatory response
RIP	Receptor interacting protein	Involved in signaling transduction in cells
RLRs	Retinoic acid-inducible gene I (RIG-I)-like helicases receptors	Intracellular PRRs, mainly sensing viral dsRNA in cells
SLAM	Signaling lymphocytic activation molecule	Transduction of activated signal
SLP-76	SH2-containing leukocyte-specific protein of 76 kd	An adaptor protein, involved in immune signaling transduction
STAT	Signal transducer and activator of transcription	A transcriptional factor, transducing signals and initiating downstream gene expression
TACI	Transmembrane activator and CAML interactor	A receptor for APRIL, belonging to TNFR family, involved in tumor formation and autoimmune diseases
TAK1	TGF-β activated kinase 1	A kinase, phosphorylating adaptor protein, involved in signaling transduction

Abbreviation	Full name	Functions
TANK	TRAF family member-associated NF-kB activator	Involved in signaling transduction in cells
TBK	TANK-binding kinase	A kinase, phosphorylating adaptor protein, involved in signaling transduction
TIGIT	T-cell immunoglobulin and ITIM domain	A cell membrane receptors serving as stimulatory molecules
Tim-3	T-cell immunoglobulin and mucin containing protein-3	Immunosuppression
TIMP-1	Tissue inhibitor of metalloproteinase 1	Inhibition of protease activity
TIRAP	TIR domain-containing adapter protein	Involved in signaling transduction in cells
TIR domain	Toll/IL-1 receptor (TIR) domain	An intracellular domain of TLR, recruiting adaptor protein and transducing downstream signal
TLRs	Toll-like receptors	Membrane-bound PRRs, sensing PAMPs
TNF	Tumor necrosis factor	A proinflammatory cytokine, involved in a variety of inflammation
TRAIL	TNF-related apoptosis-inducing ligand	A member of TNF family, inducing apoptosis
TRAF6	Tumor necrosis factor receptor-associated factor 6	Involved in signaling transduction in cells
TRAM	TRIF-related adapter molecule	Involved in signaling transduction in cells
TREM	Triggering receptors expressed on myeloid cells	Cell activation
TRIF	TIR domain-containing adapter-inducing interferon-β	Involved in signaling transduction in cells
TWEAK	TNF-related weak inducer of apoptosis	A member of TNF family, inducing apoptosis
Tyk2	Tyrosine kinase 2	A kinase, phosphorylating adaptor protein, involved in signaling transduction
ZAP-70	zeta associated protein of 70 kDa	A protein binding to intracellular domain of CD3, phosphorylating adaptor protein

Appendix B: CD Antigens

CD antigen	Other names	Molecular weight (kDa)	Expression cell	Function
CD1a, b, c, d, e, etc.		43–49	Dendritic cells, B cells, Langerhans cells, certain thymocytes, etc.	Presentation of lipid and glycolipid antigens
CD2	LFA-2, E-rosette receptor	45–58	Thymocytes, T cells, NK cells, etc.	An adhesion molecule that is involved in T-cell activation
CD3		γ: 25–28 δ: 20 ε: 20	Thymocytes, T cells, NKT cells, etc.	TCR complex molecules involved in TCR-mediated signal transduction
CD4	L3T4	55	Th cells, thymocyte subpopulations, monocytes/macrophages, granulocytes, etc.	Co-receptor for MHC-II molecule, acting as a thymocyte differentiation marker and as HIV receptor
CD5	Leu1, T1, Ly-1	58	Mature T cells, early thymocytes, a few mature B cells, etc.	Regulation of TCR- and BCR-mediated signaling
CD6	T12	105/130	Thymocytes, most of cells with CD6 are peripheral T cells, B-cell subsets, etc.	An adhesion molecule that allows developing thymocytes to bind to thymic epithelial cells, potentially acting as a co-stimulatory molecule for mature T cells

S. J. Zheng, *Veterinary Molecular Immunology*,
https://doi.org/10.1007/978-981-99-8929-4

CD antigen	Other names	Molecular weight (kDa)	Expression cell	Function
CD7	gp40	40	Pluripotent hematopoietic stem cells, thymocytes, T cells, etc.	Primarily affects T-cell development
CD8	T8, Leu-2, Lty-2	64	CTL, thymocyte subpopulations, etc.	Co-receptors for MHC-I molecules
CD9	MRP-1, p24	24	Platelets, early B cells, activated B cells, activated T cells, eosinophils, basophils, etc.	Regulates cell adhesion, migration and promotes platelet activation
CD10	CALLA	100	Precursor B cells, precursor T cells, bone marrow stromal cells, etc.	Cell membrane-bound endopeptidase (endopeptidase) with currently unknown function as a tumor prognostic marker
CD11a	LFA-1, αL integrin	180	Monocytes, macrophages, granulocytes, lymphocytes, etc.	A subunit of the LFA-1 molecule, being a cell membrane glycoprotein that acts as a cell-to-cell adhesion agent by binding to ICAM-1
CD11b	MAC-1, αM integrin, CR3	165	Monocytes, macrophages, granulocytes, NK cells, and T- and B-cell subsets	A subunit of the MAC-1 molecule that primarily involved in interaction of neutrophil and monocyte with endothelial cells, mediates phagocytosis of particles containing iC3b or IgG on the surface, and mediates chemotaxis of inflammatory cells
CD11c	αX integrin, p150	145	Dendritic cells, monocytes, macrophages, granulocytes, NK cells, and T- and B-cell subsets	Is a subunit of the CD18 molecular complex that has a common role with macrophage CD11b in tissues
CD12		90–120	Monocytes, macrophages, platelets, etc.	Currently unknown

CD antigen	Other names	Molecular weight (kDa)	Expression cell	Function
CD13	Aminopeptidase N, EC3.4.11.2, gp150	150	Monocyte/ macrophage precursor cells, mature cytosolic granulocytes, monocytes, bone marrow stromal cells, osteoblasts, few lymphocytes, etc.	Associated with cell adhesion, and inflammatory response
CD14	LPS-R	53	Monocytes, macrophages, granulocytes, etc.	A receptor for endotoxin (LPS) involved in cell activation
CD15	Lewis X/Le(x)	59	Granulocytes, monocytes, embryonic tissue cells, etc.	A CD2 ligand, its function currently unknown, possibly associated with reproduction
CD16	FcRIIIA	50–80	Macrophages, NK cells, neutrophils, etc.	A subunit of the Fc receptor involved in phagocytosis of antigen-antibody complexes as well as antibody-dependent cell-mediated cytotoxicity (ADCC)
CDw17	Lactosylceramide	Currently unknown?	Monocytes, granulocytes, basal granulocytes, platelets, B-cell subsets, tonsillar dendritic cells, etc.	Involved in binding bacteria, probably related to phagocytosis
CD18	β2 integrin	95	All white blood cells	A subunit of integrin-β_2 chain that binds to CD11a, b, or c to form an integrin, a cell adhesion molecule
CD19	B4	90	Follicular dendritic cells, early developing B-cell subpopulations, and others	A receptor like CD21 and CD81, serving as co-receptors of the BCR, and playing an important role in B-cell development, activation, and differentiation

CD antigen	Other names	Molecular weight (kDa)	Expression cell	Function
CD20	B1, Bp35	33	B cells, certain T cells	?
CD21	C3d receptor, CR2, EBV-R	140	B cells, follicular dendritic cells, immature thymocyte subpopulations, etc.	A receptor for EbV, C3d, C3dg, and iC3b, together with CD19 and CD81 acting as co-receptors for the BCR and is involved in BCR-mediated cell signal transduction
CD22	BL-CAM, Lyb8	140	Mature B cell, pre-B and pro-B cell	An adhesion molecule that is involved in the regulation of cell signaling
CD23	B6, BLAST-2, Fcε RII	45	Human B cells as well as follicular dendritic cells	Involved in the synthesis of IgE, promotes the release of cytokines such as TNF, IL-1, IL-6, GM-CSF by monocytes
CD24	BA-1, HSA	35–45	B cells (but not expressed in plasma cells) and granulocytes	Little is known; experimental evidence suggests that CD24 belongs to a sialic acid glycoprotein in the lipid rafts of cell membrane and serves as a prognostic marker for tumors (regarding tumor formation and progression)
CD25	IL-2, Tac Ag	55	Activated T cells, activated monocytes/ macrophages	A receptor for IL-2
CD26	EC3.4.14.5, ADA-binding protein	110/120	Mature or activated T cells, B cells, NK cells, macrophages	A co-stimulatory molecule in the process of T-cell activation
CD27	S152, T14	55	Mature thymocytes, activated T and B cells, NK cells, macrophages, most peripheral T cells, memory B cells, etc.	Regulating co-stimulatory signals during T- and B-cell activation, involved in thymocyte development
CD28	T44, Tp44	90	Mature $CD3^+$ thymocytes, most peripheral T cells, plasma cells, etc.	A co-stimulatory molecule involved in T-cell activation

CD antigen	Other names	Molecular weight (kDa)	Expression cell	Function
CD29	β1 integrin chain, GP, Platelet GPIIa	130	White blood cell	The β subunit of the VLA-1 integrin
CD30	Ber-H2 Ag, Ki-1Ag	120	Activated T cells, B cells, NK cells, monocytes, etc.	Associated with cell proliferation and cell death
CD31	PECAM-1	130–140	Endothelial cells, platelets, NK cells, monocytes, neutrophils, T-cell subsets, etc.	An adhesion molecule, involved in cell adhesion
CD32	FCRII, Fcγ RII	40	B cells, granulocytes, monocytes, etc.	An Fc receptor, binding to Ab or Ab-Ag complex by Fc fragment of Ab
CD33	gp67, p67	67	Monocytes, precursor myeloid cells, etc.	A sialic acid adhesin acting as a sialic acid-dependent cell adhesion agent
CD34		40	Early lymphoid stem and precursor cells, small blood vessel endothelial cells, embryonic fibroblasts, certain neural cells, etc.	A cell adhesion molecule, involved in cell adhesion
CD35	CR1, C3bR, C4bR	160/190/220/250/165/195/225/255	Erythrocytes, neutrophils, B cells, monocytes	A receptor for complement C4b/C3b-coated particles, mediating adhesion and phagocytosis of granular material, and promoting C4b/C3b cleavage with an inhibitory effect on complement activation
CD36	GPIV(IIIb)	88–113	Platelets, mature monocytes/macrophages, endothelial cells, etc.	An adhesion molecule, involved in platelet adhesion and aggregation, in platelet adhesion to monocytes or to tumor cells, in recognition of apoptotic neutrophils, and as a scavenger receptor for oxidized LDL
CD37		40–52	B cells, neutrophils, monocytes, granulocytes, etc.	Involved in signal transduction
CD38	T10	45	Early B cells, germinal center B cells, plasma cells, early T cells, etc.	NAD glycohydrolase

CD antigen	Other names	Molecular weight (kDa)	Expression cell	Function
CD39	Bp50	40–52	B cells, T cells, neutrophils, granulocytes, monocytes, etc.	Involved in signal transduction
CD40	Bp50	48	B cells, macrophages, follicular dendritic cells, endothelial cells, fibroblasts, keratinocytes, $CD34^+$ hematopoietic stem cell precursor cells, etc.	A receptor involved in B-cell activation, growth, differentiation, and antibody isotype switching, in rescue of apoptotic B cells in the germinal center, in promotion of cytokine production by macrophages, and in upregulation of dendritic cell adhesion molecules
CD41	GPIIb/IIIa, α IIb integrin	CD41α: 122 CD41β: 125	Platelets and megakaryocytes	A platelet fibrinogen receptor subunit that aggregates platelets by binding fibrinogen, etc.
CD42a	GPIX	23	Platelets and megakaryocytes	A platelet adhesion molecule that allows platelets to adhere to the vascular endothelium
CD42b	GPIB-a	135/23	Platelets and megakaryocytes	It is a platelet adhesion molecule that allows platelets to adhere to the vascular endothelium
CD42c	GPIB-b	22	Platelets and megakaryocytes	It is a platelet adhesion molecule that allows platelets to adhere to the vascular endothelium
CD42d	GPV	85	Platelets and megakaryocytes	It is a platelet adhesion molecule that allows platelets to adhere to the vascular endothelium
CD43	gpL115, leukosialin	T cells: 95–115 Neutrophils: 115–135	Leukocytes (except mature B cells)	Probably involved in inhibition of cell adhesion
CD44	H-CAM, Pgp-1	85	Most cells	An adhesion molecule that mediates leukocyte adsorption, homing, and inflammatory chemotaxis
CD44R		Epithelial cells: 85–250 Lymphocytes: 85–200	Epithelial cells, monocytes, and activated leukocytes	Possibly related to leukocyte adsorption, rolling, homing, and inflammatory chemotaxis

CD antigen	Other names	Molecular weight (kDa)	Expression cell	Function
CD45	T200, EC3.1.3.4 Different molecular weight isomers include: B220, CD45R, CD45RA, CD45RB, CD45RC, CD45RO	210/220/ 180/200	All leukocytes	Mediates T- and B-cell activation
CD46	MCP	52–58, 35–40	Peripheral blood lymphocytes, non-blood tissues (salivary glands, renal ducts, etc.)	Prevention of complement activation and deposition on the plasma membrane
CD47	Integrin associated protein (IAP)	45–60, 47–55	Most cells	An adhesion molecule and vasoreactive receptor, acting as a marker of self on cell to avoid being phagocytosed by macrophage. Its ligand is signal-regulatory protein (SIRP)α.
CD48		45	All leukocytes (except neutrophils)	An adhesion molecule whose receptor is CD2
CD49a	VLA-1, α1 integrin	200	Activated T cells, monocytes, neuronal cells, smooth muscle cells, etc.	A CD29-associated integrin that binds collagen and laminin-1
CD49b	VLA-2, α2 integrin	160	B cells, monocytes, platelets, megakaryocytes, neuronal cells, epithelial cells, endothelial cells, osteoblasts, etc.	A CD29-associated integrin that binds collagen and laminin-1
CD49c	VLA-3, α3 integrin	125/30	B cells and most adherent cells	A CD29-associated integrin that binds collagen, laminin-5, adhesin, entactin, and invasin
CD49d	VLA-4, α4 integrin	145	Multiple cell types, including T cells, B cells, monocytes, platelets, eosinophils, basophils, mast cells, thymocytes, NK cells, dendritic cells, etc. (except neutrophils)	A cell adhesion molecule that binds to the cell surface ligands VCAM-1 and MADVCAM-1, as well as adhesion proteins, and is associated with lymphocyte migration and homing, as well as being a co-stimulatory molecule for T-cell activation

CD antigen	Other names	Molecular weight (kDa)	Expression cell	Function
CD49e	VLA-5, α5 integrin	135	Multiple cell types, including memory T cells, monocytes, platelets, etc.	A CD29-associated integrin that binds fibronectin, invasins, and others
CD49f	VLA-6, α6 integrin	125/25	T cells, monocytes, platelets, megakaryocytes, trophoblasts, etc.	A CD29-associated integrin that binds fibronectin, invasins, and others
CD50	ICAM-3	130	All leukocytes	A co-stimulatory molecule of the immune response that regulates LFA-1/ICAM-1 as well as integrin-dependent cell adhesion
CD51	αV integrin	125	Platelets, megakaryocytes, etc.	A subunit of α-V integrin that binds vitronectin, adhesin, vasoreactin, and may also adhere to apoptotic cells
CD52	CAMPATH-1, HE5	25–29	Thymocytes, lymphocytes, monocytes, macrophages, epithelial cells of the germinal tract in male animals, etc.	Not very clear, experimental evidence suggests that CD52 is a negative regulator of T-cell activation
CD53	MRC OX44	32–42	White blood cell	Involved in signal transduction, BCR cross-linking for B-cell activation
CD54	ICAM-1	90, 75–115	Activated T cells, activated B cells, monocytes, endothelial cells, etc.	A ligand for CD11a/CD18 or CD11b/CD18, a receptor for rhinovirus, and a receptor for infected Plasmodium erythrocytes
CD55	DAF	70/55	Most cells	Prevention of complement activation and deposition on the plasma membrane
CD56	NCAM	175–220	Human NK cells, $CD4^+$ and $CD8^+$ cell subsets	A cell adhesion molecule in the nerve system
CD57	HNK1, leu-7	110	NK cells, $CD4^+$ and $CD8^+$ cell subsets, monocytes, etc.	A characteristic surface antigen of NK cells

CD antigen	Other names	Molecular weight (kDa)	Expression cell	Function
CD58	LFA-3	55–70	Multiple types of cells	An adhesion molecule that makes killer T cells contact target cells, and APCs contact T cells, and thymocytes contact thymic epithelial cells
CD59	IF-5Ag, H19, HRF20	18–25	Multiple types of cells	Inhibitory effect on complement-mediated membrane perforation, associated with T-cell activation and adhesion
CDw60	GD3	130	T-cell subsets, platelets, thymic epithelial cells, activated keratinocytes, synovial fibroblasts, glomerular cells, smooth muscle cells, glial cells, etc.	Possibly acts as a co-stimulatory molecule for mitogen-activated T cells
CD61	β3 integrin, GPIIIa	110	Platelets, megakaryocytes, macrophages, etc.	An integrin subunit associated with CD41 or CD51
CD62E	E-Selectin, ELAM-1	140	Acutely activated endothelial cells, skin and synovial cells of chronically inflamed lesions	An adhesion molecule that mediates the migration of leukocytes to endothelial cells at inflammatory site
CD62L	L-Selectin, LECAM-1	65	Peripheral B cells, NK cells, T cells, monocytes, granulocytes, thymocytes, etc.	An adhesion molecule that mediates the homing of lymphocytes and the migration of inflammatory cells to the inflammatory sites, etc.
CD62P	P-Selectin, LECAM-3	120	Endothelial cells, platelets, megakaryocytes, etc.	Binding to PSGL-1, mediating leukocyte migration along activated endothelial cells in inflammatory areas toward inflammatory site
CD63	LAMP-3	40–60	Endothelial cells, platelets, detached neutrophils, megakaryocytes, etc.	Involved in cell signaling transduction, cell adhesion, migration, and survival, serving as an important marker for exosomes

CD antigen	Other names	Molecular weight (kDa)	Expression cell	Function
CD64	FcRI	72	Monocytes, macrophages, blood and germinal center dendritic cells, polymorphonuclear cells, myeloid cells, etc.	Involved in phagocytosis, Fc receptor-mediated endocytosis of antigen complexes, antigen capture, presentation, mediating ADCC and release of cytokines and reactive oxygen intermediates
CD65	VIM-2		Myeloid cell	Associated with cell adhesion and migration
CD65s	Sialylated-CD65		Monocytes and granulocytes	Involved in phagocytosis and calcium influx
CD66a	Biliary Glycoprotein	160–180	Granulocytes and epithelial cells	An adhesion molecule that promotes neutrophil activation
CD66b	CGM6, NCA-95	95–100	Granulocyte	An adhesion molecule that promotes neutrophil activation
CD66c	NCA	90	Granulocytes and epithelial cells	An adhesion molecule that promotes neutrophil activation
CD66d	CEACAM3	30	Neutrophils	Function unknown, probably related to cell adhesion and migration
CD66e	CEA	180–200	Epithelial cell	Function unknown, possibly related to cell adhesion
CD66f	PSG	54–72	Embryonic liver cells	Function unknown, possibly involved in immune regulation, protecting the fetus from immune damage caused by the mother
CD68	gp110, macrosialin	110	Monocytes, macrophages, dendritic cells, neutrophils, basophils, mast cells, myeloid precursor cells, CD34+ hematopoietic stem cell subsets, activated T cells, certain peripheral B cells, etc.	A highly glycosylated protein that binds a variety of ligands, acts as a receptor for Malaria sporozoites, and as a scavenger receptor

CD antigen	Other names	Molecular weight (kDa)	Expression cell	Function
CD69	AIM (activation inducer molecule), VEA (very early activation)	60	Activated leukocytes (including T cells), thymocytes, B cells, NK cells, neutrophils, eosinophils, Langerhans cells, B cells in germinal centers, $CD4^+$ T cells, etc.	Participates in the early activation of lymphocytes, monocytes, and platelets, promotes calcium influx and synthesis of cytokines and their receptors, and promotes the lysis of target cells by NK cells
CD70	CD27-ligand, Ki-24	75/95/170	Activated T and B lymphocytes and macrophages	A ligand for CD27, probably involved in the activation of B and T lymphocytes
CD71	T9, transferring receptor	190	All replicating cells	Involved in the process of iron ion intake
CD72	Ly-19.2, Ly-32.2, Lyb-2	42	B cells (except plasma cells)	Negative regulatory molecules of B-cell immune response
CD73	ECTO-5′nucleotidase	69/70/72	T and B lymphocyte subsets, peripheral B cells, follicular dendritic cells, epithelial cells, endothelial cells, etc.	Involved in purine and pyrimidine dephosphorylation in the formation of RNA and DNA
CD74	MHC-II invariant chain/li	33/35/41	B cells, activated T cells, macrophages, activated endothelial and epithelial cells, etc.	Initiating signaling transduction pathways in multiple cells, associated with cell migration and autoimmune disease
CDw75	Lactosamines	53	B cells, peripheral T-cell subsets, red blood cells, etc.	A ligand for CD22, involved in cell adhesion
CD76		32	B-cell, T-cell subsets	Associated with tumor formation
CD77	Gb3	40	Germinal center B cells, endothelial cells	A receptor for Shiga toxin, associated with apoptosis
CD79a	MB-1	40–45	B cells	Components of the BCR, related to signal transduction
CD79b	B29	37	B cells	Components of the BCR, related to signal transduction

CD antigen	Other names	Molecular weight (kDa)	Expression cell	Function
CD80	B7-1	60	Activated B cells, macrophages, dendritic cells etc.	Involved in T-cell activation, interacting with CD28 or CTLA-4
CD81	TAPA-1	26	Hematopoietic stem cells, epithelial cells, endothelial cells, etc.	A member of the CD19/21/leu-13 signaling complex, acting as a B-cell co-receptor
CD82	4f9, c33, ia4, kai1, r2	45–90	Activated and differentiated hematopoietic stem cells	Involved in signal transduction
CD83	HB15	43	B, T lymphocytes, monocytes, dendritic cells, microglia	Involved in immunosuppression (immune checkpoints), associated with homeostasis and immune tolerance
CD84	GR6, SLAMF5	73	Monocytes, platelets, circulating B cells, etc.	Involved in leukocyte adhesion, activation, and differentiation, associated with autoimmunity and tumors
CD85	GR4	110	Plasma cells, B cells, NK cells, etc.	Probably related to T-cell activation
CD86	B7.2	80	Dendritic cells, memory B cells, germinal center B cells, monocytes, etc.	Involved in T-cell activation, interacting with CD28 or CTLA-4
CD87	UPA-R	41–68, 35–59	T cells, NK cells, monocytes, macrophages, vascular endothelial cells, fibroblasts, smooth muscle cells, keratinocytes, placental trophoblasts, hepatocytes, etc.	A receptor for UPA (UPA converts plasminogen to plasmin), probably acting as an integrin (associated with cell adhesion)
CD88	C5a receptor	43	Granulocytes, monocytes, dendritic cells, etc.	Involved in activation of granulocytes and in C5a-mediated inflammatory responses

CD antigen	Other names	Molecular weight (kDa)	Expression cell	Function
CD89	Fcα receptor	45–70, 55–75, 70–100, 50–65	Myeloid cells, activated acidophilic cells, dust cells, and splenic macrophages at all developmental stages	Induction of phagocytosis, degranulation and killing of pathogenic microorganisms
CD90	THY-1	25–35	Hematopoietic stem cells, thymocytes, T cells, lymph node tubular endothelial cells, etc.	Involved in costimulation of lymphocyte and inhibition of hematopoietic stem cell proliferation and differentiation
CD91	α2 Macroglobulin receptor	515/85	Monocytes and various non-hematopoietic cells	β_2 microglobulin receptor
CD92	Choline transporter-like protein 1 (CTL1)	73	Multiple tissues and cells	Transportation of choline (choline)
CD93	GR11	110	Myeloid cells, hematopoietic stem cells, NK cells, platelets, neurons, microglia, endothelial cells	Associated with innate immunity, inflammatory response, and regulation of phagocytosis of apoptotic cells
CD94	Kp43	70	NK cells, $CD8^+$ T cells	A subunit of the NKG2-A complex that is associated with inhibition of NK cell function
CD95	APO-1, Fas	45/90	Activated T and B lymphocytes	A receptor for FasL (CD178), mediating apoptosis
CD96	TACTILE	64	Activated T, NK, and NKT cells	Binding to CD155 as well as CD111, generating immunosuppressive signals
CD97	BL-KDD/F12, CD55 ligand	75–85	Activated T and B lymphocytes, monocytes, granulocytes, etc.	Combined with CD55
CD98	4F2, FRP-1, RL-388	125	Monocytes, T cells, B cells, etc.	Cell activation
CD99	CD99R, E2	32	Multiple cells, including endothelial cells, hematopoietic stem cells	Associated with cell adhesion, migration, death, differentiation, etc.

CD antigen	Other names	Molecular weight (kDa)	Expression cell	Function
CD100	SEMA4D	30/120	A variety of cells, including: red blood cells, platelets, activated T cells, germinal center B cells, etc.	Promotes T-cell activation and adhesion
CD101	p126, v7	240	Monocytes, granulocytes, dendritic cells, activated T cells, etc.	Involved in T-cell activation
CD102	Intracellular adhesion molecule (ICAM)-2	55–65	Vascular endothelial cells, monocytes, platelets, certain mature lymphocytes, etc.	Involved in immune response and lymphocyte circulation
CD103	HML-1, αE integrin	175	Lymphocytes in peripheral tissues	Probably related to the tissue specificity of lymphocytes
CD104	β4 integrin	220	Thymocytes, endothelial cells, epithelial cells, Schwann cells, nurse cells, etc.	An integrin associated with CD49f that binds fibronectin
CD105	Endoglin	180	Endothelial cells, monocytes/tissue macrophages, etc.	Regulating cell response to TGF-β 1
CD106	VCAM-1	100–110	Endothelial cell	An adhesion molecule, acting as a ligand for VLA-4
CD107a	LAMP-1	100–120	Activated platelets	A lysosome-associated protein, related to protein degradation and autophagy
CD107b	LAMP-2	100–120	Activated platelets	A lysosome-associated protein, related to protein degradation and autophagy
CDw108	JMH human-blood-antigen	76	Lymphocytes, lymphoblast cells	Probably acting as an adhesion molecule
CD109	Platelet activation factor	170	Activated T cells, platelets, endothelial cells, multiple tumor cells, etc.	Interacting with GRP78 to inhibit TGF-β signaling

CD antigen	Other names	Molecular weight (kDa)	Expression cell	Function
CD110	Myeloproliferative leukemia protein (MPL)	82–84	Stem cells, megakaryocytes, platelets, etc.	Associated with platelet formation, tumorigenesis, and as a prognostic marker for tumors
CD111	PRR/nectin-1	64–72	Stem cells, endothelial cells, macrophages, neutrophils, etc.	Involved in hematopoietic cell development and differentiation, related to cell adhesion
CD112	PRR2/nectin-2	64–72	Bone marrow hematopoietic stem cells, endothelial cells, monocytes, neutrophils, etc.	Involved in hematopoietic cell development and differentiation, related to cell adhesion
CD113	PRR3/nectin-3	83	T cells, vascular endothelial cells, placental tissue cells	Interacting with CD112 to promote T-cell migration
CD114	G-CSF-receptor	95/139	Granulocytes, monocytes, etc.	Mediating myeloid cell proliferation and differentiation
CD115	M-CSF receptor	150	Monocytes, macrophages	An M-CSF receptor
CD116	GM-CSF receptor α chain	80	Monocytes, neutrophils, eosinophils, etc.	A receptor that primarily binds GM-CSF subunits
CD117	SCF-receptor, c-kit	145	Hematopoietic stem cells (in bone marrow)	A growth factor receptor
CD118	LIF receptor	190	Epithelial cell	Associated with interferon signaling transduction
CD119	IFN-γ receptor	90	Monocytes, B cells, epithelial cells, etc.	An interferon receptor
CD120a	TNF receptor type 1	50–60	Monocytes, neutrophils, etc.	A TNF receptor
CD120b	TNF receptor type 2	50–60	Monocytes, neutrophils, activated lymphocytes, etc.	A TNF receptor
CD121a	IL-1 receptor type 1	80	Thymocytes, T cells, etc.	An interleukin-1 receptor
CD121b	IL-1 receptor type 2	60–70	B cells, macrophages, monocytes, etc.	An interleukin-1 receptor
CD122	IL-2 receptor β chain	70–75	T cells, B cells, NK cells, monocytes/macrophages, etc.	Involved in IL-2 and IL-15-mediated signal transduction

CD antigen	Other names	Molecular weight (kDa)	Expression cell	Function
CD123	IL-3 receptor α chain	70	Bone marrow stem cells, granulocytes, monocytes, megakaryocytes, etc.	Part of the IL-3 receptor
CD124	IL-4/IL-13 receptor	140	B cells, T cells, precursor cells	A subunit of the IL-4 and IL-13 receptors
CD125	IL-5 receptor α chain	60	Eosinophils, basophils, etc.	An IL-5 receptor
CD126	IL-6 receptor α chain	80	B cells, T cells, monocytes, hepatocytes, etc.	An IL-6 receptor
CD127	IL-7 receptor α chain	65–75, 90	Precursors of B cells (such as Pro-B cells), T cells, etc.	An IL-7 receptor
CDw128a	CXCR1 IL-8 receptor A	55–67	Multiple tissue cells	Chemokine receptors
CDw128b	CXCR2 IL-8 receptor B	55–67	Multiple tissue cells	Chemokine receptors
CD130	gp130 receptor subunit for IL-6, IL-11, LIF, OSM, LNTF, CT-1	130–140	All cells	A subunit of receptor for signal transduction mediated by IL-6, IL-11, leukemia inhibitory factors, and so on
CD131	Common β chain for IL-3, IL-5, GM-CSF receptor	120	Myeloid cell	A receptor subunit associated with IL-3, GM-CSF, and IL-5 receptor-mediated signaling
CD132	Common γ chain for IL-2, IL-4, IL-7, IL-9, and IL-15 receptors	64, 65–70	T cells, B cells, NK cells, monocytes/ macrophages, neutrophils, etc.	A subunit of IL-2, IL-4, IL-7, IL-9, IL-15 and CD360 receptors
CD133	AC-133	120	Stem cells, tumor cells	Associated with cell signaling and tumor formation
CD134	OX40	50	Activated T cells	Involved in regulation of cytokine expression, antigen presentation, signal transduction
CD135	Flt3/Flk2, STK-1	130	Pluripotent stem cell	A receptor for hematopoietic growth factor
CD136	Macrophage stimulating protein receptor	180	Epithelial cell	Induced cell migration, proliferation, and morphological changes
CD137	4-1BB	30	T cells, B cells, monocytes, etc.	A co-stimulatory molecule for T-cell proliferation

CD antigen	Other names	Molecular weight (kDa)	Expression cell	Function
CD138	SYNDECAN-1	20	Plasma cells and epithelial cells	Binding type I collagen
CD139	Podocalyxin	209/228	B cells, monocytes, granulocytes, follicular dendritic cells, etc.	Probably related to cell activation, graft rejection
CD140a	PDGF receptor α	180	Endothelial cell	α receptor for platelet growth factor
CD140b	PDGF receptor β	180	Endothelial cell	β receptor for platelet growth factor
CD141	Thrombomodulin	75	Endothelial cells, megakaryocytes, platelets, monocytes, neutrophils, etc.	Involved in the activation of protein C
CD142	Tissue factor	45	Monocytes, vascular endothelial cells at inflammatory sites, etc.	Promoting blood clotting in the microenvironment
CD143	Angiotensin converting enzyme	170/180/ 90/110	Endothelial cells, activated megakaryocytes, certain T cells, etc.	It is an enzyme with blood pressure regulation
CD144	VE-CADHERIN	135	Endothelial cell	Regulation of endothelial cell adhesion, permeability and migration
CD145	CD40L/gp39	16.3	Activated T cells, antigen-presenting cells, endothelial cells, certain stromal cells, etc.	Promoting B-cell activation and signal transduction
CD146	MUC18/S-endo	118	Endothelial cells, activated T cells, etc.	An adhesion molecule
CD147	Neurothelin/Basigin	50–60	Endothelial cells, leukocytes, red blood cells, platelets	An adhesion molecule
CD148	HPTP-ETA/DEP-1	240–260, 200–250	Granulocytes, monocytes, T cells, dendritic cells, platelets	Inhibition of cell growth

CD antigen	Other names	Molecular weight (kDa)	Expression cell	Function
CD149	MEM-133	120	Peripheral blood lymphocytes, monocytes, neutrophils, eosinophils, platelets, etc.	Involved in cell adhension and migration, signal transduction
CD150	SLAM	70	T cells, B cells, thymocytes, etc.	B-cell co-stimulatory molecules
CD151	PETA-3	32	Platelets, megakaryocytes, immature hematopoietic cells, endothelial cells, etc.	An adhesion molecule
CD152	CTLA-4	44	Activated T cells	Negative regulator for T-cell activation
CD153	CD30 ligand	40	Activated T cells	A ligand for CD30, exerting a co-stimulatory effect on T cells
CD154	CD40 ligand, T-BAM, TRAP, gp39	33	Activated CD4 T cells	A ligand for CD40, promoting B-cell activation and proliferation
CD155	Poliovirus receptor, nectin-like 5	80–90	Monocytes, macrophages, thymocytes, neuronal cells, etc.	Binding to CD226, TIGIT, CD96, etc. suppressing immune response, acting as a receptor for poliovirus
CD156	TACE/ADAM17, MS2	69	Neutrophils, monocytes, etc.	Involved in leukocyte migration
CD157	BP-3/IF-7, BST-1	42–45	Granulocytes, monocytes, macrophages, endothelial cells, etc.	Promoting lymphocyte growth
CD158a	EB6, MHC-I specific receptor, p50.1, p58.1	58/50	NK cells, T cells, etc.	Regulating NK cell-mediated cell lysis
CD158b	GL183, MHC-I specific receptor, p50.2, p58.2	58/50	NK cells, T cells, etc.	Regulating NK cell-mediated cell lysis
CD159a	NKG2a	43	NK cell	Associated with NK cell action
CD159c	NKG2c	40	NK cell	Associated with NK cell action
CD160	By55	26	T cells, NK cells, etc.	Related to the role of NK cells and T cells

CD antigen	Other names	Molecular weight (kDa)	Expression cell	Function
CD161	NKR-P1	60	NK cells, T cells, etc.	Regulating NK cell-mediated cell lysis and inducing thymocyte proliferation
CD162	PSGL-1	250/160/220	Peripheral blood T cells, monocytes, granulocytes, B cells	Mediating leukocyte migration
CD163	GHI/61, M130	130	Monocyte-macrophage	Associated with cell activation and inflammatory response, acting as a receptor for porcine reproductive and respiratory syndrome virus (PRRSV)
CD164	MGC-24	80	T cells, monocytes, granulocytes, B cells, precursor cells, etc.	Regulation of cell proliferation and apoptosis
CD165	AD2/gp37	37	Peripheral lymphocytes, thymocytes, monocytes, platelets	An adhesion molecule
CD166	ALCAM, CD6 ligand	100	Activated T cells, monocytes, epithelial cells, neuronal cells, fibroblasts, etc.	An adhesion molecule that binds CD6
CD167a	Discoidin domain receptor, DDR1	120	Epithelial cells, myoblasts	A tyrosine kinase receptor
CD168	RHAMM	85–95	Fibroblasts, lymphocytes, hematopoietic cells, malignant B cells	Associated with cell migration
CD169	Sialoadhesin	220	Macrophage	Involved in interaction between cells
CD170	Siglec-5	140	Neutrophil (type of white blood cell)	Involved in sialic acid-dependent intercellular interactions
CD171	L1	200	Monocytes, T cells, B cells	A cell adhesion molecule that maintains the structure of lymph nodes
CD172	SIRPα/SHPS-1/CD172a	110	Monocytes, granulocytes, dendritic cells, etc.	Associated with suppression of immune response, T-cell activation, neutrophil migration, and local inflammatory response

CD antigen	Other names	Molecular weight (kDa)	Expression cell	Function
CD173	H2, blood group O antigen	Carbohydrate epitopes	Red blood cells, stem cells, platelets	Blood group antigen
CD174	Lewis Y blood group antigen	Carbohydrate epitopes	Hematopoietic stem cells, epithelial cells	A blood group antigen
CD175	TN	Carbohydrate epitopes	Stem cells, certain tumor cells	Function unknown, possibly related to apoptosis
CD175s	Sialyl-Tn	Carbohydrate epitopes	Erythropoietic cells	Function unknown, possibly related to tumor formation and apoptosis
CD176	Thomson-Friedenreich antigen/TF/core-1	Carbohydrate epitopes	Stem cells, leukemia cells	Function unknown, possibly related to tumor formation and apoptosis
CD177	NB1	58–64	Neutrophils	A glycoprotein, function unknown
CD178	Fas ligand	38–42	Activated T cells	An apoptosis-inducing molecule that binds Fas (CD95)
CD179a	V pre beta	16–18	Pro-B cells and early B cells	Formation of subunits of the pre-B-cell surrogate receptor
CD179b	Lambda5	22	Pro-B cells and early B cells	Formation of subunits of the pre-B-cell surrogate receptor
CD180	Rp105/Bgp95	95–105	Monocytes, dendritic cells, B cells	A pattern recognition molecule (PRR), but lacking the TIR domain and associated with cell activation signaling
CD181	CXCR1	39	Neutrophils, T cells	Associated with cell migration
CD182	CXCR2	40	Monocytes, granulocytes, T cells	Associated with cell migration
CD183	CXCR3	40	Activated T cells and NK cells	A G protein-coupled chemokine receptor associated with cell migration
CD184	CXCR4	45	Stem cells, B cells, monocytes, dendritic cells, endothelial cells	A G protein-coupled chemokine receptor and co-receptor for HIV, associated with cell migration
CD185	CXCR5	45	B cells, T cells	Associated with cell chemotaxis

CD antigen	Other names	Molecular weight (kDa)	Expression cell	Function
CD186	CXCR6, BONZO	40	T cell	Associated with cell chemotaxis
CD191	CCR1	39	T cells, monocytes, stem cells	Associated with cell chemotaxis
CD192	CCR2	40	Monocytes, T cells, B cells, basophils	Associated with cell chemotaxis
CD193	CCR3	45	Eosinophils, basophils, T cells, dendritic cells, etc.	Associated with cell chemotaxis
CD194	CCR4	41	T cells, basophils, platelets, monocytes, etc.	Associated with cell chemotaxis
CD195	CCR5	45	Monocytes, T cells	Associated with cell chemotaxis
CD196	CCR6	45	T cells, B cells, dendritic cells, etc.	Associated with cell chemotaxis
CD197	CCR7	45	T cell	Associated with cell chemotaxis
CDw198	CCR8	50	T cells, monocytes, etc.	Associated with cell chemotaxis
CDw199	CCR9	50	T cell	Associated with cell chemotaxis
CD200	OX-2	45–50	Thymocytes, B cells, endothelial cells, activated T cells, etc.	Related to cell signaling
CD201	EPCR/PROCR	50	Stem cells, endothelial cells	Involved in signaling between cells, immunoregulation, immune recognition
CD202b	Tie2/Tek/ angiopoietin-1 receptor TEK tyrosine kinase	150	Endothelial cells, hematopoietic stem cells	Involved in cell adhension, immunoregulation and signaling
CD203c	NPP3 ENPP3, PD-beta	130–150	Basophils, mast cells, etc.	Regulation of the inflammatory response
CD204	Macrophage scavenger receptor 1/MSR1	220	Monocyte-macrophage, breast tumor cells	Function unknown, associated with tumor cell infiltration as a tumor marker
CD205	DEC-205	205	Dendritic cells, thymic epithelial cells, etc.	A mannose receptor, a member of the CLR family, primarily recognizing fungal mannose

CD antigen	Other names	Molecular weight (kDa)	Expression cell	Function
CD206	Macrophage mannose receptor	180	Dendritic cells, macrophages, monocytes, etc.	A mannose receptor that acts as a pattern recognition molecule, primarily recognizing fungal mannose
CD207	Langerin	40	Langerhans cell	The pattern recognition molecule CLR, which recognizes mannose
CD208	DC-LAMP/LAMP3	70–90	Inter-digital dendritic cell	Function unknown, possibly related to immune regulation
CD209	DC-SIGN	44	Dendritic cell	Binding to HIV
CDw210	IL-10R	90–110	T cells, B cells, NK cells, monocytes, macrophages	The receptor for IL-10
CD212	IL-12Rβ 1	100	Activated T cells, NK cells	A subunit of the IL-12 receptor
CD213a1	IL-13Rα 1	65	B cells, monocytes, fibroblasts, endothelial cells, etc.	A subunit of the IL-13 receptor
CD213a2	IL-13Rα 2	65	B cells, monocytes, etc.	A subunit of the IL-13 receptor
CD217	IL-17R	120	Multiple types of cells	A receptor for IL-17
CD218a	IL-18Rα	70	B cells, neutrophils, dendritic cells	A subunit of the IL-18 receptor
CD218b	IL-18Rβ	70	B cells, neutrophils, dendritic cells	A subunit of the IL-18 receptor
CD220	Insulin receptor	140/70	Multiple types of cells	The insulin receptor
CD221	IGF-1 receptor	140/70	Multiple types of cells	The IGF-1 receptor
CD222	IGF2 receptor/ Mannose-6-phosphate receptor	250	Multiple types of cells	The IGF-2 receptor, and also the mannose-6-phosphate receptor
CD223	LAG-3	70	Activated T cells and NK cells	A cytokine receptor
CD224	Gamma glutamyl transferase (GGT)	100	Leukocytes and stem cells	Prevention of antioxidant stress by mediating GSH synthesis

CD antigen	Other names	Molecular weight (kDa)	Expression cell	Function
CD225	Leu-13/IFITM1	14	Multiple types of cells	Associated with cell adhesion, migration, and antiviral immune signaling, forming complexes with CD81, CD19, and CD21
CD226	DNAM-1	65	T cells, NK cells, monocytes, platelets	Binds to CD155, inhibiting the immune response
CD227	MUC-1	120–225	Stem cells, epithelial cells, T cells	Protecting cell surface, exerting anti-adhesive effects, associated with tumor cell infiltration
CD228	Melanotransferrin	80–95	Stem cells and melanoma cells	Involved in transportation of Fe, metabolism and immune response
CD229	Ly-9/SLAMF3	95/110	T cells and B cells	Involved in cell activation and homeostasis
CD230	Prion protein	35	Multiple types of cells	A prion protein
CD231	Tetraspanin-7/ TALLA-1/A15/ CCG-B7/TM4SF2	27.9	Brain tissue cells, cancerous T cells	Interacting with prion proteins, associated with brain retardation and brain cell degeneration
CD232	VESPR (plexin C1)	200	Monocytes, dendritic cells, and many other cells	Associated with cell migration
CD233	Band 3/AE1/ SLC4A1	90	Red blood cell	?
CD234	Duffy/DARC	35–45	Erythrocytes, endothelial cells	Interacting with CD82, associated with cell migration
CD235a	Glycophorin A	36	Red blood cell	Blood group glycoprotein A
CD235b	Glycophorin B	20	Red blood cell	Blood group glycoprotein B
CD236	Glycophorin C/D	32/23	Red blood cells and stem cells	Blood group glycoprotein C/D
CD236R	Glycophorin C	32	Red blood cells and stem cells	Blood group glycoprotein C
CD238	Kell	93	Red blood cells and stem cells	Blood group-related antigen
CD240CE	Rhesus 30 CE	30–32	Red blood cell	?
CD240D	Rhesus 30 D	30–32	Red blood cell	?
CD241	Rhesus 50 glycoprotein/Rh type A	50	Red blood cell	?

CD antigen	Other names	Molecular weight (kDa)	Expression cell	Function
CD242	ICAM-4	42	Red blood cell	Probably related to cell adhesion
CD243	ABCB1/ MDR-1/P-gp	180	Stem cells as well as other cells	Transportation of proteins
CD244	2B4	70	NK cells vs. T cells	A receptor for NK cell activation
CD245	P220/240	220–240	Monocytes, B and T lymphocytes, NK cells	A receptor for NK cell activation
CD246	Anaplastic lymphoma kinase (ALK)	80	Cancerous lymphocytes	Associated with tumor cell proliferation
CD247	TCR zeta chain	16	NK cells and T cells	A subunit of T-cell receptor
CD248	Tumor endothelial marker 1 (TEM1)/ endosialin	175	Epithelial cells, fibroblasts, and tumor endothelial cells	Associated with atherosclerosis, pulmonary fibrosis, cancer, etc.
CD249	Aminopeptidase A	160	Endothelial and epithelial cells, activated T cells	?
CD250	Reserved for TNF	?	T cell	TNF family molecules
CD251	Reserved for lymphotoxin	To be determined	T cell	Lymphotoxin
CD252	OX40 Ligand/ CD134L	34	Dendritic cells, endothelial cells, activated B cells	Regulating the expression of cytokines, antigen presentation, signal transduction, etc.
CD253	TRAIL	33–34	Activated T cells, NK cells, activated B cells and monocytes	Induction of apoptosis and immune regulation
CD254	TRANCE, RANKL	35	Activated T cells, stromal cells, osteoblasts	Involved in cell activation, signal transduction
CD255	TWEAK/TNFSF12	31	T cell	A TNF family member, inducing apoptosis
CD256	APRIL	16	Myeloid cell	A TNF family members, associated with signaling
CD257	Blys, BAFF	45	Myeloid cell	A TNF family members, associated with signaling
CD258	LIGHT	28	Activated T cells and immature dendritic cells	A TNF family members, associated with signaling
CD259	NGF	?	?	?
CD260	Lymphotoxin-β receptors	?	?	?

CD antigen	Other names	Molecular weight (kDa)	Expression cell	Function
CD261	TRAIL-R1, APO2	57	Leukocytes, tumor cells	A membrane protein molecule that is associated with signaling
CD262	TRAIL-R2, DR5	60	Multiple types of cells	A membrane protein molecule that is associated with apoptotic signaling
CD263	TRAIL-R3, DcR1	65	Multiple types of cells	A membrane protein molecule that is implicated in regulating apoptotic signaling
CD264	TRAIL-R4, DcR2	35	Multiple types of cells	A membrane protein molecule that is implicated in regulating apoptotic signaling
CD265	RANK, TRANCE-R	97	Multiple types of cells	A membrane protein molecule that is associated with cell activation, signaling
CD266	TWEAK-R	14	HUVEC	A membrane protein molecule that is associated with signaling
CD267	TACI	32	B cells, activated T cells, etc.	A membrane protein molecule that is associated with signaling
CD268	BAFF-R	25	B cells, T cells, etc.	A membrane protein molecule that is associated with signaling
CD269	BCMA	27	B cell	A membrane protein molecule that is associated with signaling
CD270	Herpes virus entry mediator (HVEM)/ LIGHT-R/ TNFRSF14	26	Multiple cell expression	A member of the TNF receptor family, associated with immune signaling, binds to the receptor CD272 to inhibit T-cell activation
CD271	p75 neurotrophin receptor/NGFR	75	Nerve cells, stromal cells, follicular dendritic cells, etc.	Associated with cell proliferation, as a marker for tumor prognosis
CD272	BTLA	31	Th1 cells	A immunosuppressive receptor, acting as an immune checkpoints, binding to HEVM (CD270) to inhibit T-cell activation

CD antigen	Other names	Molecular weight (kDa)	Expression cell	Function
CD273	B7-DC, PD-L2	25	Dendritic cells, activated T cells, activated monocytes	A membrane protein associated with signaling
CD274	B7-H1, PD-L1	40	Dendritic cells, activated T cells, activated monocytes	A membrane protein molecule associated with signaling
CD275	ICOSL	60	Dendritic cells, B cells, T cells, activated monocytes, etc.	A membrane protein associated with signaling
CD276	B7-H3	40–45	Dendritic cells, activated monocytes, etc.	A membrane protein associated with signaling
CD277	BTN3A/BT3.1/BTF5	56	T cells, B cells, NK cells, monocytes, dendritic cells, stem cells, etc.	Associated with tumor immunosuppression
CD278	ICOS, AILIM	56	Activated T cells	Binding to B7H2 promotes cell activation
CD279	PD1	55	Activated T cells and B cells	Associated with immune signaling
CD280	MRC2/Endo180/TEM22/urokinase-type plasminogen activator receptor-associated protein	180	Myeloid precursor cells, fibroblasts, endothelial cells, macrophages, etc.	Associated with cell migration, protein degradation
CD281	TLR1, TIL	90	Monocytes, neutrophils, etc.	Involved in identification of pathogens and transduction of immune signals
CD282	TLR2, TIL4	85	Monocytes, neutrophils, etc.	Involved in identification of pathogens and transduction of immune signals
CD283	TLR3	100	Fibroblasts, dendritic cells, etc.	Involved in identification of pathogens and transduction of immune signals
CD284	TLR4	85	Monocytes, dendritic cells, endothelial cells, etc.	Involved in identification of pathogens and transduction of immune signals

CD antigen	Other names	Molecular weight (kDa)	Expression cell	Function
CD285	TLR5	?	Monocytes, dendritic cells, etc.	Involved in identification of pathogens and transduction of immune signals
CD286	TLR6	92	Monocytes, dendritic cells, etc.	Involved in identification of pathogens and transduction of immune signals
CD287	TLR7	?	Monocytes, dendritic cells, endothelial cells, etc.	Involved in identification of pathogens and transduction of immune signals
CD288	TLR8	120	Leukocytes, monocytes, dendritic cells, etc.	Involved in identification of pathogens and transduction of immune signals
CD289	TLR9	115–120	Dendritic cells, activated B cells, etc.	Involved in identification of pathogens and transduction of immune signals
CD290	TLR10	95	Dendritic cells, B cells, etc.	Involved in identification of pathogens and transduction of immune signals
CD291	TLR11	?	Innate immune cells	Involved in identification of pathogens and transduction of immune signals
CD292	BMPR1A	50–58	Bone marrow precursor cell (in cell biology)	Involved in immune response, cell adhesion, tumor surveillance
CDw293	BMPR1B	50–58	Bone marrow precursor cell (in cell biology)	Involved in immune recognition, signaling, immunoregulation
CD294	CRTH2, GPR44	55–70	Th2 cells, eosinophils, basophils, etc.	Associated with Th2 immune response
CD295	LEPR, OBR	130–150	Multiple types of cells (including stem cells)	A leptin (leptin) receptor that regulates reproduction, glucose stabilization, bone formation, wound healing, and the immune system
CD296	ART1	37	Epithelial cells, muscle cells, T cells, myeloid cells, etc.	Ribosyltransferases, associated with amino acid modifications of target proteins

CD antigen	Other names	Molecular weight (kDa)	Expression cell	Function
CD297	ART4, Dombrock Blood Group	38	Erythrocytes, erythropoietic cells, activated monocytes, etc.	Ribosyltransferases, associated with amino acid modifications of target proteins
CD298	Na, K ATPase beta3 subunit	52	Multiple types of cells	Sodium, potassium ATPase β3 subunits, associated with cell activation
CD299	DC-SIGNR/L-SIGN	45	Endothelial cells, respiratory epithelial cells	Containing CLR recognition domain, recognizing pathogenic glycosides, involved in the recognition of syncytial viruses, coronaviruses, Tubercle bacilli, etc.
CD300a	IRp60/CMRFH	60	Dendritic cells, monocytes, lymphocytes, etc.	Recognition of phosphatidyle-thanolamine (PE), phosphatidylserine (PS), associated with tumor formation, cell activation, and apoptosis
CD300c	CMRF35A	To be determined	Dendritic cells, monocytes, lymphocytes, etc.	Recognition of phosphatidylethano-lamine (PE), phosphatidylserine (PS), associated with cell activation and apoptosis
CD300e	CMRF35L1	To be determined	Dendritic cells, monocytes, lymphocytes, etc.	Binding to ceramide and sphingomyelin, associated with cell activation signaling
CD301	mgl, clec10a, clecsf14	38	Mononuclear cell	Belonging to the pattern recognition molecule CLR, recognizing mannose
CD302	DCL1	30	Granulocytes, monocytes, dendritic cells, hepatocytes, hepatic sinusoidal endothelial cells, etc.	Belonging to the pattern recognition molecule CLR, associated with innate immunity

CD antigen	Other names	Molecular weight (kDa)	Expression cell	Function
CD303	BDCA2, CLEC4E	38	Dendritic cell	Belonging to the pattern recognition molecule CLR, recognizing pathogenic non-glycosylated oligosaccharides and associated with cell signaling
CD304	BDCA4, Neuropilin-1 (NRP1)	140	Dendritic cells, B cells, respiratory epithelial cells	Binding to the COVID-19 virus receptor ACE, promoting viral infection and associated with immune recognition
CD305	LAIR-1	40	NK cells, T cells, B cells, monocytes, etc.	An immunosuppressive receptor that binds to Collagen, C1q, Protein D, etc., inhibiting cell activation, and associated with inflammatory responses
CD306	LAIR-2	20	A soluble LAIR-1 inhibitor	Competes with CD305 for ligands and regulates CD305-mediated cell functions
CD307	IRTA2, FcRH5	100	B cell	Function unknown, can be used as a leukemia marker
CD307a	FCRL1	45	B cells (including immature B cells, mature B cells, memory B cells), B lymphocytoma	Promoting $Ca2^{+}$ signaling and CD86 expression
CD307b	FCRL2	53	B cells (including immature B cells, mature B cells, memory B cells), B-cell tumors	Promoting $Ca2^{+}$ signaling, can be used as a prognostic marker for B-cell tumors
CD307c	FCRL3	80	B cells (including immature B cells, mature B cells, plasma cells, memory B cells), T cells (mainly Treg), B lymphocytoma	Associated with immune tolerance and autoimmune disease

CD antigen	Other names	Molecular weight (kDa)	Expression cell	Function
CD307d	FCRL4	55	B cells (predominantly expressed in memory B cells, with lower expression in mature B cells and germinal center B cells), B lymphocytoma	Primarily inhibiting BCR signaling
CD308	VEGFR1	?	To be determined	?
CD309	VEGFR2, FLK1, KDR	230	Endothelial cells, stem cells, monocytes, etc.	Function unknown, probably related to angiogenesis, endocytosis
CD310	VEGFR2	?	Thymocyte	Involved in T-cell development and differentiation
CD311	EMR1	?	Microvascular endothelial cell	?
CD312	EMR2	90	Neutrophils, monocytes, etc.	Myeloid cell differentiation antigen
CD313	EMR3	?	Granulocyte	Possibly related to immune regulation
CD314	NKG2D	42	NK cells, NKT cells, T cells, etc.	A membrane molecule that binds to a ligand, initiating a signal to kill the target cell
CD315	CD9P1, SMAP6	135	B cells, activated monocytes, etc.	?
CD316	EWI-2, IgSF8	63	B cells, T cells, NK, dendritic cells, etc.	A receptor for HSPA8
CD317	BST2, HM1.24	29–33	B cells, T cells, monocytes, NK cells, etc.	Associated with cell migration and tumor formation
CD318	CDCP1, TRASK, SIMA135, gp140	~140	Epithelial cells, hematopoietic cells, bone marrow stromal stem cells, certain tumor cells	A ligand for CD6, associated with immune signaling
CD319	CRACC, SLAMF7, CS1	66	NK cells, B cells, T cells, dendritic cells, etc.	A NK cell receptor, associated with NK cell cytotoxicity and the regulation of the immune response

CD antigen	Other names	Molecular weight (kDa)	Expression cell	Function
CD320	8D6, TCblR	26	A variety of cells, including epithelial cells, follicular dendritic cells, etc.	Vitamin B12 receptor, uptake of vitamin B12
CD321	JAM-1, F-11R	32~35	Multiple types of cells	A cell adhesion molecule, associated with platelet aggregation
CD323	JAM-3	To be determined	Multiple types of cells	A cell adhesion molecule
CD324	E-Cadherin, Cadherin 1	120	Epithelial cells, stem cells, erythropoietic cells, etc.	Associated with cell activation, stem cell proliferation and differentiation
CD325	N-Cadherin, Cadherin 2	140	Nerve cells, endothelial cells, stem cells, etc.	A cell adhesion molecule
CD326	EpCAM	40	Epithelial cells	Associated with cell adhesion, activation, proliferation, and differentiation
CD327	Siglec6, OB-BP1	50	Dendritic cells, B cells, trophoblasts, etc.	Inhibition of cell activation and proliferation
CD328	Siglec7, AIRM1	75	NK cells, monocytes, T cells, etc.	Associated with NK cell immunosurveillance, antiviral infection, and immunosuppressive effects
CD329	Siglec9	50	NK cells, monocytes, granulocytes, T cells, etc.	Associated with NK cell immunosurveillance, antiviral infection, and immunosuppressive effects
CD330	Siglec10	What?	Macrophage	Interacting with CD24, associated with tumor immune evasion
CD331	FGFR1, FLT2	130	Fibroblasts, epithelial cells, etc.	Associated with tumor formation and signal transduction
CD332	FGFR2, KGFR	115–135	Fibroblasts, epithelial cells, etc.	Associated with tumor formation and signal transduction
CD333	FGFR3, JTK4	115–135	Fibroblasts, epithelial cells, etc.	Associated with tumor formation and signal transduction
CD335	nkp46, ncr1, ly94	46	NK cell	An NK cell activation receptor

CD antigen	Other names	Molecular weight (kDa)	Expression cell	Function
CD336	NKp44, NCR2, Ly95	44	NK cell	Associated with NK cell activation or inhibition
CD337	nkp30, ncr3, ly117	30	NK cell	An NK cell activation receptor
CD338	ABCG2, BCRP	72	Stem cells	A stem cell surface marker
CD339	Jagged1, JAG1	150	Stromal cells, epithelial cells	A receptor for notch
CD340	HER-2	185	Stem cells	Associated with tumor formation
CD344	frizzled-4/FZD4	60	Stem cells, many other kinds of cells	Stem cell surface markers, other functions not known
CD349	FZD9	64	Stem cells, many other kinds of cells	Involved in immunoregulation and immune signal transduction
CD350	FZD10	65	Stem cells, many other kinds of cells	Involved in immune response, recognition, and adhension
CD351	FCAMR, FCA/MR	30	B cells (mainly mature B cells, memory B cells), and tissues of kidney, small intestine, mucosal lamina propria, lymph nodes, and lymphoid follicles	A member of the Fc receptor (FCR) family (part of the immunoglobulin superfamily), containing multiple IgM and IgA binding sites, mediating endocytosis, associated with systemic as well as mucosal immunity
CD352	SLAMF6, NTBA, KALI	23–35	NK cells, NKT cells, T cells, B cells, eosinophils, blood monocytes, dendritic cells	A member of the SLAM family (belonging to the immunoglobulin superfamily), promoting cytokine secretion and killing by NK cells, promoting CTL cytotoxicity
CD353	SLAMF8, BLAME	29	B cells, blood mononuclear cells, dendritic cells	A SLAM family member (belonging to the immunoglobulin superfamily)
CD354	TREM1, TREM-1	15–24	Expressed mainly in monocytes, macrophages, neutrophils, dendritic cells, NK cells, and weakly in T and B cells	A TREM family member (part of the immunoglobulin superfamily), promoting inflammatory response

CD antigen	Other names	Molecular weight (kDa)	Expression cell	Function
CD355	CRTAM	43	NK cells, T cells, B cells	A member of the nectin family (part of the immunoglobulin superfamily), promoting the cytotoxic effects of NK cells and CTL, the secretion of gamma interferon, and antitumor effects
CD356	NKG2D ligand	24	Tumor cell, virus-infected cells	A ligand for NKG2D receptor of NK cells, involved in immunoregulation and tumor cell surveillance
CD357	TNFRSF18, GITR, AITR	24	Monocytes, NK cells, and T and B cells	A member of the TNF receptor superfamily, binding to GITRL to generate co-stimulatory signals, involved in T-cell activation, apoptosis, autoimmune disease/ inflammatory responses, and leukocyte migration
CD358	TNFRSF21, DR6	68	Monocytes, bone marrow immature B cells	A member of the TNF receptor superfamily, associated with early apoptosis in B-cell development and with NF-κB and JNK signaling pathways
CD360	IL21RA, IL-21R	57	NK cells, monocytes, granulocytes, T and B lymphocytes	An IL-21 receptor subunit that forms IL-21 receptor (heterodimer) with CD132, promoting B-cell activation, proliferation, and differentiation into plasma and memory cells, promoting NK, NKT, and CTL proliferation, virulence, and antitumor effects
CD361	EVI2B, EVDB	46	Widely expressed in a variety of cells, monocytes, macrophages, neutrophils, dendritic cells, NK cells, T and B cells, etc.	A member of the EVI2B family

CD antigen	Other names	Molecular weight (kDa)	Expression cell	Function
CD362	Syndecan-2 (SDC2), HSPG, SYND2	20	B cells, activated T cells, monocytes, granulocytes	A transmembrane protein receptor proteoglycan, involved in cell adhesion, migration, growth signaling, etc., one of its ligands is CD267 (TACI), which is associated with cell motion, and tumor formation
CD363	s1pr1, edg1, s1p1	43	NK cells, T and B lymphocytes	A member of the G-protein binding receptor superfamily, primarily associated with NK, T- and B-cell migration
CD365	HAVCR1, TIM-1	42	Hepatocytes, epithelial cells, macrophages, dendritic cells, T and B lymphocytes	A cell phospholipid membrane receptor, viral (e.g., hepatitis A virus, hepatitis C virus) receptors
CD366	Tim-3	31	White blood cells, leukemic cancer cells	To be determined
CD367	CLEC4A/DCIR	19–27	Monocytes, dendritic cells	A C-type lectin receptor (CLR), which primarily recognizes fungi, producing immunosuppressive signals
CD368	CLEC4D/Dectin-3	25	Monocytes, dendritic cells	A CLR family member, which primarily recognizes mycobacteria, initiating the immune response
CD369	DECTIN1/ CLEC7A	8–27	Monocytes, dendritic cells	A CLR family member, which primarily recognizes fungal cell wall beta glycosides, initiating the immune response
CD370	Clec9A/DNGR1	27	Intermediate transition mononuclear cells	A CLR family member, which primarily recognizes fungi, initiating the immune response
CD371	CLEC12A/MICL/ CCL-1	24~32	Peripheral mononuclear cells, dendritic cells	A CLR family member, which primarily recognizes fungi, producing immunosuppressive signals

Note: ? indicates currently unknown

Appendix C: Overview of Basic Theories for Immune Control and Control Measures for Infectious Diseases of Animals

The immune system of animals, a defense system that has evolved over time, includes immune organs, immune cells, and immune molecules. The various components of the immune system work together exquisitely to provide for the normal physiological functions of the body by means of immune responses. Innate immunity is an important part of the body's immune defense that is inherent in animals and consists of four types of defense barriers: mechanical barriers, physiological barriers, phagocytic barriers, and inflammatory response barriers. Innate immunity not only blocks, inhibits, and kills pathogens early during pathogenic infection, but also serves as a bridge to acquired immunity. Innate immunity and acquired immunity work closely with each other at different levels to completely eliminate the infectious pathogens to restore and maintain the normal physiological conditions of the body. Therefore, innate immunity plays a vital role in the host defense against infectious diseases.

In the early stage of pathogenic infection, innate immunity plays a major role, and phagocytes in particular play a key role in the phagocytosis, killing, and clearance of pathogens. In addition, phagocytes destroy pathogens while presenting their antigenic epitopes to specific T lymphocytes (including $CD4^+$ and $CD8^+$ T cells) for activation and further differentiation into effector T cells and memory T cells for cell-mediated immunity based on their ability to parasitize intracellularly. $CD4^+$ T cells play an important supportive role in B-cell activation, prompting B cells to activate and differentiate into plasma cells to produce antibodies, and thus play a crucial role in humoral immunity, while some activated B cells differentiate into memory B cells. The generation of high-quality functional memory cells is the theoretical basis for vaccination against infectious diseases.

The host produces an immune response after initial exposure to an antigen (vaccination or infection), a process that involves the phagocytosis, clearance, and processing of the antigen by innate immunity and a transition to acquired immunity, where specific B and T cells are activated against the antigen and differentiate into plasma cells to produce antibodies and effector T cells. As pathogens are

S. J. Zheng, *Veterinary Molecular Immunology*,
https://doi.org/10.1007/978-981-99-8929-4

eliminated, the effector lymphocytes (cytotoxic T cells and plasma) gradually undergo apoptosis and are cleared while memory T cells and B cells persist for a long time. These memory lymphocytes are like trained soldiers, and when the same antigen enters the host again, they are immediately activated to produce a large number of effector cells to attack (i.e., recall immune response) and destroy it. For example, plasma cells produce antibodies to neutralize the pathogen, blocking the pathogen from infecting the target cells, and effector T cells destroy the infected target cells or cells harboring the pathogen. Thus, the main purpose of vaccination is to generate functional memory lymphocytes and eliminate pathogenic infections. However, the immune response of the organism to different antigens varies, e.g., for pathogens that can proliferate intracellularly (viruses, intracellular parasitic bacteria such as *Listeria monocytogenes*, *Salmonella*, *Brucella*, *Mycobacterium tuberculosis*, etc.), the main approach is cell-mediated immunity, with cytotoxic T cells (CTL) destroying the environment (pathogen-infected cells) in which they live, releasing the pathogen and, with the cooperation of humoral immunity, ultimately removing the pathogen completely. For extracellular parasitic bacteria (e.g., *E. coli*, *Staphylococcus*, *Streptococcus*, etc.) and inactivated vaccines, humoral immune response is primarily elicited because such antigens do not parasitize within the host cells. Since antibodies can neutralize and prevent pathogens from invading target cells and also provide early protection to piglets through colostrum, but cannot enter cells to neutralize pathogens, successful prevention and control against viruses and intracellular parasites by vaccination requires, in addition to high-affinity antibodies, cell-mediated immune response to destroy the cells in which pathogens survive and proliferate, thus causing pathogens released from the hiding cells and subsequently neutralized by antibodies. Therefore, elimination and clearance of intracellular pathogens require the process in which the pathogen is released from the cells, neutralized by antibodies, and finally phagocytosed, killed, and cleared by phagocytes. Since only mild virulent live vaccines can effectively elicit a strong cell-mediated immunity in vaccinated animals, the inoculation of mild virulent live vaccines is essential in the prevention and control of infectious diseases caused by intracellular pathogens.

Of note, vaccination is only an auxiliary tool in the prevention and control of infectious diseases, and is by no means sufficient. Clinical evidence from the practice indicates that no matter how strong the immunity of the host is, it can only resist a certain degree of pathogenic infection, but not strong attack by a large number of virulent pathogens, so environmental disinfection of pathogens is fundamental to disease control. Through appropriate control measures of sanitary disinfection and management, the contents of pathogens in the shed or house could be effectively reduced, thereby reducing the exposure of animals to large numbers of pathogens. Thus, with an appropriate immunization measure and effective sanitary disinfection, animals could be effectively protected, giving the best performance and achieving maximum economic benefits.

The following factors should be considered when taking immunization measures.

1. Vaccination is only an auxiliary means for disease control, and immunization does not always produce an expected preventive effect, which is related to many factors, such as vaccine quality, vaccine serotype, timing and dose of immunization, immunosuppressive diseases, genetic characteristics, etc. However, sometimes animals are afflicted with a disease because a large number of pathogenic attacks overcome the protection provided by vaccination, so it is important to protect animals by immunization with appropriate vaccines under an improved environmental condition to control the number of pathogens.
2. The decision to vaccinate against a particular vaccine should take into account the economic benefits and assess the risks involved in not vaccinating. The risk assessment should consider the farm's previous disease history, current incidence, and animal safety. Vaccine efficacy, side effects, and risks also need to be considered, especially for breeding animals, immunization with attenuated live vaccines should be avoided as much as possible. A comprehensive risk assessment should be conducted to prevent human-mediated introduction of vertically transmitted diseases.
3. Except for special cases, vaccination should be carried out according to the manufacturer's instructions and the immunization program should be developed according to the target animals of protection. For example, if the target animals of protection are newborn piglets, immunization should be carried out 3–5 weeks before farrowing to stimulate the sow to produce antibodies, subsequently protecting the piglets through maternal antibodies in colostrum. If the target of protection is chicks, the breeder should be boosted with inactivated vaccine 2 weeks before producing eggs to obtain high titers of yolk antibodies to protect the chicks after hatching. If the vaccination is intended to defend against reproductive disorders, it is more reasonable to vaccinate mammals before breeding or poultry before egg production.
4. If the situation of an epidemic is relatively stable, either a single vaccine or combined vaccine can be used. However, if a certain epidemic is prevalent on the farm, the use of a single vaccine is more effective. Experimental evidence have confirmed that if the interval of vaccination between two different vaccines is less than 3 days, the effect of the later vaccination will be affected to some extent.
5. Types of commonly seen vaccines and the principles of use.

   The two main types of vaccines commonly used in prevention of swine diseases are live and inactivated vaccines.

   (a) Live vaccines. Live vaccines include attenuated live vaccines, live vector recombinant vaccines, and gene-deficient vaccines. Attenuated live vaccines are made by passing the pathogen through a certain host system (e.g., cell culture, inoculation of chicken embryos or animals) to reduce the virulence of pathogen, and there are also vaccines made by isolation of natural avirulent or attenuated pathogen. Live vector recombinant vaccines are vaccines made by inserting a gene encoding a protective antigen into a live vector (viral or bacterial vector). Gene-deficient vaccines are vaccines made by knocking out (deleting) the virulence gene of the pathogen. Live vaccines require certain storage and use conditions to ensure that a sufficient amount of viable vaccine enter the animal at the time of vaccination. If the vaccine strain is not sufficiently viable (i.e., most died) at the time of vaccination, which is below effective dose, it is difficult to induce the animal to produce protective immunity. In general, mild (or attenuated) virulent vaccines have a full range of antigenic epitopes, and are more effective to elicit strong immunity in animals when used appropriately. The immunizing effect of a gene-deficient vaccine is dependent on the segment of the deleted gene. If the deleted segment of the gene has a suppressive effect on host immunity, the vaccine is more effective than the attenuated vaccine. However, if the deleted virulence gene has no immune suppressive effect and is only related to replication, the vaccine is less effective than the attenuated vaccine because the pathogen needs to replicate in the host to elicit strong cell-mediated immunity. In addition, certain toxins have better immunogenicity, and knockout of immunogenicity-related toxin inevitably affects the immune response in host, making it less effective than the attenuated vaccine. The live vector recombinant vaccine is relatively weak in protection because the number of antigens inserted into the expression vector is limited and the inserted antigen is not dominantly expressed compared with the vector antigens, thus can only induce weak immune response and provide protection to a certain extent, but the advantage is that the live vector itself is a attenuated or gene-deficient vaccine against another disease, so that one shot can prevent two or more diseases. In general, the live vector recombinant vaccine elicits a strong immunity against the vector homologous wild strain, but the immune response induced by inserted antigen is relatively week. Therefore, the choice for the use of live vaccines should take into account the different environmental and feeding conditions. For example, for a severely contaminated farm, a mild virulent live vaccine or a gene-deficient vaccine with good efficacy should be chosen, while for a farm with a better environment, high level of feeding management and no epidemics, a live vector recombinant vaccine can be used as needed to reduce the number of immunizations, thus reducing animal stress and improving the performance of livestock and poultry.

(b) Inactivated vaccines. Inactivated vaccines are often mixed with adjuvants, such as oil-adjuvant inactivated vaccines and aluminum hydroxide gel vaccines, etc. Inactivated vaccines induce a strong humoral immune response and produce a high titer of antibodies, but the cell-mediated immune response is weak. Circulating antibodies are able to neutralize pathogens in body fluid to block them from infecting cells, but are unable to enter the cells with or without pathogenic infection. Cell-mediated immunity specifically helps to eliminate intracellular pathogens by destroying the pathogen-infected cells through cytotoxic effect of effector T cells, causing the release of intracellular pathogens, which are subsequently cleared by direct neutralization, opsonin-mediated phagocytosis, and complement cascade responses in the presence of antibodies, complement, and other inflammatory factors.

Therefore, inactivated vaccines should preferably be used alternately with live vaccines. Inoculation of animals with inactivated vaccines 2–3 weeks before parturition increases the antibody content in the colostrum and is very helpful to protect the newborn animals. Similarly, booster immunization with inactivated vaccine 2 weeks before the initiation of the egg-laying period in breeding hens provides better protection for offspring chicks.

6. The use of combined or single vaccines should be adapted to the epidemic situation of animals. Generally speaking, single vaccine is more effective than the combined vaccines. However, combined vaccines can prevent several diseases at one shot, saving time and efforts and reducing animal stress. Whether you should choose a combined vaccine or a single vaccine depends on the disease control situation of the farm. The basic principle is: if the farm has good disease control, good environment, and feeding management in place, the combined vaccine can be used to reduce stress and protect the animal to give the best performance. If the farm has poor environmental and epidemic situations, it is recommended to use single vaccine to maximize the effect of immunizing animals with the vaccine. The two single vaccines should be used at least 3 days apart, otherwise the vaccine given first will interfere with the effect of the one given second.
7. Basic principles of vaccination and control of breeding livestock

   Seed is the "chip" of agricultural production, and so is the breeding livestock and poultry in farming. The cultivation and development of good livestock and poultry breeders require decades of persistent efforts and hard work, and livestock and poultry breeder play a key role in the development of the farming industry. The prevention and control of breeder-transmitted diseases is the top priority. In view of the past practical experience and lessons learned, disease prevention and control should start from the starting breeder herd as well as the grandparent breeder. The original breeding stock farms usually have strict production management and biosecurity measures, and the vertically transmitted severe diseases in breeders are controlled by quarantine and eradication (such as avian leukemia, Pullorum disease, etc.), but the diseases that can be controlled by quarantine and eradication at present are after all very limited, and most diseases still need to be prevented and controlled by vaccination. The most easily neglected vulnerability is the introduction of pathogens to breeders by vaccination with contaminated vaccines. To avoid such risks, the basic principles should be considered as follows:

   (a) Choosing inactivated vaccines whenever possible and avoiding live vaccines, i.e., "inactivated rather than live."
   (b) Avoiding the use of attenuated live vaccines or genetically deficient vaccines for control of vertically transmitted epidemics (avoiding revert of virulence of vaccine or recombination with wild virus).
   (c) If the use of a particular live vaccine is required, a thorough examination of exogenous pathogenic contamination must be performed to prevent the infection of the breeders by exogenous pathogens.

Appendix D: Factors Affecting Herd Immunity in Animals

Immune response is a normal physiological reaction process that is influenced by a variety of factors. In vaccinated animals, the immune response varies from one animal to another, with some being stronger and some weaker, but the degree of immune response in the whole animal population is normally distributed as a bell curve, so that the majority of animals can produce strong immunity after vaccination, but due to individual differences, there are still few numbers of animals with poor immune response or even no immune response, and thus cannot resist pathogenic attack so as to become afflicted with a disease when infected by a virulent pathogen. If herd immunity is strong, no epidemics of pathogenic transmission will occur. If herd resistance is weak, larger epidemics will occur. Thus, herd immunity in animals is a determining factor in preventing the prevalence of epidemics, and the factors affecting herd immunity are as follows:

1. **Genetic factors**. The immune response of animals to the inoculated antigen is to some extent genetically controlled. Therefore, the strength of the immune response to the same vaccine varies from breed to breed, and even from one individual to another of the same breed.
2. **Nutritional status**. The nutritional status of the animal is one of the factors that affect the immune response. The deficiency of vitamins, trace elements, and amino acids reduces the immune response of the animal. For example, vitamin A deficiency will lead to atrophy of lymphoid organs and affect lymphocyte activation, proliferation, differentiation, and receptor expression, resulting in a decrease in the number of T lymphocytes and NK cells in the body, a decrease in phagocytosis, and a decrease in antibody production capacity of B lymphocytes. Thus, nutritional status is a factor that cannot be ignored in the prevention and control of diseases.
3. **Environmental factors**. Environmental factors include temperature, humidity, ventilation, sanitation, and disinfection in the shed or house. The immune function of the animal is regulated to a certain extent by the nervous and endocrine

S. J. Zheng, *Veterinary Molecular Immunology*,
https://doi.org/10.1007/978-981-99-8929-4

systems. If the environment is too cold or too hot, the humidity is too high, and the ventilation is poor, the animals will experience different degrees of stress, resulting in a decrease in their immune response to antigens. In such a case, vaccination cannot elicit an expected immune response, which is manifested by low antibody level and diminished cell-mediated immunity. Good environmental hygiene and disinfection can reduce the number of pathogens and eliminate the chance of pathogenic infection, allowing animals to safely survive the immune response-induction period after vaccination.

With good environmental management, the chances of animal morbidity can be greatly reduced, and animals can be protected even if their antibody levels are not high. If the environment is poor and there are a large number of pathogens in the environment, the possibility for animals to be infected exists even in animals with high antibody levels. Although multiple immunizations can result in high antibody levels, this is not what disease control is intended to achieve because high immunity (very high antibody levels) is a kind of stress for the animal itself. It is well documented that high levels of antibodies can reduce the productivity of animals after multiple immunizations. Thus, good environmental hygiene is as important as vaccination in the prevention and control of infectious diseases.

4. **Vaccine**. Storage and transportation of vaccines are very important aspects in the prevention and control of infectious diseases by immunization. Improper storage and transportation can deteriorate the quality of vaccines or even make them ineffective. Some special vaccines should be stored frozen at low temperature. Mild virulent freeze-dried vaccines should be stored at 2–8 °C. Inactivated vaccines should be stored at 2–8 °C and freezing should be strictly forbidden, otherwise the quality of vaccine will decrease and immunization effect will be affected. In the process of using vaccines, there are many factors affecting the effect of immunization, such as vaccine dilution, vaccination route, immunization procedure, etc. Each factor should be given sufficient attention.
5. **Serotypes**. The pathogens of some diseases contain multiple serotypes, such as swine erysipelas, porcine pneumonia, porcine colibacillosis, foot-and-mouth disease, avian influenza, avian bronchitis, etc., which cause difficulties in the prevention and control of these diseases. If the vaccine strains differ from the serotypes of the disease-causative pathogens, it is difficult to achieve good preventive and control effects. Therefore, the use of polyvalent vaccines should be considered for epidemics with multiple serotypes.
6. **Maternal antibodies**. Passive immunization with maternal antibodies is very important for newborn animals, but it can also affect the effect of vaccination, especially when immunizing animals with attenuated live vaccines, and if high levels of maternal antibodies are present in young animals, it can greatly affect the effect of vaccination. Therefore, the age of animals receiving the first vaccination should be determined based on the results of maternal antibody assays.
7. **Other factors**. In addition to the factors mentioned above, other factors such as immunosuppressive diseases, interference between vaccines (especially between attenuated live vaccines), and the quality of vaccinators may affect the effect of vaccination.

Appendix E: Immunological Diagnosis and Diagnostic Techniques for Major Diseases on Pig Farms

With the development of the pig industry, there is an increasing need for a variety of specific diagnostic methods. Laboratory techniques allow for faster and more accurate diagnosis of diseases, which has been gradually adopted by production units. The major immunological techniques for the diagnosis of major diseases on pig farms are as follows.

Technology

Whole Blood Plate Agglutination Test

Materials

Swine erysipelas agglutination antigen, slide, #9 syringe needle, alcohol swab, alcohol lamp, blood sample-collecting loop.

Procedure

(1) Add one drop of antigen (shake well before use) to the slide.
(2) Disinfect the needle and the blood sample-collecting loop by wiping it with an alcohol swab, and followed by cauterizing the needle and loop on an alcohol lamp.
(3) Puncture the pig ear vein with the needle, collect a blood sample with the loop, mix loop blood thoroughly with the antigen solution on a glass plate, and observe the results within 2-3 minutes.
(4) Determination of results: The presence of granular agglutination is considered positive, otherwise negative.

S. J. Zheng, *Veterinary Molecular Immunology*,
https://doi.org/10.1007/978-981-99-8929-4

Notes

(1) Adjust the blood sampling loop so that the volume of blood taken should be approximately equal to that of antigen.
(2) The antigen should be well shaken before use.
(3) Needles and blood collection loops should be sterilized and disinfected before blood collection to prevent infection.

Applications

This test is mainly used in practice for the diagnosis of swine erysipelas. It should be noted during diagnosis that pigs that have been vaccinated against swine erysipelas can also show positive reaction, and the history regarding immunization of animals with vaccines should be obtained beforehand.

Serum Agglutination Test (in the Case of Brucellosis)

Slide Agglutination Test

Materials

Brucella plate agglutination antigen, Brucella positive serum, Brucella negative serum, serum sample to be tested, quarantine box, inoculation loop, dropper, etc.

Procedure

(1) Turn on the power of the quarantine box so that the temperature of the glass plate reaches about 30 °C.
(2) Take 0.08 mL, 0.04 mL, 0.02 mL, and 0.01 mL of the serum to be tested and add them to each of the four squares on the plate, while setting up a positive serum control.
(3) Add 0.03 mL of antigen dropwise near the serum in each square, then mix the antigen with the serum using an inoculation loop. Gently shake the quarantine box to mix the antigen with the serum thoroughly and record the results within 3–5 min.
(4) The criteria of recording results: the mixture with a large agglutination mass and the liquid completely transparent is considered "#", indicating 100% agglutination. The mixture with a visible agglutination piece and the liquid not very transparent is considered "+ +", indicating 50% agglutination. The mixture showing the outcome between the above two results is considered "+++", indicating 75% agglutination. The mixture showing cloudy liquid with small particles is considered "+", indicating 25% agglutination. The mixture of the liquid displaying uniform and cloudy is considered "–", indicating 0% agglutination.

(5) Result is considered positive if more than 50% agglutination occurs at 0.04 mL of pig serum, and suspicious if more than 50% agglutination occurs at 0.08 mL of pig serum. Pigs with suspicious reactions should be retested after 3–4 weeks. If the results of the re-examination are still suspicious and the pigs are free of clinical signs and a large number of negative reactions, they should be judged as negative.

Test Tube Agglutination Test

Materials

Brucella test tube agglutination antigen solution (1:20 dilution), Brucella (+) serum (1:25 dilution); Brucella (-) serum (1:25 dilution); diluent (0.5% carbonate saline); sterilized test tubes, pipettes and test tube holders; serum to be tested.

Operations

(1) Take 7 sterilized test tubes and place them on a test tube rack and make a dilution according to Table E.1.

Each tube should contain 1.0 mL of liquid when the above process is complete.

(2) The seven test tubes were shaken well and placed in a 37 °C incubator for 24 h to determine the results.

(3) Determination of the results: the test tube containing liquid completely transparent, with agglutinates uniformly spread in the bottom of the tube, when shaking, agglutinates shown in the form of flakes, blocks or particles, is

Table E.1 Brucella test tube agglutination reactions

Test tube	1	2	3	4	5	6	7
Serum Dilution / Dilution / Component (mL)					Control		
	1:25	1:50	1:100	1:200	Positive serum 1:25	Negative serum 1:25	Antigen control
0.5% Carbonate saline	2.4	0.5	0.5	0.5	-	-	0.5
Serum to be tested	0.1	0.5	0.5	0.5	0.5 (positive)	0.5 (negative)	-
Antigen solution (1:20 dilution)	0.5	0.5	0.5	0.5	0.5	0.5	0.5

Discard 1.5mL (from tube 1) → 0.5 transferred to tube 2 → 0.5 to tube 3 → 0.5 to tube 4; Discard 0.5mL (from tube 4)

considered "#", indicating 100% agglutination. The test tube containing liquid not very transparent, with obvious agglutinates in the bottom of the tube, when shaking, agglutinates shown in the form of blocks or small flocs, is considered "+ +", indicating 50% agglutination. The test tube showing the results in between the above is considered "+++", indicating 75% agglutination. The test tube containing liquid of poor transparency, with only a small amount of agglutination at the bottom of the tube, is considered "+", indicating 25% agglutination. The test tube containing uniform and cloudy liquid, with no agglutination at the bottom of the tube is considered "–", indicating 0% agglutination.

The highest dilution of the serum at which 50% agglutination occurs is the agglutination titer of that serum. When the agglutination titer of the pig serum is 1:50 or more, it is determined to be a positive reaction. If the agglutination titer is 1:25, it is considered suspicious. In such a case, blood samples should be collected again 3–4 weeks later and retested. If the retested serum is still suspicious and the pigs have no clinical signs and a large number of negative reactions, it can be considered a negative reaction.

If few individual swine serum samples show weak positive results, and the herd has no clinical signs, blood samples should be collected and retested again after 3–4 weeks to avoid "false positive" due to nonspecific reactions.

Examples of the Application of Serum Agglutination Tests to Other Swine Diseases

(1) Swine pasteurellosis. Detection of antibodies in serum by slide agglutination test with bacterial antigens.
(2) Swine erysipelas. Both slide agglutination test and test tube agglutination test can be used for the detection of antibodies to swine erysipelas. When the test tube agglutination test is done to test the serum, a positive reaction can be determined only if the agglutination titer is above 1:200.
(3) Infectious atrophic rhinitis in swine. The test tube agglutination test was used to test the serum of pig over 3 months of age. Serum dilution was performed at 1:10. 1:20, 1:40, 1:80...... etc. The serum is considered positive if agglutination titer at or over 1:80, the serum is considered suspicious if titer is at 1:40, and the serum is considered negative if the titer below 20.

Agar Diffusion Test

Materials

Good quality agar powder or agarose, 0.01M pH 7.2 PBS (containing 0.76% NaCl); antigen, positive serum and serum samples to be tested, slides (or culture dishes), punch, pipette, etc.

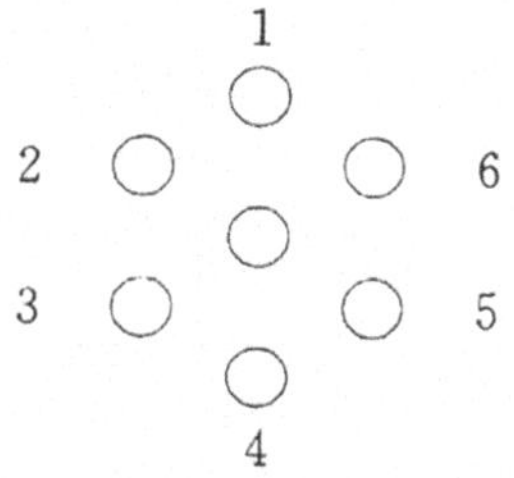

Fig. E.1 Schematic illustration of wells in the agar plate of agar diffusion test

Procedure

(1) Preparation of agar plates. Add 1.2 g of agar powder to a 150 mL triangular flask, then add 100 mL PBS solution, add 0.01% thimerosal plus cotton plugs, boil the agar solution in a water bath or in a microwave oven, and keep the flask at 60 °C in a water bath after the agar is completely dissolved. Take 3.5 mL of agar liquid, and pour it on a clean slide, or take 15 mL to pour it into a clean culture dish. After cooling down and solidified, the agar plates are stored at 4 °C till use.
(2) Punching and adding samples. An agar plate is punched to make a hole (well) of 4 mm in diameter with a punch, and holes in agar are 3 mm apart (as in Fig. E.1). The other side of the agar plate should be quickly burned to seal the bottom of agar holes to avoid sample leaking at the bottom.

    Add antigen to central well, and positive serum to peripheral wells 2 and 5, add the serum to be tested to wells 1, 3, 4, and 6.
(3) Determination of antibody titer: central well is filled with antigen, peripheral wells with twofold diluted serum, one well for each dilution.
(4) Placing the agar plates in a wet box at room temperature or 37 °C for 24–48 h.
(5) Determination of results: If a clear and dense precipitation line appears between the positive serum and the antigen wells, if the serum to be tested forms a precipitation line with the antigen well, or if the end of the precipitation line of the positive serum is curved toward the adjacent serum well, the serum to be tested is considered positive. If no precipitation line is formed between the serum to be tested and the antigen, or if the precipitation line of the positive serum extends straight toward or away from the adjacent serum well, the serum to be tested is considered negative. The titer of the antibody is determined by the highest dilution of the serum that shows a positive reaction (clear precipitation line) as the agar diffusion titer of the serum.

Notes

(1) To prevent mold contamination in agar plates, thiomersal (0.01% of the total) should be added to the agar gel for storage.
(2) Avoid the agar gel over-thin or over-thick when preparing agar plates.

(3) Do not move the agar gel when punching the hole to avoid the gap between the agar gel and the slide, otherwise the liquid will easily flow out from the bottom after adding the sample. The gap-prevented method: After punching the hole, gently heat the other side of the slide with an alcohol lamp and test the temperature with the skin on the back of your hand to feel a little bit hot. The bottom of the hole is sealed when it gets slightly hot.
(4) The sample should be filled up without overflowing, and if it overflows, the overflowing liquid can be removed by absorbing using filter paper in time.

Application

The test is easy to perform, accurate, rapid, and reproducible, thus it is widely used in clinic, e.g., it is often used on intensive pig-raising farms for the detection of serum antibodies against swine erysipelas, swine pneumonia, swine paratyphoid, hog cholera, swine vesicular disease, and toxoplasmosis.

The method for the detection of serum antibodies can also be used to detect antigens.

Indirect Hemagglutination Test (IHA)

Materials

Antigen-sensitized red blood cells (lyophilized sensitized red blood cells), positive and negative pig sera and pig sera to be tested, diluent (0.01M pH 7.2 PBS containing 2% unimmune serum and 0.05% NaN_3), 96-well "V"–shaped well microplate plate. Pipettes, drip tip, and micro-rotator.

Procedure

(1) Inactivation of the serum to be tested: the pig serum to be tested is inactivated in a water bath at 60 °C for 30 min.
(2) Add 25 μL of diluent to each well of the "V" plate using Pipettes.
(3) 25 μL of the serum to be tested is added to well 1, and mixed for 4–5 times with a pipette in a pumping up and down manner. Make serial dilutions by transferring 25 μL mixture to well 2 for mixing, sequentially to well 11, remove and discard 25 μL from well 11, and thus serum dilutions from wells 1–11 are 1:2, 1:4, 1:8, 1:16, 1;32, 1:64, 1:128, 1:256, 1:512, 1:1024, 1:2048, respectively. The last well (well 12) is used for the sensitized red blood cell control, meanwhile positive and negative serum controls should be established.
(4) Add 25 μL of sensitized erythrocytes per well, shake for 1–2 min on an rotator, and place the plate at room temperature or 37 °C for 1–2 h to observe the results.

(5) Determination of results: Reaction well with a thin layer of aggregated erythrocytes, covering the entire bottom of the well or with the edge curled in a ruffled shape, is considered "#", indicating 100% agglutination. Reaction well with a thin layer of agglutinated erythrocytes, but with a small area of aggregation, dense in the center and loose at the edge, is considered "++", indicating 50% agglutination. The agglutination in between is considered "+++", indicating 75% agglutination. Reaction well with most of the red blood cells concentrated in the center with a small amount of agglutination around is considered "+", indicating 25% agglutination. Reaction well with red blood cells dotted at the bottom, smooth around and no dispersed agglutination, is considered "–", indicating 0 agglutination.

The highest dilution of the serum at which 50% agglutination (++) occurs is determined to be the indirect hemagglutination titer of the serum.

Examples for the Application of IHA

The indirect hemagglutination test is a rapid, simple, sensitive, and specific serological technique for the detection of serum antibodies to a wide range of pathogens.

(1) Foot-and-mouth disease virus (FMDV). IHA is used to detect antibodies to infection-associated antigens. This method determines whether an animal has been infected with foot-and-mouth disease virus by detecting antibodies to FMDV antigen. In the test, the serum to be tested is diluted to be: 1:10, 1:20.... Those sera with an indirect hemagglutination titer of 1:20 or more are considered positive. Meanwhile a control group should also be set up.
(2) Mycoplasmal pneumonia of swine. This test is mainly used to detect serum antibodies to the mycoplasmal pneumonia of swine. In the experiment, after inactivation, the serum to be tested should be adsorbed by aldehydated erythrocytes before use, and the method is: taking 0.2 mL of the serum to be tested, adding 0.3 mL 2% aldehydated erythrocytes, adsorbing at 37 °C and precipitating naturally or centrifuging at 2000 rpm for 10 min to take the supernatant, which is 2.5 times diluted serum, and then performing a serial dilutions of 1:5, 1:10, 1:20, 1:40, 1:80, and 1:160. The hemagglutination titer of the positive control serum should be ≥1:40; negative control <1:5; no self-coagulation in the antigen-sensitized erythrocyte control wells. Sera to be tested with a hemagglutination titer of ≥1:10 are considered positive, and those with a hemagglutination titer <1:5 are considered negative. Sera with a hemagglutination titer between negative and positive are considered suspicious. Negative and suspicious pigs must be retested after another 4 weeks, and if both tests are negative, they are considered free of Mycoplasmal pneumonia. If both results are suspicious, they are considered positive. This test can be used for screening and diagnosis of Mycoplasmal pneumonia of swine.

(3) Hog cholera. This test is based on a twofold series of dilution of inactivated sera to be tested, and a hemagglutination titer greater than 1:16 is considered positive, mainly used for hog cholera antibody titer detection and surveillance.
(4) Toxoplasmosis. The test is performed with a twofold serial dilution of the inactivated serum to be tested. The positive control (provided by Lanzhou Veterinary Research Institute) should have a hemagglutination titer of ≥1:1024, all other controls are negative, and the serum to be tested, with a hemagglutination titer of ≥1:64, is considered positive.
(5) Chlamydia of swine. The test is performed on a twofold serial dilution of the inactivated serum to be tested. The serum with a hemagglutination titer of ≥1:64 is considered positive. The serum with a hemagglutination titer of ≤1:16 is considered negative, and between positive and negative reaction is considered suspicious. The serum with a suspicious result should be retested, and if retested result remains suspicious, the serum is considered positive.

Fluorescent Antibody Assay

Materials

Tissue sections, viral tissue cultures, bacterial suspensions, 95% ethanol, 0.01M pH 7.2 PBS, antiserum, FITC-labeled anti-pig IgG (e.g., rabbit anti-pig IgG), FITC-labeled pig antibody (specific IgG), a fluorescent microscope.

Procedure

Direct Method

(1) Bacterial suspensions or infected tissue may be used to smear or touch slides, and tissue sections can be circled with a pen and blown dry.
(2) Fixation of the specimen: place the slide in 95% ethanol for 5–10 min at room temperature, wash with PBS, and blown dry.
(3) Add FITC-labeled specific porcine IgG dropwise to the circled sample, and incubate at 37 °C for 30 min.
(4) Wash, dry, seal and examined by fluorescence microscopy.

Indirect Method

(1) Virus-infected tissue can be cryosectioned, and virus-infected cell culture slides are circled with a pen, and blown dry.
(2) Fixation of specimens is performed in the same way as the direct method as above described.

(3) Add antiserum dropwise to the specimen site on the slide, incubate at room temperature or 37 °C for 30 min, then wash with PBS, dip sequentially in 3 cups of PBS solution for 5 min, and take out to dry.
(4) Add FITC-labeled anti-pig IgG (diluted to titer) dropwise to the circled sample on the slide, and incubate at 37°C for 30 min.
(5) Wash, dry, seal, and examine the slides by fluorescence microscopy.
(6) Positive and negative controls should be set up for both the direct and indirect methods.

Determination of results

Yellow-green fluorescence can be observed in positive reactions but not in negative controls.

Notes

(1) This test is prone to nonspecific fluorescence, so the washes should be performed thoroughly according to the protocol, and negative and positive controls should be set up.
(2) Sections should be thin and not thick, and smears should not be too thick.
(3) Positive reactions are usually observed as a fluorescent antibody staining showing the feature of fixed bacteria or viruses, and determination of the results should be made with reference to relevant information about the pathogen.

Examples for the Application of This Technique

(1) Porcine rabies. The hippocampus tissue of porcine brain was taken to prepare touching slides, and made them dried and fixed in acetone, stained with rabies fluorescent antibody, washed with borate buffer (pH 9.0) and then with distilled water. After drying, the specific fluorescence was examined under a fluorescence microscope.
(2) Swine erysipelas. Tissue sections were prepared with diseased tissues (spleen, kidney, liver, lymph nodes). Examination was carried out by the direct method with fluorescent antibodies to Swine erysipelas type A and B. Bright green bacteria observed under a fluorescence microscope were considered positive.
(3) Swine dysentery. Pig feces were collected to make smears on slide, and a direct method of fluorescent antibody technique was employed to examine the slide. The samples with yellow-green fluorescent Treponema hyodysenteriae (spirochete-like bacteria) observed under a fluorescence microscope were considered as positive, by which the final diagnosis could be made.
(4) Hog cholera. The kidney or spleen or tonsil of the pig was collected to prepare tissue-touching or frozen slides. Hog cholera fluorescent antibodies were used to probe the viral antigen in the tissues on slides, which were observed under a fluorescence microscope, and would be considered positive if there were bright green or yellow-green fluorescent cells with clear outline, fluorescence-stained

cytoplasm and unstained nucleus. A hog cholera serum inhibition test should be set up to ensure the specificity of the fluorescent antibody assay.

(5) Porcine epidemic diarrhea. Fluorescent antibody assay is primarily used to detect viral antigens within the intestinal epithelial cells or cell cultures of neonatal piglets that have died naturally. Smears of intestinal epithelial cell eluates on slides were prepared and detected by the direct method. Frozen sections or cell cultures for porcine epidemic diarrhea could also be used to detect porcine epidemic diarrhea virus by fluorescent antibody assay. If stained with fluorescent antibodies diluted with 0.02% Evans blue, samples are considered positive if there is diffuse or granular bright green fluorescence within the cell plasma as observed under a fluorescence microscope, and considered negative if the cells are red-fluorescent. It should be noted that when examining the slides stained with FITC-labeled fluorescent antibodies, positive results should be bright green fluorescence. Any fluorescence that appears dark green or yellow was nonspecific.

(6) Transmissible gastroenteritis of swine. Collect tonsils and mesenteric lymph nodes samples to prepare frozen sections or tissue-touching slides, or collect the precipitates of jejunal mucosal epithelial cells to prepare smear slides. Detection of viral antigens was performed directly with fluorescent antibodies (diluted by 0.01M pH 7.2 PBS containing 0.02% Evans blue). Positive result was determined by observing specific fluorescence within the cells under a fluorescence microscope. All cells would be considered negative if no specific fluorescence was observed within the cell plasma.

Two Methods of Enzyme-Linked Immunosorbent Assay (ELISA): Indirect Method and Sandwich Method

Materials

0.05M pH 9.6 carbonate buffer, diluent: 0.01M pH 7.2 PBS (containing 0.13M NaCI, 5% fetal bovine serum); washing buffer: 0.01M pH 7.2 PBS (containing 0.13M NaCl, 0.05% Tween-20), 30% H_2O_2, o-phenylenediamine (OPD), pH 5.0 disodium hydrogen phosphate-citrate buffer, 2M H_2SO_4, antigen, positive serum, negative serum, serum to be tested, enzyme-labeled anti-PEDV IgG, enzyme-labeled rabbit anti-pig IgG, 96-well polystyrene ELISA plate, ELISA reader.

Procedure

Indirect ELISA

(1) Prepare a certain concentration of antigen-coating solution with carbonate buffer to coat the ELISA plate with 100 μL per well, and place it for 2 h at 37 °C or overnight in a refrigerator at 4 °C.

(2) Dump the liquid in the wells, fill each well with washing buffer, wash 3 times for 3 min each, and then dump the washing buffer and pat it upside down several times on water-soaking paper.
(3) Add 100 μL of the diluted serum to be tested to each well while setting up negative and positive serum controls, and incubate for 0.5 h at 37 °C. If antibody titer is detected, refer to step (4).
(4) Determination of antibody titer. The serum to be tested is serially diluted by folds, add each dilution to two wells, i.e., add the same dilution of serum to the 2 adjacent wells 100 μL each. The negative and positive serum controls should be set up. Incubate the plate for 0.5 h at 37 °C.
(5) Washing plates as in step (2).
(6) Add enzyme-labeled rabbit anti-pig IgG 100 μL to each well, and incubate for 0.5 h at 37 °C.
(7) Washing plates as in (2).
(8) Add 100 μL of freshly prepared substrate solution (dissolve 40 mg of OPD in 100 mL of pH 5.0 phosphate-citrate buffer and add 0.15 mL 30% H_20_2 before use) 100 μL per well, and observe the color development at room temperature (about 5–30 min).
(9) Add 50 μL 2 mol/L H_2SO_4 per well to terminate the reaction.
(10) The optical density (OD) of each well at OD 490 nm was measured using an ELISA reader.

Sandwich method using the detection of PEDV antigen as an example

(1) Antibody coating. ELISA plates were coated with porcine anti-PEDV antibody or PED immune serum (diluted with carbonate buffer) at a volume of 100 μL per well, and placed at 37 °C for 2 h or in a refrigerator at 4 °C overnight.
(2) Washing plates, the same as step (2) as described in indirect method.
(3) Add samples to be tested. Add 100 μL sample to each well, usually 2 wells per sample, set up negative and positive controls as well as enzyme-conjugate control (antibody + PBS + enzyme conjugate), and place the ELISA plate at 37 °C for 0.5 h.
(4) Washing plates, the same as step (2).
(5) Addition of substrate solution: the same as indirect method step (8).
(6) Same as indirect method (9).
(7) Same as indirect method (10).

Determination of results. There are three common methods for the determination of results for ELISA.

(1) Negative and positive. The sample well, with the absorption value of the tested sample ≥ the average absorption value of negative controls + 2SD (standard deviation), is considered positive, otherwise negative.
(2) P/N ratio method. The sample well, with the ratio of absorption value of sample/the mean absorption value of negative controls ≥2–3, is considered positive.
(3) Endpoint method. The highest dilution of the sample that shows a positive reaction is the titer of that sample.

Notes

(1) High pH and low ionic strength conditions were preferred during the coating process, and the optimal coating concentration of protein is selected at 1–100 μg/mL.
(2) Serum or antigen diluent should contain 5–10% serum of another species of animals or 1% BSA or 0.5% gelatin to act as a blocking agent for preventing nonspecific reactions. Otherwise, blocking step should be performed by adding a blocking solution between steps (2) and (3).
(3) Washing should be adequate, enzyme conjugates should be diluted and used as required, and substrate solutions should always be freshly prepared before use.
(4) At color development step, the reaction should be terminated immediately when the negative control just turns slightly yellow.
(5) When determining the results by P/N ratio method, the result determination may be misleading if the average OD value of negative control wells is less than 0.1.
(6) When detected by an automated ELISA reader, the index of ELISA reader should be calibrated according to the manufacturer's instructions.
(7) The average of the OD490 values of two wells of the same dilution is the OD value for that dilution, and the control wells should be treated the same way.

Examples for the Application of ELISA

Since the invention of ELISA in the 1970s, the test has been widely used in the diagnosis of diseases, epidemiological investigations, and surveillance of serum antibodies. The indirect ELISA method is mainly used for the detection and diagnosis of serum antibodies to infectious diseases. The sandwich ELISA method is used for the detection of antigenic components of a variety of pathogens (bacteria, viruses) and diagnosis of infectious diseases.

(1) Hog cholera. Hog cholera virus (HCV) antigen that had been purified by monoclonal antibodies (provided by China Veterinary Drug Inspection Institute) was used as coating antigens to develop an indirect ELISA method, which can differentiate anti-virulent HCV serum antibodies from that against attenuated live HCV vaccine (differentiate infected from vaccinated, DIVA). The method of coating antigen was performed per the user's instructions. The criteria for the result determination are: on the attenuated live HCV antigen-coated ELISA plate, the serum sample with an OD value — that of the negative control serum $\geq$0.2 is considered positive; on the virulent HCV antigen-coated ELISA plate, the serum sample with the OD value — that of negative control serum $\geq$0.5 is considered positive. This method is a very effective tool for hog cholera antibody assay, so that the status of HCV infection in pigs and its epidemiological survey could be evaluated, providing a basis for the development of an adequate

immunization program for the control of HCV infection, and ultimately eradicating HCV infection.

(2) Porcine epidemic diarrhea (PED). Double antibody sandwich ELISA was used to directly detect causative virus in the feces, and is suitable for rapid diagnosis, as reported by Changchun PLA Agricultural and Animal Husbandry University. The tested fecal samples should be processed (feces were diluted 5–10 times with diluent to make a suspension, centrifuged at 3000 rpm for 15 min, and the supernatant is used for ELISA assay). Each sample was added to two wells, and positive control (capture antibody + positive stool + enzyme-conjugated detection antibody), negative control (capture antibody + negative stool + enzyme-conjugated detection antibody), enzyme-conjugate control (capture antibody + PBS + enzyme-conjugated detection antibody). The sample with a P/N ratio of $\geq$2 was considered positive, otherwise negative.

(3) Swine trichinosis. The disease can be diagnosed by detection of antibodies against swine trichinosis using an indirect ELISA method, which is also suitable for swine trichinosis quarantine.

(4) Mycoplasmal pneumonia of swine. Mycoplasmal pneumonia of swine can be diagnosed by detection of antibodies to mycoplasmal pneumonia using an indirect ELISA method. Using micro-volume whole blood ELISA (developed by the Jilin Provincial Institute of Veterinary Science), early-infected pigs and those in the late stage of infection (lesion absorption stage) can be diagnosed by detection of specific antibodies.

Dot Enzyme-Linked Immunosorbent Assay (Dot-ELISA)

Materials

Nitrocellulose membrane (NCM) (pore size 0.45 μM), 0.01M pH 7.2 PBS, blocking solution (PBS with 10% horse serum or 0.1% gelatin), 30% H_2O_2; 3,3-diaminobenzidine (DAB), antigen solution, negative control solution, and solution to be tested (including serum to be tested, positive serum, and negative serum), enzyme-labeled rabbit anti-pig IgG, 0.05 M Tris-HCl buffer, washing buffer (PBS with 0.05% Tween-20).

Procedure

This section mainly introduces an indirect Dot-ELISA, which can detect both antigens and antibodies.

Detection of Antigens

(1) Preparation of NCM. Draw a 6 × 6 mm square on the NCM, mark a circle in the center of the square by pressing the rounded end of punch, place the NCM in distilled water to soak for 10 min, and take out the NCM and set aside to dry for storage till use.
(2) Dot NCM with sample to be tested. Add 10 μL sample solution to NCM circle, set up both positive and negative antigen controls, and dry the NCM naturally at room temperature or at 37 °C.
(3) Blocking. Place NCM in the blocking solution at 37 °C for 30 min, and wash the NCM once with washing buffer for 3 min.
(4) Add working concentrations of porcine positive serum. Place the NCM in porcine positive serum at working concentrations and incubate at 37 °C for 1 h, wash the NCM with washing buffer three times for 3 min each.
(5) Add enzyme-labeled rabbit anti-pig IgG and incubate at 37 °C for 1 h, followed by washing NCM with washing buffer 3–4 times for 3 min each.
(6) Add substrate solution (freshly prepared 0.05 MTris-HCl buffer 100 ml with 40 mg DAB and 50 μL 30% H_2O_2) and incubate for 5–20 min.
(7) Terminate the color development by immersing the NCM in distilled water for 2–3 min.
(8) Dry the NCM naturally at room temperature or at 37 °C.
(9) Observe the results.

Detection of Antibodies

(1) Preparation of NCM is made as above described step (1) in Dot-ELISA for the detection of antigen.
(2) Dot NCM with a known antigen solution, and dry the NCM naturally or at 37 °C.
(3) Blocking. Block NCM as above described step (3) in Dot-ELISA for the detection of antigen test antigen.
(4) Cut antigen-dotted NCM into small pieces, place the NCM into serially diluted serum (1:10, 1:50, 1:100, 1:200, 1:400, 1:800, 1:1600, 1:3200, 1:6400, 1:12,800), and incubate for 1 h at 37 °C; meanwhile negative and positive serum controls should be set up, followed by 3 washes with washing buffer for 3 min each time.

(5), (6), (7), (8), and (9) are the same as above described steps (5), (6), (7), (8), and (9) for the detection of antigens, respectively.

Determination of results: sample dot with the same as the negative control showing no color inside the circle is considered negative (–). Sample dot with dark brown color in clear contrast to white background (outside the dot) is considered strong positive (#). Sample dot with light brown color in clear contrast to the white background is considered positive (++). Sample dot with color change in between is considered positive (+++). Sample dot with weak spots or homogenous brown color within the spots is considered positive (+).

The highest dilution of the serum that shows a positive reaction is determined as the Dot-ELISA titer of that serum.

Notes

(1) A pilot study should be conducted to determine the dilution of the antigen, serum, and enzyme-labeled antibodies.
(2) Insoluble hydrogen donors, such as DAB and 4-chloro-l-naphthol, should be used in Dot-ELISA, soluble hydrogen donors should not be used.
(3) Determine the antigen or antibody to be tested, depending on the purpose of the test.
(4) Blocking solutions can be used such as the serum of another species of animals, BSA, or gelatin solution for blocking. Choose the best blocking conditions based on pilot studies.

Applications of Dot-ELISA

Dot-ELISA can be used as conventional ELISA for the detection of antigen or antibodies. For applications, see ELISA tests.

(H) Virus Neutralization Test

The virus neutralization test is highly specific and can detect serum neutralizing antibodies with known viruses and also use specific antisera to detect viruses, while the test is labor-intensive and difficult to carry out in a general laboratory of pig farm. Thus its application is somewhat limited in production practice. However, it is still a standard and reliable method for the detection or identification of serum antibodies or viruses.

Appendix F: Recommendation Scheme for Immunization Against Common Infectious Diseases in Animals (Trial)

This scheme was developed in accordance with the "Animal Infectious Disease Prevention and Control Law of the People's Republic of China" by the Ministry of Agriculture in order to better and more carefully implement "National Long-term Animal Infectious Disease Prevention and control Plan (2012–2020)." This book reprints it as Appendix Six as follows:

Diseases That Require Immunization

Brucellosis, Newcastle disease, rabies, sheep pox and goat pox, anthrax, porcine pseudorabies, echinococcosis, porcine reproductive and respiratory syndrome, Japanese encephalitis, swine erysipelas, porcine circovirus disease, infectious bronchitis in chickens, infectious bursal disease in chickens, duck plague, low pathogenicity (H9 subtype) avian influenza, and other animal infectious diseases.

Recommendation Immunization Program

Animal raising units, based on practical conditions, should regularly monitor the level of antibodies after immunization and adjust the immunization program in time according to the test results.

S. J. Zheng, *Veterinary Molecular Immunology*,
https://doi.org/10.1007/978-981-99-8929-4

Brucellosis

Regional Division

Category I regions refer to Beijing, Tianjin, Hebei, Inner Mongolia, Shanxi, Heilongjiang, Jilin, Liaoning, Shandong, Henan, Shaanxi, Xinjiang, Ningxia, Qinghai, Gansu and other 15 provinces and the Xinjiang Production and Construction Corps. Comprehensive immunization of cattle and sheep is implemented for three consecutive years on a county basis. Immunization of breeding bulls and rams is prohibited. Diary stock are not immunized in principle, but in counties where the individuals with pathogen positive rate over 2% could be vaccinated after approval by the provincial veterinary authority for immunization application submitted by the county veterinary authority. Pathogen-positive livestock should be eliminated before immunization. The veterinary authority of counties, where the disease control criteria have already been reached, or reached ahead of time, or stable control and eradication standards have been reached, can apply to the provincial veterinary authority for exemption from immunization. Upon approval, immunization can be no longer implemented.

After three consecutive years of immunization, counties that have reached the control standard as assessed by the provincial veterinary authority can stop immunization.

Category II regions refer to Jiangsu, Shanghai, Zhejiang, Jiangxi, Fujian, Anhui, Hunan, Hubei, Guangdong, Guangxi, Sichuan, Chongqing, Guizhou, Yunnan, and Tibet, totaling 15 provinces. Immunization is not implemented in principle. The veterinary authority of counties, where the disease control standard have not reached and need immunization, can apply to the provincial veterinary authority for immunization, and can implement immunization program after approval by the provincial veterinary authority and send files to the Ministry of Agriculture.

The purification area refers to Hainan Province, where immunization is prohibited.

Immunization Procedure

In areas where immunization against brucellosis is approved, susceptible livestock should be tested first according to the recommended procedures and methods in the vaccine-manufacturer's instructions, and only negative livestock can be immunized.

Vaccines used: Brucella live vaccine M5 or M5-90 strain is used to prevent bovine and ovine brucellosis; Brucella live vaccine S2 strain is used to prevent brucellosis in goats, sheep, pigs, and cattle; Brucella live vaccine A19 or S19 strain is used to prevent bovine brucellosis.

Newcastle Disease

Comprehensive immunization is implemented for chickens.

Commercial Broilers: At 7–10 days of age, primary immunization of chicks with Newcastle disease attenuated live vaccine and/or inactivated vaccine is carried out. Two weeks later, the Newcastle disease attenuated live vaccine is used to booster immunization once.

Breeding chickens, commercial laying hens: At 3–7 days of age, primary immunization of chicks with Newcastle disease attenuated live vaccine is performed; at 10–14 days of age, the Newcastle disease attenuated live vaccine and/or inactivated vaccine is used for secondary immunization; at 12 weeks of age, the Newcastle disease attenuated live vaccine and/or inactivated vaccine is used to booster immunization, and at 17–18 weeks of age or before egg-laying, the Newcastle disease inactivated vaccine is used for immunization once. After egg-laying, booster immunization is performed according to the antibody levels in the serum as examined by hemagglutination inhibition test.

Vaccines used: Newcastle disease inactivated vaccine or attenuated live vaccine for chickens.

Rabies

Comprehensive immunization is implemented for dogs, with a focus on immunization for dogs in rural areas and urban-rural junctions with high incidence of rabies.

Primary immunization of newborn puppies is carried out at 2 months of age, secondary immunization at 3 months of age, and thereafter immunization is carried out every 12 months.

Vaccines used: Canine rabies live vaccine or inactivated vaccine.

Sheep Pox and Goat Pox

Immunization is carried out for sheep and goat in epidemic areas.

Primary immunization is carried out at around 60 days of age, and booster immunization is carried out every 12 months thereafter.

Vaccines used: Goat pox live vaccine.

Anthrax

Susceptible livestock in townships where outbreaks have occurred in the past 3 years are immunized.

Immunization is carried out once a year. In the event of an outbreak, all susceptible livestock in the epidemic and threatened areas are immunized once as an emergency control measure.

Vaccines used: Non-capsulated anthrax spore vaccine or No. 2 anthrax spore vaccine.

Pseudorabies (Aujeszky's Disease) in Swine

Immunization is carried out for pigs in epidemic areas.

Commercial pigs: Immunization is carried out once at around 55 days of age.

Breeding sows: primary immunization is carried out at around 55 days of age; pre-breeding immunization is carried out for first-time pregnant sows, and immunization of pregnant sows is carried out again 4–6 weeks before delivery.

Breeding boars: the primary immunization is carried out at around 55 days of age, and thereafter immunization is carried out every 6 months.

Vaccines used: Pseudorabies live vaccine or inactivated vaccine for pigs.

Echinococcosis

Immunization is implemented for sheep and goat in the epidemic areas of Inner Mongolia, Sichuan, Tibet, Gansu, Qinghai, Ningxia, Xinjiang, and the Xinjiang Production and Construction Corps.

Each year, the newly born sheep and goat are vaccinated, and thereafter, the immunized sheep and goat are given a booster immunization once a year.

Vaccines used: Sheep Echinococcosis Genetic Engineering Subunit Vaccine.

Porcine Reproductive and Respiratory Syndrome (PRRS)

Immunization is implemented for pigs in the epidemic areas.

Commercial pigs: Live vaccines are used to immunize pigs before and after weaning, and a booster immunization can be given 4 months later depending on the epidemic situation.

Sows: The immunization procedure for sows before 150 days of age is the same as that for commercial pigs. Depending on the epidemic situation, an inactivated vaccine can be used to immunize sows before breeding.

Boars: Inactivated vaccines are used to immunize boars. The immunization procedure for boars before 70 days of age is the same as that for commercial pigs, and a booster immunization is given every 4–6 months thereafter.

Vaccines used: Live or inactivated vaccine for PRRS.

Japanese Encephalitis of Swine

Immunization is implemented for pigs in the epidemic areas.

Each year, 1–2 months before the appearance of mosquitoes, susceptible livestock such as pigs are immunized twice, with an interval of 1–2 months. The immunization time should be determined based on the specific epidemic situation.

Vaccines used: Inactivated or live vaccine for porcine Japanese Encephalitis.

Swine Erysipelas

Immunization is implemented for pigs in the epidemic areas.

The primary immunization of pigs is performed at 28–35 days of age, and the secondary immunization is given around 70 days of age.

Vaccines used: Inactivated vaccine for Swine Erysipelas.

Porcine Circovirus Disease

Immunization is implemented for pigs in the epidemic areas.

Immunization can be carried out according to the recommended procedure for each type of Porcine Circovirus vaccine.

Vaccines used: Inactivated vaccine for Porcine Circovirus Disease.

Infectious Bronchitis of Chickens

Immunization is implemented for chickens in the epidemic areas.

Commercial broilers: The live vaccine for Infectious Bronchitis of chickens is used for the primary, secondary, and booster immunizations of chickens at 1–7 days of age, 10–14 days of age, and 56 days of age, respectively. It is recommended that broilers for slaughter at 40–50 days of age are only immunized twice.

Breeder chickens and commercial laying hens: The immunization procedure for breeders and commercial layer before 56 days of age is the same as that for commercial broilers. The inactivated vaccine for Infectious Bronchitis of chickens is used for the fourth immunization at 110–120 days of age. After the onset of egg production, immunization is carried out based on the antibody levels in the serum of tested chicken.

Vaccines use: Inactivated or live vaccine for Infectious Bronchitis of chickens.

Infectious Bursal Disease

Immunization is implemented for chickens in the epidemic areas.

Commercial broilers: The primary immunization of chickens with live vaccines against Infectious Bursal Disease (IBD) is implemented at 10–14 days of age, and the secondary immunization is administered around 22 days of age. Broilers for slaughter at 40–50 days of age should be immunized before 24 days of age.

Breeder chickens and commercial laying hens: The primary immunization of chickens with live vaccines for IBD is implemented at 10–14 days of age, and secondary immunization at 28–35 days of age. A booster immunization of chickens with inactivated vaccine for IBD is given at 110–120 days of age. After onset of egg-laying, chickens are immunized according to the antibody levels in the serum of tested chickens.

Vaccines use: Inactivated or live vaccine for Infectious Bursal Disease.

Duck Plague

Immunization is implemented for ducks in the epidemic areas.

Commercial meat ducks: Immunization of ducks with inactivated or live vaccines against duck plague is implemented around 14 days of age.

Commercial laying ducks: The primary immunization of commercial laying ducks with inactivated or live vaccine against duck plague is carried out around 14 days of age, and secondary immunization around 60 days of age, and then booster immunization is implemented every 6 months.

Breeder ducks: The primary immunization of commercial laying ducks with inactivated or live vaccine against duck plague is carried out around 14 days of age, and secondary immunization around 60 days of age. Booster immunization of duck layers with live vaccines against duck plague is implemented one month before egg-laying, and then booster immunization is implemented every 4–6 months after the onset of egg-laying.

Vaccines used: Live or inactivated vaccine for Duck Plague.

Low pathogenicity (H9 subtype) Avian Influenza

Immunization is implemented for chickens in the epidemic areas.

Commercial broilers: the primary immunization of chicken is implemented at 7–14 days of age, and the secondary immunization at 28–35 days of age. It is recommended that broilers for slaughter at 40–50 days of age receive the primary immunization only.

Breeder chickens and commercial laying hens: The primary and secondary immunization procedures are the same as that for broilers, and the booster immunization of chickens is implemented at 110–120 days of age. After onset of egg-laying, hens are immunized according to the antibody levels in the serum of tested hens. Vaccines used: Inactivated vaccine for Avian Influenza (H9 subtype).

Other Matters for Attention

(1) The specific methods and doses of vaccination for various vaccines should be employed according to the manufacturer's instructions.

(2) Effectively monitor and evaluate the effect of the vaccine. If the level of specific antibodies does not meet the requirements, booster immunization should be implemented immediately.
(3) For breeding farms and other breeding units that perform purification of related important diseases, they can carry out the purification plan and do not need to take immunization measures.
(4) Vaccines approved for production or registered by the state must be used, and vaccine management should be strengthened to follow the requirement for the storage and transportation of vaccines. For commonly seen animal diseases such as Brucellosis, new vaccine products can be considered to immunize animals if the vaccine is approved by the state.
(5) Before using vaccines, it is necessary to carefully check for the outlook quality of the vaccine, such as expiration date, the bottle damage, etc. During the immunization process, animals should be vaccinated according to the protocol per the manufacturer's instruction, and the waste should be treated with harmless disposal.
(6) It is necessary to effectively carry out personal biosafety protection to avoid infection of pathogens through skin wounds, respiratory tract, digestive tract, and visible mucous membranes or avoid causing adverse reactions.
(7) During the immunization process, disinfection should be well performed. For the immunization of livestock such as pigs, cows, sheep, and dogs, the role of "one animal, one needle" should be followed. For the immunization of poultry such as chickens and ducks, the needles should be changed frequently to prevent cross-infection.
(8) A complete immunization record should be kept, a standard and complete immunization file should be established to ensure that the immunization time, the type of vaccine used, and other information are accurate, detailed, and traceable.

March 12, 2014

Appendix G: Recommended Program for the Prevention and Control of Swine Diseases by Immunization

Appendix G of this book includes "The recommended program of immunization against swine disease" issued by the Ministry of Agriculture on March 28, 2007, which was regarding the requirement for the better preparation for the control of swine disease of 2007.

Overall Requirements

In order to effectively prevent and control the occurrence and prevalence of pig diseases and to ensure the sustainable and healthy development of the livestock industry, the safety of livestock products, and the health of the people, this program has been established.

The State implements compulsory immunization against foot-and-mouth disease and a full cover of immunization against Hog cholera, with an immunization coverage of 100%. All local authorities, taking into account the local epidemiological situation of swine diseases, implement immunization against other pig diseases. At the same time, surveillance of the effectiveness of immunization should be carried out in a timely manner, and immunization procedures should be adjusted according to the levels of antibodies to ensure the effectiveness of immunization.

Based on this program, each local authority may develop an adequate immunization program, after fully taking into account the reality of local situation.

Diseases to be Controlled by Immunization

The diseases covered by this program are: foot-and-mouth disease (FMD), Hog cholera, highly pathogenic porcine reproductive and respiratory syndrome (HP-PRRS), porcine pseudorabies, porcine Japanese encephalitis, porcine

S. J. Zheng, *Veterinary Molecular Immunology*,
https://doi.org/10.1007/978-981-99-8929-4

Parvovirus infection, transmissible gastroenteritis of swine (TGE), porcine epidemic diarrhea (PED), porcine pasteurellosis, porcine erysipelas, porcine streptococcal disease, porcine colibacillosis, Swine paratyphoid, porcine mycoplasmal pneumonia, porcine infectious atrophic rhinitis, and porcine infectious pleuropneumonia.

Recommended Immunization Procedures

Commercial Pigs

Immunization time	Vaccines
1 day old	Attenuated hog cholera vaccine (Note 1)
7 days old	Inactivated porcine mycoplasmal pneumonia vaccine (Note 2)
20 days of age	Attenuated hog cholera vaccine
21 days of age	Inactivated porcine mycoplasmal pneumonia vaccine (Note 2)
23–25 days of age	Inactivated HP-PRRS vaccine
	Inactivated porcine infectious pleuropneumonia vaccine (Note 2)
	Inactivated Streptococcus serotype II vaccine (Note 2)
28–35 days of age	Inactivated foot-and-mouth disease vaccine
	Swine erysipelas vaccine, swine pasteurellosis or swine diphtheria-swine pneumonia diphtheria combined vaccine (Note 2)
	Attenuated swine paratyphoid vaccine (Note 2)
	Inactivated infectious atrophic rhinitis vaccine (Note 2)
55 days of age	Attenuated Porcine pseudorabies gene-deficient vaccine
	Inactivated infectious atrophic rhinitis vaccine (Note 2)
60 days of age	Inactivated foot-and-mouth disease vaccine
	Attenuated hog cholera vaccine
70 days of age	Swine erysipelas vaccine, swine pasteurellosis or swine diphtheria-swine pneumonia diphtheria combined vaccine (Note 2)

Note: 1 Attenuated hog cholera vaccine prepared with spleen-lymph tissue is recommended, in particular to the farmers with HCV carriers that usually cause Hog cholera in lactating piglets via vertical transmission of HCV

Note: 2 Vaccination could be selective to optimize the control effect, depending on the epidemics of infectious diseases

Breeding Sow

Immunization time [Note 1]	Vaccines
Every 4–6 months	Inactivated foot-and-mouth disease vaccine
Before breeding of gilts	Attenuated hog cholera vaccine
	Inactivated HP-PRRS vaccine
	Inactivated porcine parvovirus vaccine
	Attenuated porcine pseudorabies gene-deficient vaccine
Before breeding of the multiparous sows	Attenuated hog cholera vaccine
	Inactivated HP-PRRS vaccine
4–6 weeks before delivery	Attenuated porcine pseudorabies gene-deficient vaccine
	E. coli bivalent genetic-engineered vaccine [Note 2]
	Porcine transmissible gastroenteritis + porcine epidemic diarrhea combined vaccine [Note 2]

Note: 1 (A) The immunization procedure for breeding pigs before 70 days of age is the same as that for commercial pigs, (B) In areas where Japanese encephalitis is endemic or threatening, twice vaccinations against this disease per year should be conducted at one-month intervals from March to May (1–2 months before the emergence of mosquitoes), and (C) Attenuated hog cholera vaccine prepared with spleen-lymph tissue is recommended

Note: 2 Vaccination could be selective to optimize the control effect, depending on the epidemics of infectious diseases

Breeding Boar

Immunization time [Note]	Vaccines
Every 4–6 months	Inactivated foot-and-mouth disease vaccine
Every 6 months	Attenuated hog cholera vaccine
	Inactivated HP-PRRS vaccine
	Attenuated porcine pseudorabies gene-deficient vaccine

Note: (A) The immunization procedure for breeding pigs before 70 days of age is the same as that for commercial pigs, (B) In areas where Japanese encephalitis is endemic or threatening, twice vaccinations against this disease per year should be conducted at one-month intervals from March to May (1–2 months before the emergence of mosquitoes), and (C) Attenuated hog cholera vaccine prepared with spleen-lymph tissue is recommended

Technical Requirements

1. Vaccines should be manufactured with the approval from the state or officially registered, used properly, and stored and transported in accordance with the conditions of vaccine storage stipulation.
2. Vaccination should be carried out as per the manufacturer's instructions, and the waste should be harmlessly treated and disposed of.
3. During the immunization process, necessary disinfection should be performed, and at the same time, "one pig, one needle" requirement should be met to prevent cross-infection.
4. If, after measurement of antibodies post vaccination, the antibody level did not meet the stipulated requirements, a booster immunization should be implemented as soon as possible.
5. Emergency vaccination of threatened pigs should be carried out when an animal disease occurs.
6. A complete immunization file should be established for checking and reference.

Appendix H: Comparison of Amino Acid Sequence Homology of Selected Cytokines Between Avian and Mouse[a]

Cytokine type	Avian cytokine amino acid sequence ID	Mouse cytokine amino acid sequence ID	Homology comparison (%)
IL-1β	NP_989855.1	NP_032387.1	35.14
IL-2	NP_989484.1	NP_032392.1	15.33
IL-4	NP_001007080.1	NP_067258.1	22.22
IL-6	XP_015136769.1 (1); NP_989959.1 precursor) (2)	NP_112445.1	20.45 (1) 23.05 (2)
IL-10	NP_001004414.2	NP_034678.1	38.33
IL-12 P35	NP_998753.1	AAA39292.1	37.55
IL-12 P40	AAP86972.1	AAA39296.1	66.80
IL-17A	NP_989791.1	NP_034682.1	48.17
IL-18	XP_015153430.1 (284aa) (1); NP_989939.1 (198aa) (2)	NP_034682.1	19.30 (1) 32.89 (2)
IFN-α	NP_990758.1	NP_034632.2	20.71
IFN-β	NP_001020007.1	NP_034640.1	12.32
IFN-γ	NP_990480.1	NP_032363.1	23.97
TNF-α	AUG72034.1	AAB65593.1	13.33
TGF-β1	AFD30526.1	NP_035707.1	54.34

[a] Original data source: https://www.ncbi.nlm.nih.gov/

S. J. Zheng, *Veterinary Molecular Immunology*,
https://doi.org/10.1007/978-981-99-8929-4

Appendix I: Comparison of Amino Acid Sequence Homology of Selected Cytokines Between Avian and Human[a]

Cytokine type	Avian cytokine amino acid sequence ID	Human cytokine amino acid sequence ID	Homology comparison (%)
IL-1β	NP_989855.1	NP_000567.1	33.45
IL-2	NP_989484.1	NP_000577.2	17.11
IL-4	NP_001007080.1	Isoform 1: NP_000580.1 Isoform 2:NP_758858.1 Isoform 3: NP_001341919.1	22.73 18.12 20.86
IL-6	XP_015136769.1 (1); NP_989959.1 (precursor) (2)	NP_000591.1	29.48 (1) 31.95 (2)
IL-10	NP_001004414.2	NP_000563.1	43.89
IL-12 P35	NP_998753.1	AAD56385.1	12.82
IL-12 P40	AAP86972.1	AAD56386.1	43.39
IL-17A	NP_989791.1	NP_002181.1	43.75
IL-18	XP_015153430.1 (284aa) (1); NP_989939.1 (198aa) (2)	NP_002181.1	20.07 (1) 29.09 (2)
IFN-α	NP_990758.1	NP_076918.1	19.7
IFN-β	NP_001020007.1	NP_002167.1	19.9
IFN-γ	NP_990480.1	NP_000610.2	29.45
TNF-α	AUG72034.1	AAA61198.1	15.42
TGF-β1	AFD30526.1	NP_000651.3	54.84

[a] Original data source: https://www.ncbi.nlm.nih.gov/

S. J. Zheng, *Veterinary Molecular Immunology*,
https://doi.org/10.1007/978-981-99-8929-4

Appendix J: Comparison of Amino Acid Sequence Homology of Selected Cytokines Between Swine and Mice[a]

Cytokine type	Porcine cytokine amino acid sequence ID	Mouse cytokine amino acid sequence ID	Homology comparison (%)
IL-2	NP_999026.1	NP_032392.1	42.01
IL-4	NP_999288.1	NP_067258.1	41.67
IL-6	NP_999564.1	NP_112445.1	42.25
IL-10	NP_999206.1	NP_034678.1	68.33
IL-12 P35	NP_999158.1	AAA39292.1	54.35
IL-12 P40	NP_999178.1	AAA39296.1	31.50
IL-17A	NP_001005729.1	NP_034682.1	67.98
IL-18	NP_999162.1	NP_032386.1	57.48
IFN-α	NP_999558.1	NP_034632.2	55.56
IFN-β	NP_001003923.1	NP_034640.1	46.24
IFN-γ	AAO34390	NP_032363.1	45.21
TNF-α	NP_999187.1	AAB65593.1	75.97
TGF-β1	AAL57902.1	NP_035707.1	89.49

[a] Original data source: https://www.ncbi.nlm.nih.gov/

S. J. Zheng, *Veterinary Molecular Immunology*,
https://doi.org/10.1007/978-981-99-8929-4

Appendix K: Comparison of Amino Acid Sequence Homology of Selected Cytokines Between Swine and Human[a]

Cytokine type	Porcine cytokine amino acid sequence ID	Human cytokine amino acid sequence ID	Homology comparison (%)
IL-2	NP_999026.1	NP_000577.2	69.48
IL-4	NP_999288.1	Isoform 1: NP_000580.1 Isoform 2: NP_758858.1 Isoform 3: NP_001341919.1	60.78 53.24 38.97
IL-6	NP_999564.1	NP_000591.1	61.32
IL-10	NP_999206.1	NP_000563.1	74.44
IL-12 P35	NP_999158.1	AAD56385.1	76.81
IL-12 P40	NP_999178.1	AAD56386.1	83.62
IL-17A	NP_001005729.1	NP_002181.1	76
IL-18	NP_999162.1	NP_032386.1	76.68
IFN-α	NP_999558.1	NP_076918.1	65.08
IFN-β	NP_001003923.1	NP_002167.1	64.71
IFN-γ	AAO34390	NP_000610.2	61.70
TNF-α	NP_999187.1	AAA61198.1	84.12
TGF-β1	AAL57902.1	NP_000651.3	94.36

[a] Original data source: https://www.ncbi.nlm.nih.gov/

S. J. Zheng, *Veterinary Molecular Immunology*,
https://doi.org/10.1007/978-981-99-8929-4

Appendix L: Nobel Prizes in Medicine in the Field of Immunology Research[a]

Year of award	Recipients	Key findings
1901	Emil Adolf von Behring (1854–1917, Marburg University, Germany)	Serotherapy, humoral immunity
1908	Ilya Ilyich Mechnikov (1845–1916, Institut Pasteur, Paris, France) Paul Ehrlich (1854–1915, Göttingen University, Göttingen, Germany)	Phagocytosis, the theory of cellular immunity Serotherapy and the side chain theory
1913	Charles R. Richet (1850–1935, Sorbonne University, Paris, France)	Anaphylaxis, immune damage
1919	Jules Bordet (1870–1961, Brussels University, Brussels, Belgium)	Complement action
1930	Kar Landsteiner (1868–1943, Rockefeller Institute for Medical Research, New York, NY, USA)	Human blood groups and blood typing, hapten
1960	Sir Frank Macfarlane Burnet (1899–1985, Walter and Eliza Hall Institute for Medical Research, Melbourne, VIC, Australia) Peter Brian Medawar (1915–1987, University College, London, UK)	Acquired immunological tolerance theory Proving the immunological tolerance theory with organ transplants
1972	Gerald M. Edelman (1929–2014, Rockefeller University, New York, NY, USA) Rodney R. Porter (1917–1985, University of Oxford, Oxford, UK)	Resolving the basic Y-shaped structure of antibodies Resolving the basic Y-shaped structure of antibodies
1980	Baruj Benacerraf (1920–2011, Harvard University, Boston, MA, USA) Jean Dausset (1916–2009, Université de Paris, Paris, France) George D. Snell (1903–1996, Jackson Laboratory, Bar Harbor, ME, USA)	Immune-related gene Human MHC (HLA) Mouse (H-2)

S. J. Zheng, *Veterinary Molecular Immunology*,
https://doi.org/10.1007/978-981-99-8929-4

Year of award	Recipients	Key findings
1984	Niels K. Jerne (1911–1994, Basel Institute for Immunology, Basel, Switzerland) Georges J.F. Köhler (1946–1995, Basel Institute for Immunology, Basel, Switzerland) César Milstein (1927–2002, MRC Laboratory of Molecular Biology, Cambridge, UK)	Theory of clonal selection Development of the monoclonal antibody hybridoma technique Development of the monoclonal antibody hybridoma technique
1987	Susumu Tonegawa (1939–, Basel Institute for Immunology, Basel, Switzerland)	Genetic principle for generation of antibody diversity
1996	Peter C. Doherty (1940–, The Australian National University, Canberra, ACT, Australia) Rolf M. Zinkernagel (1944–, The Australian National University, Canberra, ACT, Australia)	Specificity of the cell-mediated immune defense Specificity of the cell-mediated immune defense
2011	Bruce A. Beutler (1957–, University of Texas Southwestern Medical Center at Dallas, Dallas, TX, USA; The Scripps Research Institute, La Jolla, CA, USA) Jules A. Hoffmann (1941–, University of Strasbourg, Strasbourg, France) Ralph M. Steinman (1943–2011, Rockefeller University, NY, USA) (dendritic cells)	Mechanisms of innate immunity Mechanisms of innate immunity Dendritic cells
2016	Yoshinori Ohsumi (Yoshinori Ohsumi) (1945–, Tokyo Institute of Technology, Tokyo, Japan)	Mechanism of autophagy
2018	James P. Allison (1948–, University of Texas MD Anderson Cancer Center, Houston, TX, USA; Parker Institute for Cancer Immunotherapy, San Francisco, CA, USA) Tasuku Honjo (1942–, Kyoto University, Kyoto, Japan)	Cancer therapy by inhibition of negative immune regulation Cancer therapy by inhibition of negative immune regulation
2023	Katalin Karikó (1955–, Szeged University, Szeged, Hungary; University of Pennsylvania, Philadelphia, PA, USA) Drew Weissman (1959–, Penn Institute for RNA Innovations, University of Pennsylvania, Philadelphia, PA, USA)	Nucleoside base modifications that enabled the development of effective mRNA vaccines against COVID-19 Nucleoside base modifications that enabled the development of effective mRNA vaccines against COVID-19

[a] Note: Information from https://www.nobelprize.org/prizes

The manufacturer's authorised representative in the EU is Springer
Nature Customer Service Centre GmbH, Europaplatz 3, 69115 Heidelberg,
Germany. If you have any concerns regarding our products, please
contact ProductSafety@springernature.com

Printed and bound by CPI Group (UK) Ltd, Croydon, CR0 4YY
01/10/2026
02225863-0011